T0260312

Renault
Owners
Workshop
Manual

John Fowler

Models covered
All Renault 18 Saloon & Estate models including
special editions, Deauville and Turbo
1397 cc, 1565 cc, 1647 cc & 1995 cc petrol engines

Does not cover Diesel engine versions

(598-6T6)

ABCDE
FGHIJ
KLMNO
PQR

2

Haynes Publishing
Sparkford Nr Yeovil
Somerset BA22 7JJ England

Haynes Publications, Inc
861 Lawrence Drive
Newbury Park
California 91320 USA

Acknowledgements

Thanks are due to the Champion Sparking Plug Company Limited who supplied the illustrations showing spark plug conditions, to Holt Lloyd Limited who supplied the illustrations showing bodywork repair, and to Duckhams Oils who provided lubrication data. Thanks are also due to Regie Renault, particularly Renault Limited (UK) for their assistance with technical information and the provision of certain illustrations. Sykes-Pickavant provided some of the workshop tools. Lastly, thanks are due to all those people at Sparkford who assisted in the production of this manual.

© **Haynes Publishing Group 1991**

A book in the **Haynes Owners Workshop Manual Series**

Printed by J. H. Haynes & Co. Ltd, Sparkford, Nr Yeovil, Somerset BA22 7JJ, England

All rights reserved. No part of this book may be reproduced or transmitted in any form or by any means, electronic or mechanical, including photocopying, recording or by any information storage or retrieval system, without permission in writing from the copyright holder.

ISBN 1 85010 281 3

British Library Cataloguing in Publication Data
Fowler, John. *1930–*
    Renault 18 owner's workshop manual. –
(Owner's Workshop Manuals)
    1. Renault automobile
    I. Title   II. Series
629.28'722   TL215.R4
    ISBN 1-85010-281-3

Whilst every care is taken to ensure that the information in this manual is correct, no liability can be accepted by the authors or publishers for loss, damage or injury caused by any errors in, or omissions from, the information given.

Restoring and Preserving our Motoring Heritage

Few people can have had the luck to realise their dreams to quite the same extent and in such a remarkable fashion as John Haynes, Founder and Chairman of the Haynes Publishing Group.

Since 1965 his unique approach to workshop manual publishing has proved so successful that millions of Haynes Manuals are now sold every year throughout the world, covering literally thousands of different makes and models of cars, vans and motorcycles.

A continuing passion for cars and motoring led to the founding in 1985 of a Charitable Trust dedicated to the restoration and preservation of our motoring heritage. To inaugurate the new Museum, John Haynes donated virtually his entire private collection of 52 cars.

Now with an unrivalled international collection of over 210 veteran, vintage and classic cars and motorcycles, the Haynes Motor Museum in Somerset is well on the way to becoming one of the most interesting Motor Museums in the world.

A 70 seat video cinema, a cafe and an extensive motoring bookshop, together with a specially constructed one kilometre motor circuit, make a visit to the Haynes Motor Museum a truly unforgettable experience.

Every vehicle in the museum is preserved in as near as possible mint condition and each car is run every six months on the motor circuit.

Enjoy the picnic area set amongst the rolling Somerset hills. Peer through the William Morris workshop windows at cars being restored, and browse through the extensive displays of fascinating motoring memorabilia.

From the 1903 Oldsmobile through such classics as an MG Midget to the mighty 'E' type Jaguar, Lamborghini, Ferrari Berlinetta Boxer, and Graham Hill's Lola Cosworth, there is something for everyone, young and old alike, at this Somerset Museum.

Haynes Motor Museum

Situated mid-way between London and Penzance, the Haynes Motor Museum is located just off the A303 at Sparkford, Somerset (home of the Haynes Manual) and is open to the public 7 days a week all year round, except Christmas Day and Boxing Day.

Telephone 01963 440804.

Contents

Spark plug condition and bodywork repair colour section between pages 32 and 33

Renault 18 Saloon

Renault 18 Estate

About this manual

Its aim

The aim of this manual is to help you get the best from your car. It can do so in several ways. It can help you decide what work must be done (even should you choose to get it done by a garage), provide information on routine maintenance and servicing, and give a logical course of action and diagnosis when random faults occur. However, it is hoped that you will use the manual by tackling the work yourself. On simpler jobs it may even be quicker than booking the car into a garage and going there twice to leave and collect it. Perhaps most important, a lot of money can be saved by avoiding the costs the garage must charge to cover its labour and overheads.

The manual has drawings and descriptions to show the function of the various components so that their layout can be understood. Then the tasks are described and photographed in a step-by-step sequence so that even a novice can do the work.

Its arrangement

The manual is divided into thirteen Chapters, each covering a logical sub-division of the vehicle. The Chapters are each divided into Sections, numbered with single figures, eg 5; and the Sections into paragraphs (or sub-sections), with decimal numbers following on from the Section they are in, eg 5.1, 5.2 etc.

It is freely illustrated, especially in those parts where there is a detailed sequence of operations to be carried out. There are two forms of illustration: figures and photographs. The figures are numbered in sequence with decimal numbers, according to their position in the Chapter – Fig. 6.4 is the fourth drawing/illustration in Chapter 6. Photographs carry the same number (either individually or in related groups) as the Section or sub-section to which they relate.

There is an alphabetical index at the back of the manual as well as a contents list at the front. Each Chapter is also preceded by its own individual contents list.

References to the 'left' or 'right' of the vehicle are in the sense of a person in the driver's seat facing forwards.

Unless otherwise stated, nuts and bolts are removed by turning anti-clockwise, and tightened by turning clockwise.

Vehicle manufacturers continually make changes to specifications and recommendations, and these, when notified, are incorporated into our manuals at the earliest opportunity.

Whilst every care is taken to ensure that the information in this manual is correct, no liability can be accepted by the authors or publishers for loss, damage or injury caused by any errors in, or omissions from, the information given.

Introduction to the Renault 18

The Renault 18 was first introduced to the French market in April 1978, and became available in the UK in February 1979, and in the USA in 1981. It combines a spacious and aerodynamically designed bodyshell with thoroughly proven power units and transmissions. Both overhead valve and overhead camshaft engines are fitted to the range and in the UK a 1565 cc Turbo version is available.

Manual and automatic transmissions are available; mounted in-line with, and behind, the engine. The differential unit is built into the transmission and transmits drive through the driveshafts to the front wheels.

Front suspension is a double wishbone independent system,

whilst at the rear a rigid axle with coil springs is employed. Telescopic shock absorbers and anti-roll bars are fitted front and rear.

Rack-and-pinion steering is fitted, and braking is by a double hydraulic circuit system, with a pressure limiting valve in the circuit to the rear wheels to prevent locking of the wheels in an emergency stop situation. On early models discs are fitted at the front and drums at the rear, but on later models discs are fitted all round. A servo unit provides assistance to the braking system.

The body incorporates features developed in the Renault Basic Research Vehicle, and the result is a very strong body structure incorporating many safety features.

Dimensions and weights

UK models
Dimensions
Overall length:

R1340 TL and R1341 TS	172.1 in (4.37 m)
R1340 GTL and R1341 GTS	172.5 in (4.38 m)
R1342 and R1345	172.8 in (4.39 m)
R1350	175.2 in (4.45 m)
R1351	175.6 in (4.46 m)
R1352, R1353, R1355 and R2350	176.4 in (4.48 m)

Overall width:

R1340 TL, R1341 TS and R1350	66.2 in (1.68 m)
R1340 GTL, R1341 GTS, R1342, R1351, R1352, R1353, R1355 and R2350	66.5 in (1.69 m)
Overall height (unladen): All models	55.1 in (1.4 m)
Wheelbase: All models	96.1 in (2.44 m)
Rear track: All models	53.2 in (1.35 m)

Front track:

R1340, R1341, R1342, R1350 and R1351	55.5 in (1.41 m)
R1345, R1352, R1353, R1355 and R2350	55.9 in (1.42 m)
Ground clearance: All models	4.7 in (120 mm)

Turning circle, between walls:

With manual gearbox	36.1 ft (11.00 m)
With automatic transmission	38.4 ft (11.70 m)

Weights
Kerb weight:

R1340	2007 lb (910 kg)
R1341	2139 lb (970 kg)
R1345	2293 lb (1040 kg)
R1350	2194 lb (995 kg)
R1351	2260 lb (1025 kg)
R1355	2459 lb (1115 kg)
R2350	2172 lb (985 kg)

Maximum permissible all-up weight:

R1340	2955 lb (1340 kg)
R1341	3021 lb (1370 kg)
R1345	3175 lb (1440 kg)
R1350	3186 lb (1445 kg)
R1351	3252 lb (1475 kg)
R1355	3440 lb (1560 kg)
R2350	3186 lb (1445 kg)

USA models
Dimensions
Overall length:

R1348 and R1341	178.9 in (4.54 m)
R1358 and R1351	181.6 in (4.61 m)
Overall width: All models	66.6 in (1.69 m)
Overall height (unladen): All models	55.2 in (1.40 m)
Wheelbase: All models	96.1 in (2.44 m)
Front track: All models	55.9 in (1.42 m)
Rear track: All models	52.8 in (1.34 m)
Ground clearance: All models	4.7 in (120 mm)

Turning circle, between kerbs:

With manual gearbox	33.8 ft (10.3 m)
With automatic transmission	36.1 ft (11.00 m)

Weights
Kerb weight:

R1348 and R1341 with manual gearbox	2430 lb (1090 kg)
R1348 and R1341 with automatic transmission	2425 lb (1100 kg)
R1358 and R1351 with manual gearbox	2550 lb (1157 kg)
R1358 and R1351 with automatic transmission	2577 lb (1169 kg)

Maximum permissible all-up weight:

R1348 and R1341 with manual gearbox	3230 lb (1465 kg)
R1348 and R1341 with automatic transmission	3252 lb (1475 kg)
R1358 and R1351 with manual gearbox	3422 lb (1552 kg)
R1358 and R1351 with automatic transmission	3448 lb (1564 kg)

Buying spare parts and vehicle identification numbers

Buying spare parts

Spare parts are available from many sources, for example: Renault dealers, other garages and accessory shops, and motor factors. Our advice regarding spare part sources is as follows:

Officially appointed Renault garages – This is the best source of parts which are peculiar to your car and are otherwise not generally available (eg complete cylinder heads, internal gearbox components, badges, interior trim etc). It is also the only place at which you should have repairs carried out if your car is still under warranty – non-Renault components may invalidate the warranty. To be sure of obtaining the correct parts it will always be necessary to give the storeman your car's vehicle identification number, and if possible, to take the old part along for positive identification. It obviously makes good sense to go straight to the specialists on your car for this type of part for they are best equipped to supply you.

Other garages and accessory shops – These are often very good places to buy materials and components needed for the maintenance of your car (eg spark plugs, bulbs, fanbelts, oils and greases, filler paste etc). They also sell general accessories, usually have convenient opening hours, charge reasonable prices and can often be found not far from home.

Motor factors – Good factors will stock all the more important components which wear out relatively quickly (eg clutch components, pistons, valves, exhaust systems, brake cylinders/pipes/hoses/seals/shoes and pads etc). Motor factors will often provide new or reconditioned components on a part exchange basis – this can save a considerable amount of money.

Vehicle identification numbers

Modifications are a continuous and unpublicised process carried out by the vehicle manufacturers, so accept the advice of the parts storeman when purchasing a component. Spare parts lists and manuals are compiled upon a numerical basis and individual vehicle numbers are essential to the supply of the correct component.

Vehicle identification plates

On UK models a rectangular plate is secured to the heating/ventilating block on the RH side of the engine compartment or to the RH inner wing, and an oval plate is similarly secured (photo).

The vehicle identification plates

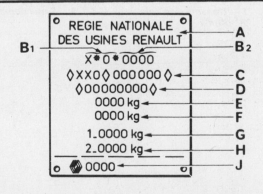

UK vehicle identification plate (rectangular)

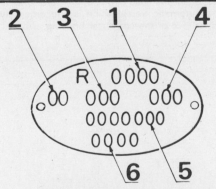

Vehicle identification plate (oval)

A	Maker's name
B1	Country number
B2	Reception number
C	French Ministry of Mines number
D	Chassis number
E	Maximum all-up weight
F	Maximum gross train weight
G	Maximum front axle load
H	Maximum rear axle load
J	Model year

1	Vehicle type
2	Transmission and appearance code
3	Basic equipment code
4	Optional equipment code
5	Fabrication number
6	Model year (not always present)

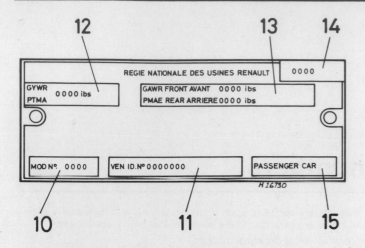

USA/Canada vehicle identification plate (rectangular)

10 Model number
11 Serial number
12 Total maximum weight

13 Maximum weight for front and rear axles
14 Date of manufacture
15 Class of vehicle

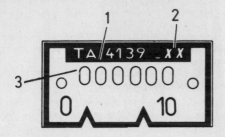

Automatic transmission identification plate

1 Transmission type   3 Serial number
2 Suffix

On USA/Canada models a rectangular plate is secured to the LH door pillar and an oval plate to the RH side of the engine compartment.

The engine identification plate will be found riveted to the cylinder block, the position depending upon the space available. Two types of plate may be employed.

The transmission identification plate will be found in the case of manual gearboxes, under the head of one of the bolts which secure the end cover. Where automatic transmission is concerned, the plate is riveted to the upper face of the converter housing.

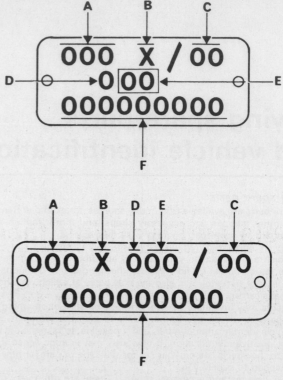

Two types of engine identification plate

A Engine type
B French Ministry of Mines letter
C Equipment code

D Renault identity code
E Engine suffix
F Serial number

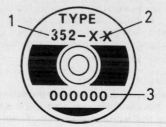

Manual transmission identification plate

1 Transmission type   3 Serial number
2 Suffix

Tools and working facilities

Introduction

A selection of good tools is a fundamental requirement for anyone contemplating the maintenance and repair of a motor vehicle. For the owner who does not possess any, their purchase will prove a considerable expense, offsetting some of the savings made by doing-it-yourself. However, provided that the tools purchased meet the relevant national safety standards and are of good quality, they will last for many years and prove an extremely worthwhile investment.

To help the average owner to decide which tools are needed to carry out the various tasks detailed in this manual, we have compiled three lists of tools under the following headings: *Maintenance and minor repair, Repair and overhaul,* and *Special.* The newcomer to practical mechanics should start off with the *Maintenance and minor repair* tool kit and confine himself to the simpler jobs around the vehicle. Then, as his confidence and experience grow, he can undertake more difficult tasks, buying extra tools as, and when, they are needed. In this way, a *Maintenance and minor repair* tool kit can be built-up into a *Repair and overhaul* tool kit over a considerable period of time without any major cash outlays. The experienced do-it-yourselfer will have a tool kit good enough for most repair and overhaul procedures and will add tools from the *Special* category when he feels the expense is justified by the amount of use these tools will be put to.

It is obviously not possible to cover the subject of tools fully here. For those who wish to learn more about tools and their use there is a book entitled *How to Choose and Use Car Tools* available from the publishers of this manual.

Maintenance and minor repair tool kit

The tools given in this list should be considered as a minimum requirement if routine maintenance, servicing and minor repair operations are to be undertaken. We recommend the purchase of combination spanners (ring one end, open-ended the other); although more expensive than open-ended ones, they do give the advantages of both types of spanner.

Combination spanners - 9, 10, 11, 12, 13, 14 & 17 mm
Adjustable spanner - 9 inch
Engine sump drain plug key
Spark plug spanner (with rubber insert)
Spark plug gap adjustment tool
Set of feeler gauges
Brake bleed nipple spanner
Screwdriver - 4 in long x $\frac{1}{4}$ in dia (flat blade)
Screwdriver - 4 in long x $\frac{1}{4}$ in dia (cross blade)
Combination pliers - 6 inch
Hacksaw (junior)
Tyre pump
Tyre pressure gauge
Oil can
Fine emery cloth (1 sheet)
Wire brush (small)
Funnel (medium size)

Repair and overhaul tool kit

These tools are virtually essential for anyone undertaking any major repairs to a motor vehicle, and are additional to those given in the *Maintenance and minor repair* list. Included in this list is a comprehensive set of sockets. Although these are expensive they will be found invaluable as they are so versatile - particularly if various drives are included in the set. We recommend the $\frac{1}{2}$ in square-drive type, as this can be used with most proprietary torque spanners. If you cannot afford a socket set, even bought piecemeal, then inexpensive tubular box wrenches are a useful alternative.

The tools in this list will occasionally need to be supplemented by tools from the *Special* list.

Sockets (or box spanners) to cover range in previous list
Reversible ratchet drive (for use with sockets)
Extension piece, 10 inch (for use with sockets)
Universal joint (for use with sockets)
Torque wrench (for use with sockets)
Mole wrench - 8 inch
Ball pein hammer
Soft-faced hammer, plastic or rubber
Screwdriver - 6 in long x $\frac{5}{16}$ in dia (flat blade)
Screwdriver - 2 in long x $\frac{5}{16}$ in square (flat blade)
Screwdriver - 1$\frac{1}{2}$ in long x $\frac{1}{4}$ in dia (cross blade)
Screwdriver - 3 in long x $\frac{1}{8}$ in dia (electricians)
Pliers - electricians side cutters
Pliers - needle nosed
Pliers - circlip (internal and external)
Cold chisel - $\frac{1}{2}$ inch
Scriber
Scraper
Centre punch
Pin punch
Hacksaw
Valve grinding tool
Steel rule/straight-edge
Allen keys
Selection of files
Wire brush (large)
Axle-stands
Jack (strong scissor or hydraulic type)

Special tools

The tools in this list are those which are not used regularly, are expensive to buy, or which need to be used in accordance with their manufacturers' instructions. Unless relatively difficult mechanical jobs are undertaken frequently, it will not be economic to buy many of these tools. Where this is the case, you could consider clubbing together with friends (or joining a motorists' club) to make a joint purchase, or borrowing the tools against a deposit from a local garage or tool hire specialist.

The following list contains only those tools and instruments freely available to the public, and not those special tools produced by the vehicle manufacturer specifically for its dealer network. You will find occasional references to these manufacturers' special tools in the text of this manual. Generally, an alternative method of doing the job without the vehicle manufacturers' special tool is given. However, sometimes, there is no alternative to using them. Where this is the case and the relevant tool cannot be bought or borrowed you will have to entrust the work to a franchised garage.

Valve spring compressor
Piston ring compressor
Balljoint separator
Universal hub/bearing puller
Impact screwdriver
Micrometer and/or vernier gauge
Dial gauge
Stroboscopic timing light
Dwell angle meter/tachometer
Universal electrical multi-meter
Cylinder compression gauge
Lifting tackle
Trolley jack
Light with extension lead

Buying tools

For practically all tools, a tool dealer is the best source since he will have a very comprehensive range compared with the average garage or accessory shop. Having said that, accessory shops often offer excellent quality tools at discount prices, so it pays to shop around.

There are plenty of good tools around at reasonable prices, but always aim to purchase items which meet the relevant national safety standards. If in doubt, ask the proprietor or manager of the shop for advice before making a purchase.

Care and maintenance of tools

Having purchased a reasonable tool kit, it is necessary to keep the tools in a clean serviceable condition. After use, always wipe off any dirt, grease and metal particles using a clean, dry cloth, before putting the tools away. Never leave them lying around after they have been used. A simple tool rack on the garage or workshop wall, for items such as screwdrivers and pliers is a good idea. Store all normal spanners and sockets in a metal box. Any measuring instruments, gauges, meters, etc, must be carefully stored where they cannot be damaged or become rusty.

Take a little care when tools are used. Hammer heads inevitably become marked and screwdrivers lose the keen edge on their blades from time to time. A little timely attention with emery cloth or a file will soon restore items like this to a good serviceable finish.

Working facilities

Not to be forgotten when discussing tools, is the workshop itself. If anything more than routine maintenance is to be carried out, some form of suitable working area becomes essential.

It is appreciated that many an owner mechanic is forced by circumstances to remove an engine or similar item, without the benefit of a garage or workshop. Having done this, any repairs should always be done under the cover of a roof.

Wherever possible, any dismantling should be done on a clean flat workbench or table at a suitable working height.

Any workbench needs a vice: one with a jaw opening of 4 in (100 mm) is suitable for most jobs. As mentioned previously, some clean dry storage space is also required for tools, as well as the lubricants, cleaning fluids, touch-up paints and so on which become necessary.

Another item which may be required, and which has a much more general usage, is an electric drill with a chuck capacity of at least $\frac{5}{16}$ in (8 mm). This, together with a good range of twist drills, is virtually essential for fitting accessories such as wing mirrors and reversing lights.

Last, but not least, always keep a supply of old newspapers and clean, lint-free rags available, and try to keep any working area as clean as possible.

Spanner jaw gap comparison table

Jaw gap (in)	Spanner size
0.250	$\frac{1}{4}$ in AF
0.276	7 mm
0.313	$\frac{5}{16}$ in AF
0.315	8 mm
0.344	$\frac{11}{32}$ in AF; $\frac{1}{8}$ in Whitworth
0.354	9 mm
0.375	$\frac{3}{8}$ in AF
0.394	10 mm
0.433	11 mm
0.438	$\frac{7}{16}$ in AF
0.445	$\frac{3}{16}$ in Whitworth; $\frac{1}{4}$ in BSF
0.472	12 mm
0.500	$\frac{1}{2}$ in AF
0.512	13 mm
0.525	$\frac{1}{4}$ in Whitworth; $\frac{5}{16}$ in BSF
0.551	14 mm
0.563	$\frac{9}{16}$ in AF
0.591	15 mm
0.600	$\frac{5}{16}$ in Whitworth; $\frac{3}{8}$ in BSF
0.625	$\frac{5}{8}$ in AF
0.630	16 mm
0.669	17 mm
0.686	$\frac{11}{16}$ in AF
0.709	18 mm
0.710	$\frac{3}{8}$ in Whitworth, $\frac{7}{16}$ in BSF
0.748	19 mm
0.750	$\frac{3}{4}$ in AF
0.813	$\frac{13}{16}$ in AF
0.820	$\frac{7}{16}$ in Whitworth; $\frac{1}{2}$ in BSF
0.866	22 mm
0.875	$\frac{7}{8}$ in AF
0.920	$\frac{1}{2}$ in Whitworth; $\frac{9}{16}$ in BSF
0.938	$\frac{15}{16}$ in AF
0.945	24 mm
1.000	1 in AF
1.010	$\frac{9}{16}$ in Whitworth; $\frac{5}{8}$ in BSF
1.024	26 mm
1.063	$1\frac{1}{16}$ in AF; 27 mm
1.100	$\frac{5}{8}$ in Whitworth; $\frac{11}{16}$ in BSF
1.125	$1\frac{1}{8}$ in AF
1.181	30 mm
1.200	$\frac{11}{16}$ in Whitworth; $\frac{3}{4}$ in BSF
1.250	$1\frac{1}{4}$ in AF
1.260	32 mm
1.300	$\frac{3}{4}$ in Whitworth; $\frac{7}{8}$ in BSF
1.313	$1\frac{5}{16}$ in AF
1.390	$\frac{13}{16}$ in Whitworth; $\frac{15}{16}$ in BSF
1.417	36 mm
1.438	$1\frac{7}{16}$ in AF
1.480	$\frac{7}{8}$ in Whitworth; 1 in BSF
1.500	$1\frac{1}{2}$ in AF
1.575	40 mm; $\frac{15}{16}$ in Whitworth
1.614	41 mm
1.625	$1\frac{5}{8}$ in AF
1.670	1 in Whitworth; $1\frac{1}{8}$ in BSF
1.688	$1\frac{11}{16}$ in AF
1.811	46 mm
1.813	$1\frac{13}{16}$ in AF
1.860	$1\frac{1}{8}$ in Whitworth; $1\frac{1}{4}$ in BSF
1.875	$1\frac{7}{8}$ in AF
1.969	50 mm
2.000	2 in AF
2.050	$1\frac{1}{4}$ in Whitworth; $1\frac{3}{8}$ in BSF
2.165	55 mm
2.362	60 mm

Jacking and towing

Jacking

The jack supplied with the vehicle is not designed for service or repair operations, but purely for changing a wheel in the event of a puncture. A strong pillar or trolley jack should be employed for maintenance and repair tasks requiring the vehicle to be raised. The jacking point is of particular importance.

At the front use a block to take the load under the side-members behind the undertray. Do not bear on the exhaust pipe, and before lifting ensure that the handbrake is applied and the rear wheels are chocked.

At the side, take the load under the sill in the centre of the front door, using a wooden block to spread the load. Ensure that the wheels are chocked.

At the rear, position the jack head under the centre of the rear axle. Ensure that the front wheels are chocked.

When using stands, do not omit to use a reinforcement block between the vehicle and the stand, such as those supplied with the vehicle jack.

Do not lift the vehicle by the towing attachment points.

When jacking the vehicle in an emergency situation, such as changing a wheel at the roadside, the jack should be found stowed in the boot on the LH side on Saloon cars, and in the rear of the Estate car behind the spare wheel. The jack has a clip on the top lifting pad, and when in use care must be taken that the clip is properly inserted into the jacking slot on the car body, from the inside with the pad beneath the sill and the brace adapter facing outwards. The body jacking points are to be found one behind each front wheel, and one in front of each rear wheel (photos).

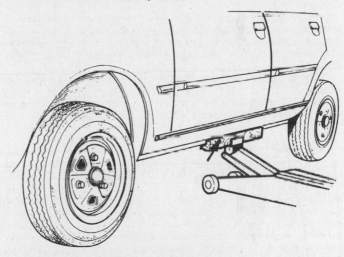

Lifting the car at the side. A specially shaped block of wood (arrowed) is best used here

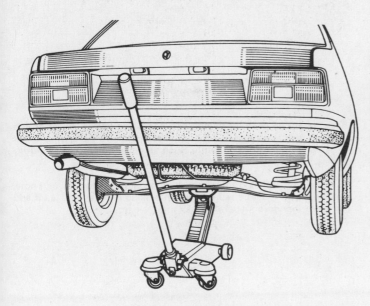

Lifting the car at the front. A block of wood is used to spread the load

Lifting the car at the rear

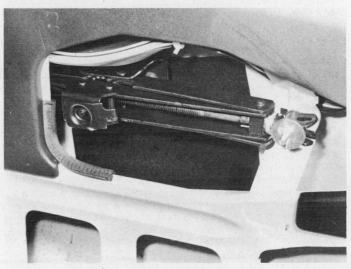

The location of the vehicle jack (Estate)

The jack in use

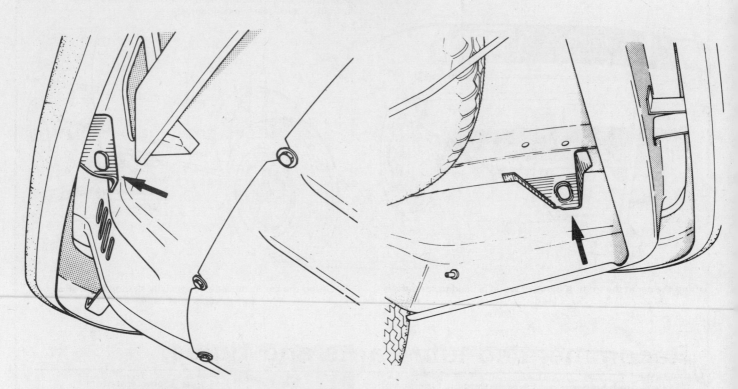

Front towing eye (arrowed)   Rear towing eye (arrowed)

Towing

If the vehicle should need to be towed, ensure that the correct procedure is followed.

Where the vehicle is fitted with automatic transmission, refer first to Chapter 7, Section 2.

Where possible, use a tow rope (in preference to chains) and ensure that it is of sufficient length. Towing hooks are provided at the front and rear of the vehicle as shown in the illustrations, and only these should be used for towing or when being towed. Never attach the tow rope to suspension or body parts. Do not use the towing hooks for operations of a heavy nature, such as pulling other vehicles from ditches.

Ensure when being towed that the steering is unlocked, and note that more effort will be required when braking if the engine is off and the servo therefore inoperative.

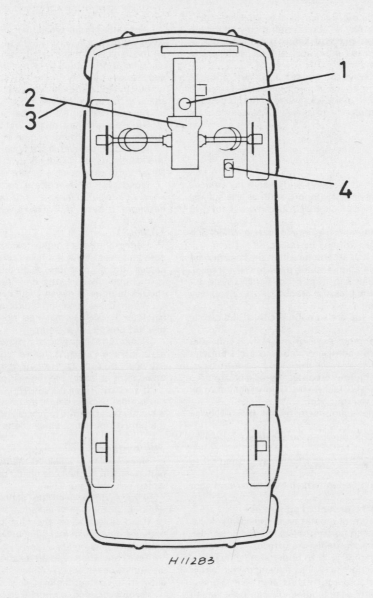

H11283

Recommended lubricants and fluids

Component or system	Lubricant type/specification	Duckhams recommendation
1 Engine	Multigrade engine oil, viscosity SAE 15W/40, 20W/40 or 20W/50, to API SE	Duckhams Hypergrade
2 Manual transmission	Hypoid gear oil, viscosity SAE 80EP to API GL5	Duckhams Hypoid 80S
3 Automatic transmission	Dexron type ATF	Duckhams Uni-Matic or D-Matic
4 Brake fluid reservoir	Hydraulic fluid to SAE J1703	Duckhams Universal Brake and Clutch Fluid
Power steering (when fitted)	Dexron type ATF	Duckhams Uni-Matic or D-Matic

Safety first!

Professional motor mechanics are trained in safe working procedures. However enthusiastic you may be about getting on with the job in hand, do take the time to ensure that your safety is not put at risk. A moment's lack of attention can result in an accident, as can failure to observe certain elementary precautions.

There will always be new ways of having accidents, and the following points do not pretend to be a comprehensive list of all dangers; they are intended rather to make you aware of the risks and to encourage a safety-conscious approach to all work you carry out on your vehicle.

Essential DOs and DON'Ts

DON'T rely on a single jack when working underneath the vehicle. Always use reliable additional means of support, such as axle stands, securely placed under a part of the vehicle that you know will not give way.

DON'T attempt to loosen or tighten high-torque nuts (e.g. wheel hub nuts) while the vehicle is on a jack; it may be pulled off.

DON'T start the engine without first ascertaining that the transmission is in neutral (or 'Park' where applicable) and the parking brake applied.

DON'T suddenly remove the filler cap from a hot cooling system – cover it with a cloth and release the pressure gradually first, or you may get scalded by escaping coolant.

DON'T attempt to drain oil until you are sure it has cooled sufficiently to avoid scalding you.

DON'T grasp any part of the engine, exhaust or catalytic converter without first ascertaining that it is sufficiently cool to avoid burning you.

DON'T allow brake fluid or antifreeze to contact vehicle paintwork.

DON'T syphon toxic liquids such as fuel, brake fluid or antifreeze by mouth, or allow them to remain on your skin.

DON'T inhale dust – it may be injurious to health (see *Asbestos* below).

DON'T allow any spilt oil or grease to remain on the floor – wipe it up straight away, before someone slips on it.

DON'T use ill-fitting spanners or other tools which may slip and cause injury.

DON'T attempt to lift a heavy component which may be beyond your capability – get assistance.

DON'T rush to finish a job, or take unverified short cuts.

DON'T allow children or animals in or around an unattended vehicle.

DO wear eye protection when using power tools such as drill, sander, bench grinder etc, and when working under the vehicle.

DO use a barrier cream on your hands prior to undertaking dirty jobs – it will protect your skin from infection as well as making the dirt easier to remove afterwards; but make sure your hands aren't left slippery. Note that long-term contact with used engine oil can be a health hazard.

DO keep loose clothing (cuffs, tie etc) and long hair well out of the way of moving mechanical parts.

DO remove rings, wristwatch etc, before working on the vehicle – especially the electrical system.

DO ensure that any lifting tackle used has a safe working load rating adequate for the job.

DO keep your work area tidy – it is only too easy to fall over articles left lying around.

DO get someone to check periodically that all is well, when working alone on the vehicle.

DO carry out work in a logical sequence and check that everything is correctly assembled and tightened afterwards.

DO remember that your vehicle's safety affects that of yourself and others. If in doubt on any point, get specialist advice.

IF, in spite of following these precautions, you are unfortunate enough to injure yourself, seek medical attention as soon as possible.

Asbestos

Certain friction, insulating, sealing, and other products – such as brake linings, brake bands, clutch linings, torque converters, gaskets, etc – contain asbestos. *Extreme care must be taken to avoid inhalation of dust from such products since it is hazardous to health.* If in doubt, assume that they *do* contain asbestos.

Fire

Remember at all times that petrol (gasoline) is highly flammable. Never smoke, or have any kind of naked flame around, when working on the vehicle. But the risk does not end there – a spark caused by an electrical short-circuit, by two metal surfaces contacting each other, by careless use of tools, or even by static electricity built up in your body under certain conditions, can ignite petrol vapour, which in a confined space is highly explosive.

Always disconnect the battery earth (ground) terminal before working on any part of the fuel or electrical system, and never risk spilling fuel on to a hot engine or exhaust.

It is recommended that a fire extinguisher of a type suitable for fuel and electrical fires is kept handy in the garage or workplace at all times. Never try to extinguish a fuel or electrical fire with water.

Note: *Any reference to a 'torch' appearing in this manual should always be taken to mean a hand-held battery-operated electric lamp or flashlight. It does NOT mean a welding/gas torch or blowlamp.*

Fumes

Certain fumes are highly toxic and can quickly cause unconsciousness and even death if inhaled to any extent. Petrol (gasoline) vapour comes into this category, as do the vapours from certain solvents such as trichloroethylene. Any draining or pouring of such volatile fluids should be done in a well ventilated area.

When using cleaning fluids and solvents, read the instructions carefully. Never use materials from unmarked containers – they may give off poisonous vapours.

Never run the engine of a motor vehicle in an enclosed space such as a garage. Exhaust fumes contain carbon monoxide which is extremely poisonous; if you need to run the engine, always do so in the open air or at least have the rear of the vehicle outside the workplace.

If you are fortunate enough to have the use of an inspection pit, never drain or pour petrol, and never run the engine, while the vehicle is standing over it; the fumes, being heavier than air, will concentrate in the pit with possibly lethal results.

The battery

Never cause a spark, or allow a naked light, near the vehicle's battery. It will normally be giving off a certain amount of hydrogen gas, which is highly explosive.

Always disconnect the battery earth (ground) terminal before working on the fuel or electrical systems.

If possible, loosen the filler plugs or cover when charging the battery from an external source. Do not charge at an excessive rate or the battery may burst.

Take care when topping up and when carrying the battery. The acid electrolyte, even when diluted, is very corrosive and should not be allowed to contact the eyes or skin.

If you ever need to prepare electrolyte yourself, always add the acid slowly to the water, and never the other way round. Protect against splashes by wearing rubber gloves and goggles.

When jump starting a car using a booster battery, for negative earth (ground) vehicles, connect the jump leads in the following sequence: First connect one jump lead between the positive (+) terminals of the two batteries. Then connect the other jump lead first to the negative (–) terminal of the booster battery, and then to a good earthing (ground) point on the vehicle to be started, at least 18 in (45 cm) from the battery if possible. Ensure that hands and jump leads are clear of any moving parts, and that the two vehicles do not touch. Disconnect the leads in the reverse order.

Mains electricity and electrical equipment

When using an electric power tool, inspection light etc, always ensure that the appliance is correctly connected to its plug and that, where necessary, it is properly earthed (grounded). Do not use such appliances in damp conditions and, again, beware of creating a spark or applying excessive heat in the vicinity of fuel or fuel vapour. Also ensure that the appliances meet the relevant national safety standards.

Ignition HT voltage

A severe electric shock can result from touching certain parts of the ignition system, such as the HT leads, when the engine is running or being cranked, particularly if components are damp or the insulation is defective. Where an electronic ignition system is fitted, the HT voltage is much higher and could prove fatal.

Routine maintenance

Maintenance is essential for ensuring safety and desirable for the purpose of getting the best in terms of performance and economy from your car. Over the years the need for periodic lubrication – oiling, greasing and so on – has been drastically reduced if not totally eliminated. This has unfortunately tended to lead some owners to think that because no such action is required, components either no longer exist, or will last forever. This is a serious delusion. It follows therefore that the largest initial element of maintenance is visual examination. This may lead to repairs or renewals.

The summary below gives a schedule of routine maintenance operations. More detailed information on the respective items is given in the Chapter concerned. Before starting on any maintenance procedures, make a list and obtain any items or parts that may be required. Make sure you have the necessary tools to complete the servicing requirements. Where the vehicle has to be raised clear of the ground pay particular attention to safety and ensure that chassis stands and/or blocks supplement the jack. Do not rely on the jack supplied with the car – it was designed purely to raise the car for changing a wheel in the event of a puncture.

Every 500 miles (800 km) or weekly – whichever comes first

Check the engine oil level, and top up if necessary (photos)

The engine oil level dipstick

Topping up the engine oil

Topping up brake fluid level

Checking a tyre pressure

Topping up coolant level

Sump drain plug

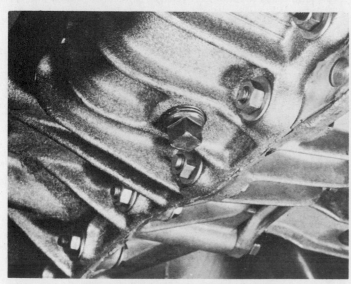

The drain plug on the manual gearbox

Filling the manual gearbox – viewed from underside. Oil level should be up to the bottom of the hole

Check the fluid level in the automatic transmission, if fitted (see Chapter 7)

Check the windscreen washer reservoir fluid level (and the tailgate washer reservoir on estate models) and top up if necessary, adding a screen wash additive such as Turtle Wax High Tech Screen Wash

Check the battery electrolyte level (see Chapter 10)

Check the brake fluid level (see Chapter 9) (photo)

Check the tyre pressures, and examine all tyres for defects (photo)

Check the coolant level (see Chapter 2) (photo)

Check the operation of the lights, screen wipers, washers and horn

Every 5000 miles (7500 km) – or six monthly, whichever comes first

Drain the engine oil (photo) and refill with new oil

Check automatic transmission fluid level and top up if necessary

Every 10 000 miles (15 000 km) or yearly, whichever comes first

Change engine oil filter and oil

Check steering gear oil level and top up if necessary

Check brake disc pads for wear and renew if necessary

Check underbody for damage and corrosion

Check and adjust all drivebelt tensions

Check and adjust contact points dwell angle and ignition timing (on conventional ignition models only)

Check and adjust or renew the spark plugs (on conventional ignition models only)

Adjust the idling speed and mixture

Check and adjust the headlight beam alignment

Check the automatic transmission fluid level and top up if necessary

Every 20 000 miles (30 000 km) or two years, whichever comes first

Visually check all mechanical units for wear, oil leakage and function

Check and adjust or renew contact points (on conventional ignition models only)

Check and adjust or renew the spark plugs (on electronic ignition models only)

Renew the fuel filter (as applicable)

Renew the air filter

Every 30 000 miles (50 000 km) or three years, whichever comes first

Drain the automatic transmission fluid and refill with new fluid (models 1985-on)

Every 40 000 miles (60 000 km) or four years, whichever comes first

Drain the manual gearbox oil and refill with new oil (photos)

Drain the automatic transmission fluid and refill with new fluid (models 1985-on)

Check rear brake lining wear on models with drum brakes and renew if necessary

Adjust the handbrake

Adjust the clutch cable (as applicable)

Check suspension and steering components for wear, damage and security

Drain, flush and refill the cooling system

Every 80 000 miles (12 000 km)

Renew the timing belt on OHC engine models

Fault diagnosis

Introduction

The car owner who does his or her own maintenance according to the recommended schedules should not have to use this section of the manual very often. Modern component reliability is such that, provided those items subject to wear or deterioration are inspected or renewed at the specified intervals, sudden failure is comparatively rare. Faults do not usually just happen as a result of sudden failure, but develop over a period of time. Major mechanical failures in particular are usually preceded by characteristic symptoms over hundreds or even thousands of miles. Those components which do occasionally fail without warning are often small and easily carried in the car.

With any fault finding, the first step is to decide where to begin investigations. Sometimes this is obvious, but on other occasions a little detective work will be necessary. The owner who makes half a dozen haphazard adjustments or replacements may be successful in curing a fault (or its symptoms), but he will be none the wiser if the fault recurs and he may well have spent more time and money than was necessary. A calm and logical approach will be found to be more satisfactory in the long run. Always take into account any warning signs or abnormalities that may have been noticed in the period preceding the fault — power loss, high or low gauge readings, unusual noises or smells, etc — and remember that failure of components such as fuses or spark plugs may only be pointers to some underlying fault.

The pages which follow here are intended to help in cases of failure to start or breakdown on the road. There is also a Fault Diagnosis Section at the end of each Chapter which should be consulted if the preliminary checks prove unfruitful. Whatever the fault, certain basic principles apply. These are as follows:

Verify the fault. This is simply a matter of being sure that you know what the symptoms are before starting work. This is particularly important if you are investigating a fault for someone else who may not have described it very accurately.

Don't overlook the obvious. For example, if the car won't start, is there petrol in the tank? (Don't take anyone else's word on this particular point, and don't trust the fuel gauge either!) If an electrical fault is indicated, look for loose or broken wires before digging out the test gear.

Cure the disease, not the symptom. Substituting a flat battery with a fully charged one will get you off the hard shoulder, but if the underlying cause is not attended to, the new battery will go the same way. Similarly, changing oil-fouled spark plugs for a new set will get you moving again, but remember that the reason for the fouling (if it wasn't simply an incorrect grade of plug) will have to be established and corrected.

Don't take anything for granted. Particularly, don't forget that a 'new' component may itself be defective (especially if it's been rattling round in the boot for months), and don't leave components out of a fault diagnosis sequence just because they are new or recently fitted. When you do finally diagnose a difficult fault, you'll probably realise that all the evidence was there from the start.

Electrical faults

Electrical faults can be more puzzling than straightforward mechanical failures, but they are no less susceptible to logical analysis if the basic principles of operation are understood. Car electrical wiring exists in extremely unfavourable conditions — heat, vibration and chemical attack — and the first things to look for are loose or corroded connections and broken or chafed wires, especially where the wires

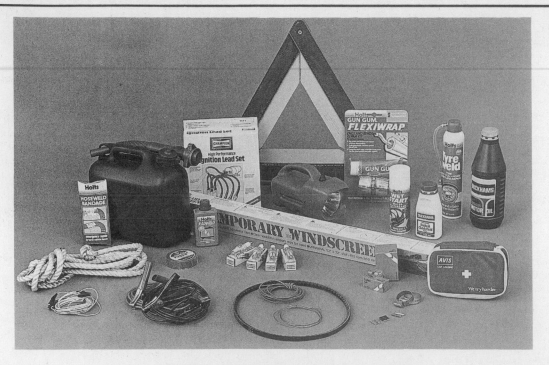

Carrying a few spares may save you a long walk

pass through holes in the bodywork or are subject to vibration.

All metal-bodied cars in current production have one terminal of the battery 'earthed', ie connected to the car bodywork, and in nearly all modern cars it is the negative (−) terminal. The various electrical components – motors, bulb holders etc – are also connected to earth, either by means of a lead or directly by their mountings. Electric current flows through the component and then back to the battery via the car bodywork. If the component mounting is loose or corroded, or if a good path back to the battery is not available, the circuit will be incomplete and malfunction will result. The engine and/or gearbox are also earthed by means of flexible metal straps to the body or subframe; if these straps are loose or missing, starter motor, generator and ignition trouble may result.

Assuming the earth return to be satisfactory, electrical faults will be due either to component malfunction or to defects in the current supply. Individual components are dealt with in Chapter 10. If supply wires are broken or cracked internally this results in an open-circuit, and the easiest way to check for this is to bypass the suspect wire temporarily with a length of wire having a crocodile clip or suitable connector at each end. Alternatively, a 12V test lamp can be used to verify the presence of supply voltage at various points along the wire and the break can be thus isolated.

If a bare portion of a live wire touches the car bodywork or other earthed metal part, the electricity will take the low-resistance path thus formed back to the battery: this is known as a short-circuit. Hopefully a short-circuit will blow a fuse, but otherwise it may cause burning of the insulation (and possibly further short-circuits) or even a fire. This is why it is inadvisable to bypass persistently blowing fuses with silver foil or wire.

Spares and tool kit

Most cars are only supplied with sufficient tools for wheel changing; the *Maintenance and minor repair* tool kit detailed in *Tools and working facilities*, with the addition of a hammer, is probably sufficient for those repairs that most motorists would consider attempting at the roadside. In addition a few items which can be fitted without too much trouble in the event of a breakdown should be carried. Experience and available space will modify the list below, but the following may save having to call on professional assistance:

Spark plugs, clean and correctly gapped
HT lead and plug cap – long enough to reach the plug furthest from the distributor
Distributor rotor, condenser and contact breaker points
Drivebelt(s) – emergency type may suffice
Spare fuses
Set of principal light bulbs
Hose clips
Tin of radiator sealer and hose bandage
Exhaust bandage
Roll of insulating tape
Length of soft iron wire
Length of electrical flex
Torch or inspection lamp (can double as test lamp)
Battery jump leads
Tow-rope
Ignition water dispersant aerosol
Litre of engine oil
Sealed can of hydraulic fluid
Emergency windscreen
Tyre valve core

If spare fuel is carried, a can designed for the purpose should be used to minimise risks of leakage and collision damage. A first aid kit and a warning triangle, whilst not at present compulsory in the UK, are obviously sensible items to carry in addition to the above.

When touring abroad it may be advisable to carry additional spares which, even if you cannot fit them yourself, could save having to wait while parts are obtained. The items below may be worth considering:

Cylinder head gasket
Alternator brushes
Throttle and clutch cables

One of the motoring organisations will be able to advise on availability of fuel etc in foreign countries.

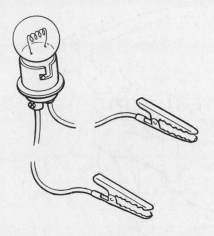

A simple test lamp is useful for investigating electrical faults

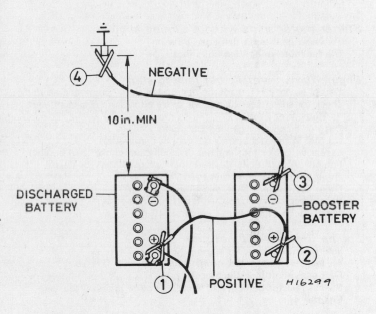

Jump start lead connections for negative earth vehicles – connect leads in order shown

---

Engine will not start

Engine fails to turn when starter operated
Flat battery (recharge, use jump leads, or push start)
Battery terminals loose or corroded
Battery earth to body defective
Engine earth strap loose or broken
Starter motor (or solenoid) wiring loose or broken
Automatic transmission selector in wrong position, or inhibitor switch faulty
Ignition/starter switch faulty
Major mechanical failure (seizure) or long disuse (piston rings rusted to bores)
Starter or solenoid internal fault (see Chapter 10)

Starter motor turns engine slowly
Partially discharged battery (recharge, use jump leads, or push start)
Battery terminals loose or corroded
Battery earth to body defective
Engine earth strap loose
Starter motor (or solenoid) wiring loose
Starter motor internal fault (see Chapter 10)

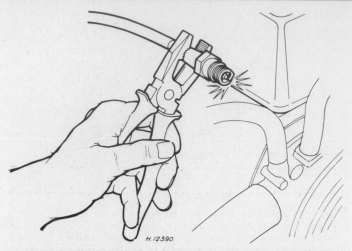

**Crank engine and check for a spark. Note use of insulated pliers –
dry cloth or a rubber glove will suffice**

Starter motor spins without turning engine
Flywheel gear teeth damaged or worn
Starter motor mounting bolts loose

Engine turns normally but fails to start
Damp or dirty HT leads or distributor cap – crank engine and
check for spark (see illustration), or try a moisture dispersant such
as Holts Wet Start
Dirty or incorrectly gapped CB points
No fuel in tank (check for delivery at carburettor)
Excessive choke (hot engine) or insufficient choke (cold engine)
Fouled or incorrectly gapped spark plugs (renew or regap)
Other ignition system fault (see Chapter 4)
Other fuel system fault (see Chapter 3)
Poor compression (see Chapter 1)
Major mechanical failure (eg camshaft drive)

Engine fires but will not run
Insufficient choke (cold engine)
Air leaks at carburettor or inlet manifold
Fuel starvation (see Chapter 3)
Ballast resistor defective (if fitted), or other ignition fault (see
Chapter 4)

Engine cuts out and will not restart

Engine cuts out suddenly – ignition fault
Loose or disconnected LT wires
Wet HT leads or distributor cap (after transversing water splash)
Coil or condenser failure (check for spark)
Other ignition fault (see Chapter 4)

Engine misfires before cutting out – fuel fault
Fuel tank empty
Fuel pump defective or filter blocked (check for delivery)
Fuel tank filler vent blocked (suction will be evident on releasing
cap)
Carburettor needle valve sticking
Other fuel system fault (see Chapter 3)

Engine cuts out – other causes
Serious overheating
Major mechanical failure (eg camshaft drive)

Engine overheats

Ignition (no-charge) warning light illuminated
Slack or broken drivebelt – retension or renew (Chapter 2)

Ignition warning light not illuminated
Coolant loss due to internal or external leakage (see Chapter 2)
Thermostat defective
Low oil level
Brakes binding
Radiator clogged externally or internally
Electric cooling fan not operating correctly (if fitted)
Engine waterways clogged
Ignition timing incorrect or automatic advance malfunctioning
Mixture too weak

Note: *Do not add cold water to an overheated engine or damage may
result*

Low engine oil pressure

Gauge reads low or warning light illuminated with engine running
Oil level low or incorrect grade
Defective gauge or sender unit
Wire to sender unit earthed
Engine overheating
Oil filter clogged or bypass valve defective
Oil pressure relief valve defective
Oil pick-up strainer clogged
Oil pump worn or mountings loose
Worn main or big-end bearings

Note: *Low oil pressure in a high-mileage engine at tickover is not
ncessarily a cause for concern. Sudden pressure loss at speed is far
more significant. In any event, check the gauge or warning light sender
before condemning the engine.*

Engine noises

Pre-ignition (pinking) on acceleration
Incorrect grade of fuel
Ignition timing incorrect
Distributor faulty or worn
Worn or maladjusted carburettor
Excessive carbon build-up in engine

Whistling or wheezing noises
Leaking vacuum hose
Leaking carburettor or manifold gasket
Blowing head gasket

Tapping or rattling
Incorrect valve clearances
Worn valve gear
Worn timing chain
Broken piston ring (ticking noise)

Knocking or thumping
Unintentional mechanical contact (eg fan blades)
Worn fanbelt
Peripheral component fault (generator, water pump etc)
Worn big-end bearings (regular heavy knocking, perhaps less
under load)
Worn main bearings (rumbling and knocking, perhaps worsening
under load)
Piston slap (most noticeable when cold)

Chapter 1 Engine

For modifications, and information applicable to later models, see Supplement at end of manual

Contents

Specifications

Engine type

Vehicle type:	Engine type
R1340 TL and GTL (manual)	847-A7-20
R1341 TS and GTS (manual)	841-C7-25
R1341 (automatic)	841-D7-26

Engine, general

	Type 847	Type 841
Number of cylinders	4	4
Bore	76 mm	79 mm
Stroke	77 mm	84 mm
Cubic capacity	1397 cm³	1647 cm³
Compression ratio	9.25	9.30
Power	64 bhp (DIN) at 5500 rpm	79 bhp (DIN) at 5500 rpm
Torque	77 lbf ft at 3500 rpm	90 lbf ft at 3000 rpm
Firing order	1-3-4-2	1-3-4-2
Location of No 1 cylinder	Flywheel end	Flywheel end
Valve gear	OHV	OHV

Valve clearances	Type 847	Type 841
Hot:		
Inlet	0.007 in (0.18 mm)	0.008 in (0.20 mm)
Exhaust	0.010 in (0.25 mm)	0.010 in (0.25 mm)
Cold:		
Inlet	0.006 in (0.15 mm)	0.008 in (0.20 mm)
Exhaust	0.008 in (0.20 mm)	0.010 in (0.25 mm)

Cylinder head		
Height, standard:		
1st model	2.827 in (71.8 mm)	3.189 in (81.0 mm)
2nd model	2.842 in (72.2 mm)	–
Height after machining:		
1st model	2.807 in (71.3 mm)	3.170 in (80.5 mm)
2nd model	2.822 in (71.7 mm)	–
Gasket face maximum bow	0.002 in (0.05 mm)	0.002 in (0.05 mm)

Valve assemblies		
Valve seat included angle:		
Inlet	120°	90°
Exhaust	90°	90°
Valve seat width:		
Inlet	0.043 to 0.055 in (1.1 to 1.4 mm)	0.051 to 0.063 in (1.3 to 1.6 mm)
Exhaust	0.043 to 0.067 in (1.4 to 1.7 mm)	0.067 to 0.079 in (1.7 to 2.0 mm)
Valve seat outside diameter:		
Inlet	1.401 in (35.6 mm)	1.460 in (37.10 mm)
Exhaust	1.204 in (30.6 mm)	1.303 in (33.10 mm)
Valve stem diameter:		
Inlet	0.276 in (7.0 mm)	0.315 in (8.0 mm)
Exhaust	0.276 in (7.0 mm)	0.315 in (8.0 mm)
Valve head diameter:		
Inlet	1.346 in (34.2 mm)	1.409 in (35.8 mm)
Exhaust	1.141 in (29.0 mm)	1.240 in (31.50 mm)
Valve guide bore	0.276 in (7.0 mm)	0.315 in (8.0 mm)
Valve guide outside diameter:		
Standard	0.433 in (11.0 mm)	0.512 in (13.0 mm)
Oversize 1	0.437 in (11.10 mm)	0.516 in (13.1 mm)
Oversize 2	0.443 in (11.25 mm)	0.522 in (13.25 mm)
Valve spring free length	1.653 in (42.0 mm)	1.905 in (48.4 mm)

Camshaft		
Number of bearings	4	5
Shaft endfloat	0.0024 to 0.0043 in (0.06 to 0.11 mm)	0.002 to 0.0047 in (0.05 to 0.12 mm)
Valve timing:		
Inlet opens	22° BTDC	22° BTDC
Inlet closes	62° ABDC	70° ABDC
Exhaust opens	65° BBDC	70° BBDC
Exhaust closes	25° ATDC	22° ATDC

Tappets		
External diameter:		
Standard	0.748 in (19.0 mm)	0.472 in (12.0 mm)
Oversize	0.756 in (19.2 mm)	0.480 in (12.2 mm)

Pushrods		
Length, 1st model	6.764 in (171.8 mm)	3.464 in (88.0 mm)
Length, 2nd model	6.921 in (175.8 mm)	–

Cylinder block liners		
Bore	2.992 in (76.0 mm)	3.110 in (79.0 mm))
Base location diameter	3.173 in (80.6 mm)	3.307 in (84.0 mm)
Liner protrusion (O-ring not fitted)	0.0008 to 0.0035 in (0.02 to 0.09 mm)	0.004 to 0.0067 in (0.10 to 0.17 mm)

Pistons and connecting rods		
Type	Alloy with 3 rings	Alloy with 3 rings
Ring widths, standard:		
Top	0.069 in (1.75 mm)	0.069 in (1.75 mm))
2nd (taper)	0.079 in (2.0 mm)	0.079 in (2.0 mm)
Oil control	0.158 in (4.0 mm)	0.158 in (4.0 mm)

	Type 847	Type 841
Gudgeon pin:		
Outside diameter	0.787 in (20.0 mm)	0.827 in (21.0 mm)
Length	2.520 in (64.0 mm)	2.716 in (69.0 mm)
Connecting rods:		
Shell bearing material	Aluminium-tin	Aluminium-tin
Endfloat, small-end	0.012 to 0.022 in (0.31 to 0.57 mm)	0.012 to 0.022 in (0.31 to 0.57 mm)

Crankshaft

	Type 847	Type 841
Number of bearings	5	5
Main bearing material	Aluminium-tin	Aluminium-tin
Endfloat	0.002 to 0.009 in (0.05 to 0.23 mm)	0.002 to 0.009 in (0.05 to 0.23 mm)
Available thrust washers	0.110 in (2.80 mm) 0.112 in (2.85 mm) 0.114 in (2.90 mm) 0.116 in (2.95 mm)	0.110 in (2.80 mm) 0.112 in (2.85 mm) 0.114 in (2.90 mm) 0.116 in (2.95 mm)
Main bearing journal diameter:		
Standard	2.157 in (54.80 mm)	2.157 in (54.80 mm)
Regrind	2.1477 in (54.55 mm)	2.1477 in (54.55 mm)
Crankpin diameter:		
Standard	1.731 in (43.98 mm)	1.890 in (48.0 mm)
Regrind	1.721 in (43.73 mm)	1.880 in (47.75 mm)

Lubrication system

	Type 847	Type 841
Oil type/specification	Multigrade engine oil, viscosity SAE 15W/40, 20W/40 or 20W/50, to API SE (Duckhams Hypergrade)	
Oil capacity (including filter)	5¾ Imp pints (3.25 litres)	7½ Imp pints (4.5 US qts, 4.25 litres)
Oil filter	Champion C107	Champion C107 (up to June '84)
Oil pressure (hot):		
Idle	10 lbf/in² (0.7 bar)	29 lbf/in² (2.0 bar)
4000 rpm	50 lbf/in² (3.5 bar)	58 lbf/in² (4.0 bar)

Torque wrench settings

	Type 847		Type 841	
	lbf ft	Nm	lbf ft	Nm
Cylinder head bolts:				
Initial tightening-cold	43	58	30	40.5
Final tightening-cold	43	69	55	74.5
Final tightening-hot	43	58	59	80
Flywheel bolts (manual)	37.5	51	37.5	51
Driveplate bolts (automatic)	–	–	50	68
Main bearing cap bolts	45	61	49	66
Crankshaft pulley bolt	Not specified		52.5	71
Big-end bearing cap nuts	34	46	34	46
Main oilway plugs:				
End	–	–	60	81
Side	–	–	33¾	45.5
Camshaft sprocket bolt	22.5	30.5	–	–
Rocker shaft pillar bolts	13	17.5	18	24.5

PART A: ALL ENGINES

1 General description

Both engines are four-cylinder in-line units, mounted at the front of the vehicle.

The 1647 cc engine has an aluminium cylinder head and block, and a high camshaft, whilst the 1397 cc version has a cast iron cylinder block and lower camshaft positioning. Camshaft drive in both cases is by chain.

The crankshaft runs in five main bearings, with thrust taken by washers situated at the centre bearing location.

Where an operation is possible with the engine still in the vehicle, it has been described from this point of view. If the operation is to be carried out with the engine on the bench, it should be clear which operations are unnecessary.

2 Major operations possible with the engine in the vehicle

The following operations are possible with the engine still in place (both engine types):

(a) *Removal of the cylinder head*
(b) *Removal of the sump and oil pump*
(c) *Removal of the pistons and liners*
(d) *Removal of the timing cover, timing chain and sprockets, tensioner and camshaft*
(e) *Removal of the flywheel*

3 Oil filter, all engines – removal and refitting

1 It may be possible to unscrew the oil filter by hand, but if not, a strap wrench of suitable design will be necessary. Be prepared for some oil spillage as the filter is removed.
2 Before fitting the new filter, lubricate the seal. Screw the filter on hand tight only (photo).
3 Once the engine is restarted, check for leaks.
4 Top up the engine oil as necessary. Normally, filter renewal will be carried out in conjunction with an oil change. If the filter alone is being renewed, expect it to absorb about half a pint (0.25 litres).

4 Valve clearances, all engines – adjustment

Correct valve clearances are vital for the correct functioning of the engine. If the clearances are too big, noisy operation and reduced efficiency will result. If the clearances are set too small, the valve may not seat properly when hot, and poor compression and burning of the valve and seat may result. Refer to the Specifications for the clearances.
1 Final adjustment of the valve clearances should be carried out once the engine has run (ie hot). When the engine is in the vehicle and a manual gearbox is fitted, the engine may be rotated by engaging top gear and either pushing the vehicle to and fro or jacking up one front wheel and turning it by hand. If automatic transmission is fitted, the engine will have to be turned a little at a time, using the starter motor,

or with a spanner on the crankshaft pulley.

2   Commence by removing the rocker cover. Turn the engine until the exhaust valve on No 1 cylinder is fully open, when the inlet valve on No 3 cylinder and exhaust valve on No 4 cylinder may be adjusted to the appropriate clearance. (Numbering from the flywheel end, exhaust valves are 1-4-5-8, inlet valves are 2-3-6-7).

3   To adjust, loosen the locknut on the rocker arm adjusting screw, place a feeler gauge of the specified thickness between the valve and rocker arm, and turn the adjusting screw until a close sliding fit is obtained on the feeler. Hold the adjusting screw still and tighten the

locknut, then recheck the clearance (photo).

4   Proceed with the remaining valves, so that all of them are adjusted in accordance with the table which follows:

Valve fully open	Check and adjust
1 (EX)	6 (IN), 8 (EX)
5 (EX)	4 (EX), 7 (IN)
8 (EX)	1 (EX), 3 (IN)
4 (EX)	2 (IN), 5 (EX)

5   Refit the rocker cover, using a new gasket if necessary.

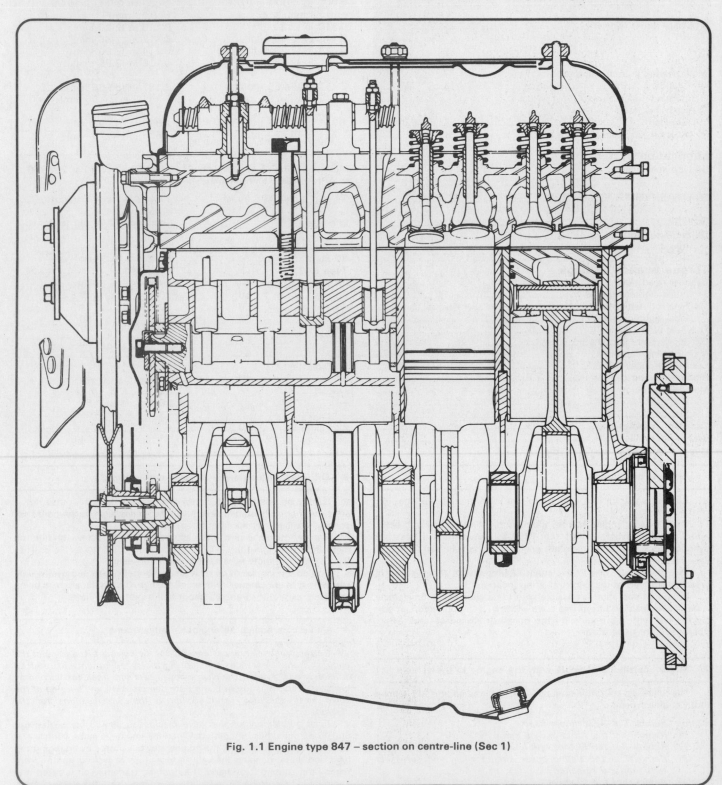

Fig. 1.1 Engine type 847 – section on centre-line (Sec 1)

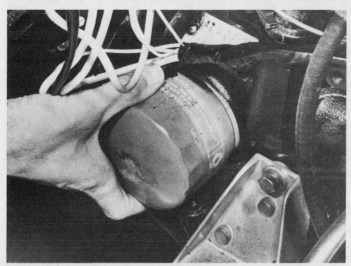

3.2 Screwing on the oil filter

4.3 Adjusting a valve clearance

6.26 The exhaust pipe, disconnected

PART B: TYPE 847 ENGINE (1397 cc)

5   Engine type 847 and gearbox – removal

1   Proceed as described in Section 6, paragraphs 1 to 26.
2   Disconnect the driveshafts from the transmission as described in Chapter 8.
3   Disconnect the speedometer cable from the transmission.
4   Disconnect the gear selector mechanism (see Chapter 6).
5   Disconnect the wiring from the reversing light switch.
6   Support the rear of the gearbox on a jack, and remove the rubber mountings.
7   Connect suitable lifting tackle to the engine lifting eyes, and take the weight.
8   Remove the nuts and cup washers under the engine mountings, one on each side.
9   Raise the engine, lower the gearbox as required, and lift out the combined unit.

6   Engine, type 847 – removal

1   Disconnect the battery.
2   Protect the wings with a suitable non-abrasive material, to prevent accidental scratching.
3   Remove the engine undertray, by taking out the three securing bolts.
4   Remove the sump plug using the correct sized square key, and allow the oil to drain.
5   Drain the cooling system (see Chapter 2).
6   Remove the air cleaner assembly (see Chapter 3).
7   Disconnect the accelerator cable at the engine end (see Chapter 3). We discovered that the tongues on the compensator could be closed by pressing a 12 mm socket over them. The compensator can then be pulled out.
8   Disconnect the choke cable at the carburettor.
9   Disconnect the bonnet release cable.
10   Remove the bonnet release platform, by taking out the 6 bolts, two at each end and two in the front centre. Take away the two top radiator mountings thereby released.
11   Remove the front grille, by taking out the screws in the top corners.
12   Remove the radiator (see Chapter 2).
13   Remove the top and bottom radiator hoses.
14   Pull the fuel pipes off the pump, and plug or clamp them to prevent fuel loss.
15   Remove the distributor vacuum pipe, and the pump-to-carburettor fuel pipe.
16   Disconnect the clutch cable at the bellhousing lever.
17   Remove the distributor cap and leads.
18   Disconnect the two wiring sockets at the bulkhead.
19   Disconnect the wiring at the reversing light switch, at the coil, and at the white connector.
20   Disconnect the wiring to the distributor.
21   Remove the three hoses at the water pump, and undo the plastic clips at the bracket on the RH side of the engine.
22   Disconnect the vacuum pipe from the servo, at the carburettor. Tuck the pipe out of the way.
23   Disconnect the engine earthing strap.
24   Disconnect all wiring to the starter motor solenoid, noting the colour coding.
25   Remove the battery (not strictly necessary, but recommended). Put it on charge if necessary.
26   Disconnect the exhaust pipe from the inlet manifold, by removing the clamp (photo).
27   Remove the starter motor (see Chapter 10).
28   Take out the clutch shield fixing bolts (photo).
29   Connect suitable lifting tackle to the engine lifting eyes.
30   Remove the upper and lower clutch housing bolts.
31   Remove the nuts under the engine mountings, one on each side, complete with the cup washers (photo).
32   Ease the engine slightly forward and up to free the mountings and clutch shield, removing this complete with the TDC sensor.
33   Pull the engine well forward, to clear the clutch shaft, and lift it up and out (photo).

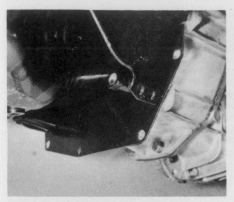

6.28 The clutch shield, bolts removed

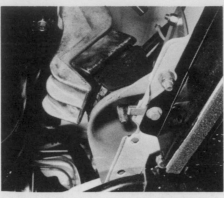

6.31 An engine mounting, with nut and cup washer removed

6.33 Lifting out the engine

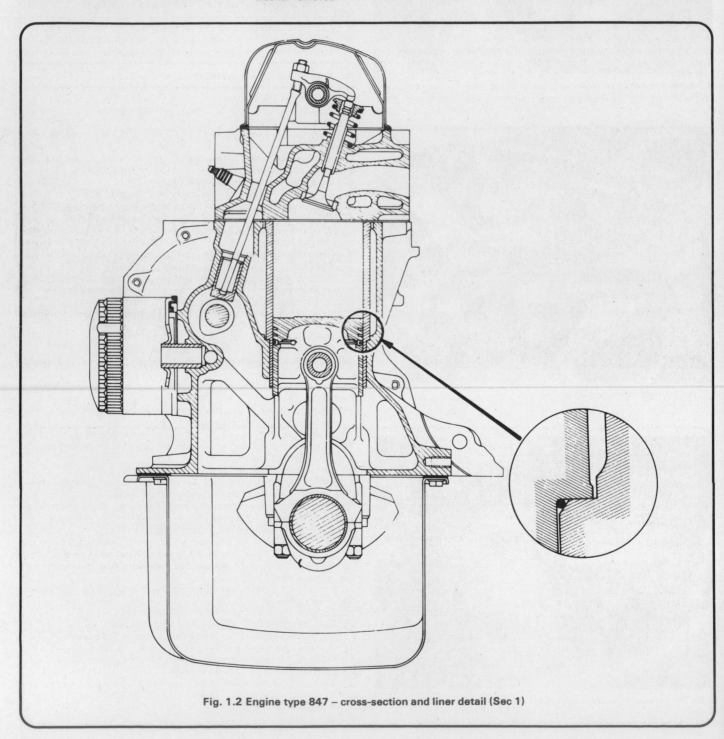

Fig. 1.2 Engine type 847 – cross-section and liner detail (Sec 1)

7 Cylinder head, engine type 847 – removal

1   If the engine is in the vehicle, proceed as follows:

(a) Disconnect the battery
(b) Remove the front top crossmember (see Section 6, paragraph 10)
(c) Remove the air filter assembly (see Chapter 3)
(d) Drain the cooling system (see Chapter 2)
(e) Remove the distributor (see Chapter 4)
(f) Remove the drivebelt
(g) Remove the alternator
(h) Disconnect all wiring, cables and hoses, as described in Section 6
(i) Disconnect the exhaust downpipe at the manifold clamp

2   Continue as follows, whether or not the engine is in the vehicle.
3   Remove the nuts, and then the rocker cover (photo).
4   Free the valve clearance adjusting screws by loosening the locknuts, remove the screws and then the pushrods. Ensure that the components are identified so that they may be refitted in the same positions (photo).
5   Loosen all the cylinder head bolts a little at a time in the order shown in Fig. 1.13. Remove all bolts except that which passes through the locating dowel near the distributor (Fig. 1.3). Leave this one remaining bolt slackened but in contact with the cylinder head. **Do not** lift the cylinder head at this point unless complete dismantling of the engine block and cylinder liners is envisaged, or the watertight seal at the bottom of the liners will be broken, thereby permitting foreign matter to enter the sump. Instead, tap the head with a suitable soft hammer on each side to unstick it, and pivot it on the locating dowel with the bolt still in place to free the cylinder head surface.
6   Remove the remaining bolt, and lift the head off.
7   Clamp the liners down to the cylinder head surface unless they are to be removed. We employed suitable bolts, nuts and plain washers (photo).

8 Sump and oil pump, type 847 engine – removal

1   Disconnect the battery and drain the engine oil.
2   Remove the engine undertray, and withdraw all the sump bolts.
3   Partially lower the sump, and turn the engine to place Nos 1 and 4 pistons at BDC.
4   Tilt the front of the sump to the left, and lower it.
5   Remove the bolts securing the oil pump, and withdraw it.

7.3 Lifting off the rocker cover

7.4 The pushrods, screws and locknuts

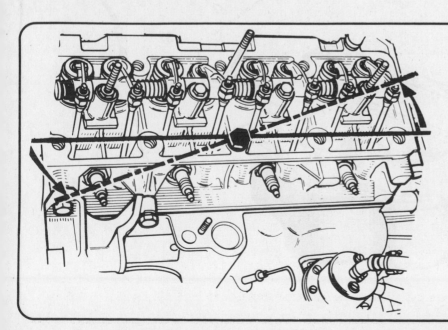

Fig. 1.3 Pivoting the cylinder head on the dowel, before lifting (Sec 7)

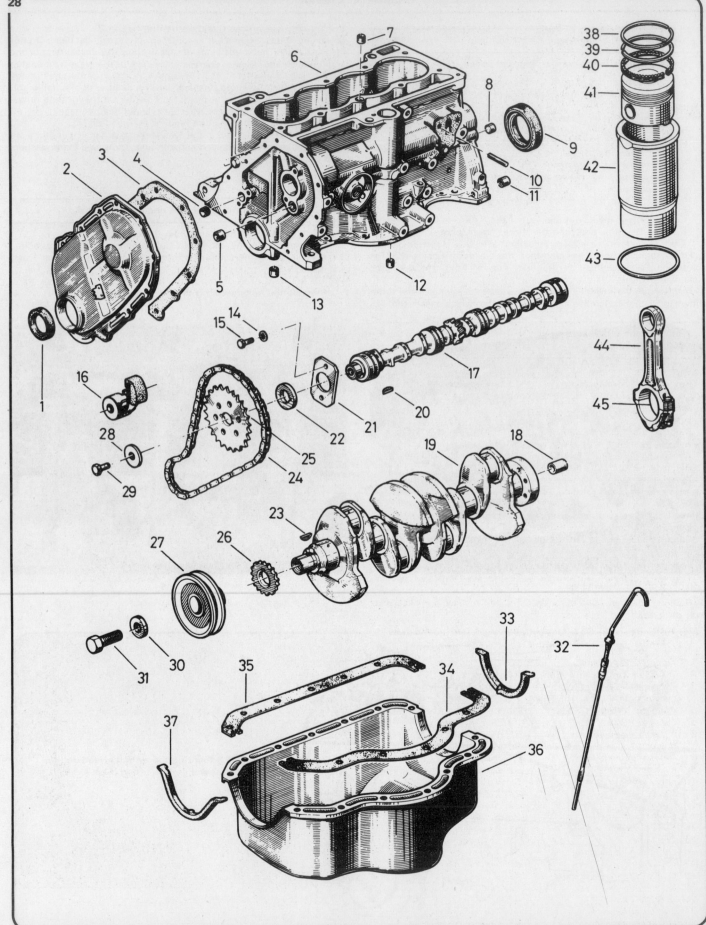

7.7 Clamping the cylinder liners

9.7 The crankshaft pulley, before removal

9 Timing cover and seal, timing chain, tensioner and sprockets, type 847 engine – removal

Timing cover

1 Remove the bonnet release platform, by disconnecting the cable and taking out the bolts, two at each end of the platform and two in the centre.
2 Disconnect the battery.
3 Remove the drivebelt (see Chapter 2).
4 Remove the alternator (see Chapter 10).
5 Take out the retaining bolts, and remove the fan and pulley.
6 Remove the sump, as described in Section 8.
7 Remove the crankshaft pulley bolt. If the flywheel is still fitted, jam it to prevent crankshaft rotation, but if not, screw two of the flywheel retaining bolts into the crankshaft, and jam the shaft by placing a suitable lever between them. Remove the crankshaft pulley (photo).
8 Remove the bolts, and withdraw the timing cover (photo).
9 Carefully prise or tap out the oil seal from the cover.

Timing chain, tensioner and sprockets

10 Take the bolt from the centre of the timing chain tensioner, and lift the tensioner off.
11 Remove the camshaft sprocket bolt, and lift the sprocket and chain off (photo).

9.8 Removing the timing cover

Fig. 1.4 Engine type 847 (less cylinder head) – exploded view (Sec 1)

1	Timing cover oil seal	23	Crankshaft key
2	Timing cover	24	Timing chain
3	Timing cover gasket	25	Camshaft sprocket
4	Oilway plug	26	Crankshaft sprocket
5	Aluminium plug	27	Crankshaft pulley
6	Cylinder block	28	Camshaft washer
7	Cylinder head locating dowel	29	Camshaft bolt
8	Aluminium plug	30	Pulley washer
9	Crankshaft oil seal	31	Pulley bolt
10	Fuel pump studs	32	Dipstick
11	Dowel	33	Rubber seal
12	Dowel	34	Sump gasket (half)
13	Dowel	35	Sump gasket (half)
14	Spring washer	36	Sump
15	Bolt	37	Rubber seal
16	Timing chain tensioner	38	Top compression ring
17	Camshaft	39	Taper compression ring
18	Crankshaft spigot bearing	40	Oil control ring
19	Crankshaft	41	Piston
20	Camshaft key	42	Liner
21	Camshaft flange	43	Liner seal
22	Distance piece	44	Connecting rod
		45	Connecting rod cap

9.11 Withdrawing the camshaft sprocket bolt

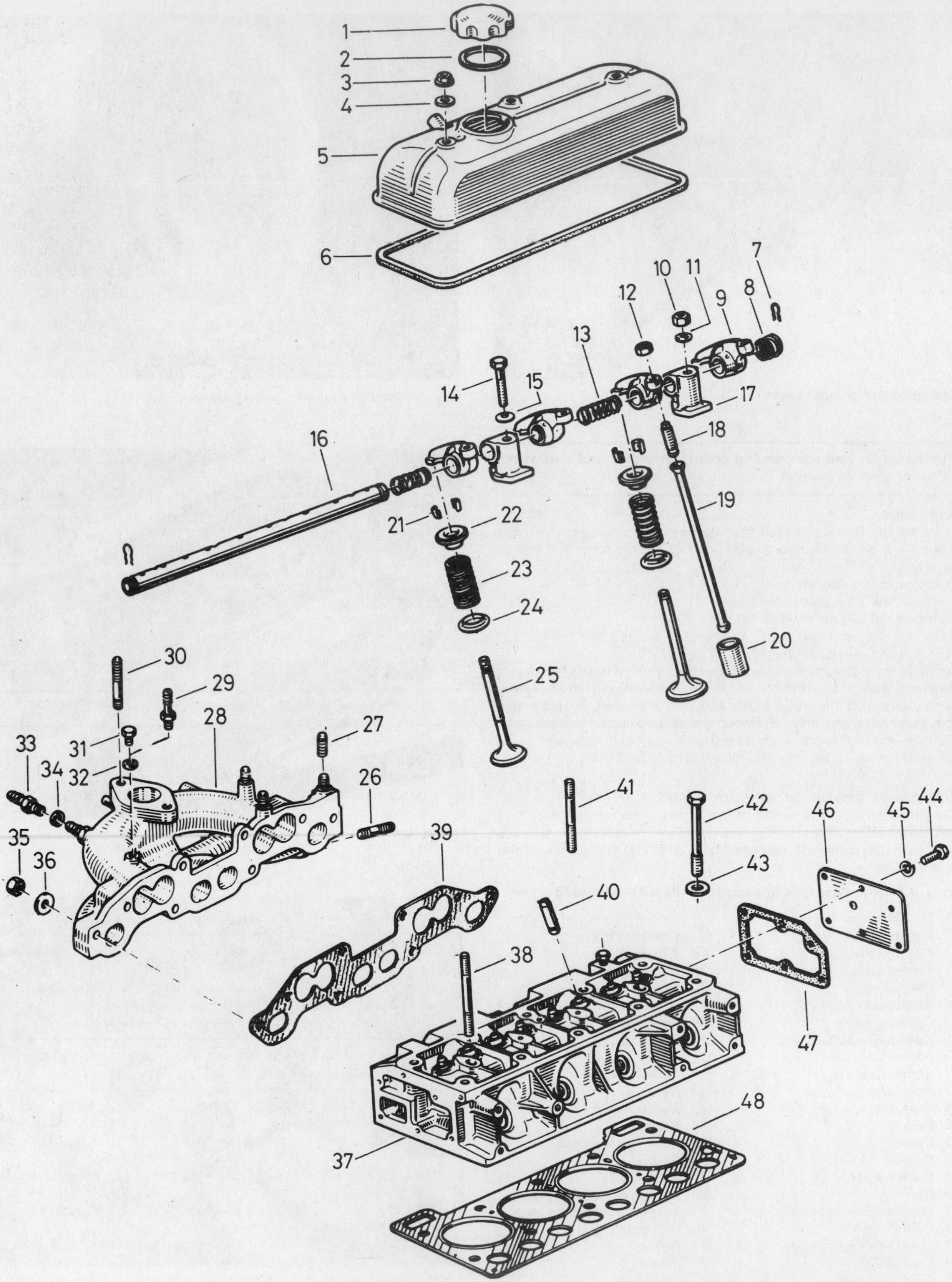
1
2
3
4
5
6
7
8
9
10
11
12
13
14
15
16
17
18
19
20
21
22
23
24
25
26
27
28
29
30
31
32
33
34
35
36
37
38
39
40
41
42
43
44
45
46
47
48

9.12 The crankshaft sprocket, before removal

10.5 Lifting out a tappet (cam follower)

10.7 The distributor drive pinion

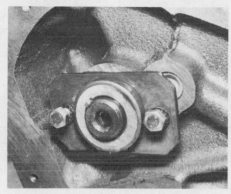

10.10a The camshaft flange and securing bolts

10.10b Withdrawing the camshaft

12 If necessary, draw off the crankshaft sprocket, using either a suitable puller, or a pair of levers carefully applied behind the sprocket (photo).

Fig. 1.5 Cylinder head, engine type 847 – exploded view (Sec 7)

1	Oil filler cap	25	Valve
2	Sealing washer	26	Stud
3	Nut	27	Stud
4	Washer	28	Manifold, inlet/exhaust
5	Rocker cover	29	Adaptor
6	Rocker cover gasket	30	Stud
7	Spring clip	31	Plug
8	End spring	32	Washer
9	Rocker arm	33	Adaptor
10	Nut	34	Washer
11	Washer	35	Nut
12	Adjuster locknut	36	Washer
13	Spring	37	Cylinder head
14	Bolt	38	Stud
15	Washer	39	Manifold gasket
16	Rocker shaft	40	Valve guide
17	Rocker bracket	41	Stud
18	Adjustment screw	42	Bolt
19	Pushrod	43	Washer
20	Tappet	44	Bolt
21	Valve collets	45	Washer
22	Top cup	46	Plate
23	Valve spring	47	Gasket
24	Base washer	48	Cylinder head gasket

10 Camshaft, type 847 engine – removal

1 Drain the sump.
2 Remove the front grille.
3 Remove the radiator (see Chapter 2).
4 Remove the cylinder head and clamp the liners (see Section 7).
5 Lift out the tappets, and ensure that they are identified correctly for refitting purposes (photo).
6 Remove the fuel pump (see Chapter 3). The pipes may be left connected.
7 Lift out the distributor drive pinion, using a bolt 12 mm dia x 1.75 mm pitch (photo).
8 Remove the sump (see Section 8).
9 Remove the camshaft sprocket (see Section 9).
10 Remove the bolts securing the camshaft flange, and draw out the shaft carefully (photos).
11 If necessary, carefully support the camshaft flange and tap the shaft through, thus releasing the collar and flange.

11 Pistons, cylinder liners, connecting rods and big-end bearings, type 847 engine – removal

1 The removal of these items is possible, if necessary, with the engine still in the vehicle.
2 Remove the cylinder head (see Section 7).
3 Remove the sump and oil pump (see Section 8).
4 Mark the connecting rods to ensure correct refitting, starting with No 1 at the clutch end, and away from the camshaft side. Mark the bearing caps also, so that they are fitted both to the correct rod and the right way round.
5 Remove the big-end nuts, followed by the caps and bearing shells (photo).
6 Identify the cylinder liners, using a quick drying paint, so that each can be refitted in the same position and the right way round.
7 Withdraw each piston and liner assembly, and temporarily refit the

11.5 Removing a big-end bearing cap

11.7 Withdrawing a piston and liner assembly

12.6 Removing a main bearing cap

12.7a Lifting out the crankshaft

12.7b Recovering a thrust washer

12.7c Lifting out a main bearing shell

13.4 The flywheel, with bolts loosened

13.6 Fitting the rear seal (sump is removed in this photograph)

13.9 The spigot bush, correctly positioned

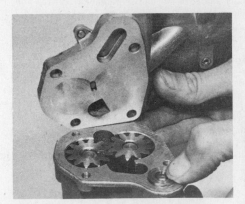

15.2a The oil pump cover being removed

15.2b The ball seat, ball and spring

15.3a Lifting out the idler gear

Are your plugs trying to tell you something?

Normal.
Grey-brown deposits, lightly coated core nose. Plugs ideally suited to engine, and engine in good condition.

Heavy Deposits.
A build up of crusty deposits, light-grey sandy colour in appearance.
Fault: Often caused by worn valve guides, excessive use of upper cylinder lubricant, or idling for long periods.

Lead Glazing.
Plug insulator firing tip appears yellow or green/yellow and shiny in appearance.
Fault. Often caused by incorrect carburation, excessive idling followed by sharp acceleration. Also check ignition timing.

Carbon fouling.
Dry, black, sooty deposits.
Fault: over-rich fuel mixture Check: carburettor mixture settings, float level, choke operation, air filter.

Oil fouling.
Wet, oily deposits. Fault: worn bores/piston rings or valve guides; sometimes occurs (temporarily) during running-in period.

Overheating.
Electrodes have glazed appearance, core nose very white – few deposits. Fault: plug overheating. Check: plug value, ignition timing, fuel octane rating (too low) and fuel mixture (too weak).

Electrode damage.
Electrodes burned away; core nose has burned, glazed appearance. Fault: pre-ignition. Check: for correct heat range and as for 'overheating'.

Split core nose.
(May appear initially as a crack). Fault: detonation or wrong gap-setting technique. Check: ignition timing, cooling system, fuel mixture (too weak).

WHY DOUBLE COPPER IS BETTER FOR YOUR ENGINE.

Unique Trapezoidal Copper Cored Earth Electrode

50% Larger Spark Area

Copper Cored Centre Electrode

Champion Double Copper plugs are the first in the world to have copper core in both centre <u>and</u> earth electrode. This innovative design means that they run cooler by up to 100°C – giving greater efficiency and longer life. These double copper cores transfer heat away from the tip of the plug faster and more efficiently Therefore, Double Copper runs at cooler temperatures than conventional plugs giving improved acceleration response and high speed performance with no fear of pre-ignition.

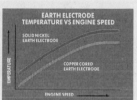

TRAPEZOIDAL COPPER CORED EARTH ELECTRODE

NEW TRAPEZOIDAL COPPER CORED EARTH ELECTRODE   CONVENTIONAL SOLID NICKEL ALLOY EARTH ELECTRODE

50% INCREASE IN SPARK AREA

EARTH ELECTRODE TEMPERATURE VS ENGINE SPEED

SOLID NICKEL EARTH ELECTRODE

COPPER CORED EARTH ELECTRODE

TEMPERATURE   ENGINE SPEED

Champion Double Copper plugs also feature a unique trapezoidal earth electrode giving a 50% increase in spark area. This, together with the double copper cores, offers greatly reduced electrode wear, so the spark stays stronger for longer.

 FASTER COLD STARTING

 FOR UNLEADED OR LEADED FUEL

 ELECTRODES UP TO 100°C COOLER

 BETTER ACCELERATION RESPONSE

 LOWER EMISSIONS

 50% BIGGER SPARK AREA

 THE LONGER LIFE PLUG

Plug Tips/Hot and Cold.
Spark plugs must operate within well-defined temperature limits to avoid cold fouling at one extreme and overheating at the other.
Champion and the car manufacturers work out the best plugs for an engine to give optimum performance under all conditions, from freezing cold starts to sustained high speed motorway cruising.
Plugs are often referred to as hot or cold. With Champion, the higher the number on its body, the hotter the plug, and the lower the number the cooler the plug.

Plug Cleaning
Modern plug design and materials mean that Champion no longer recommends periodic plug cleaning. Certainly don't clean your plugs with a wire brush as this can cause metal conductive paths across the nose of the insulator so impairing its performance and resulting in loss of acceleration and reduced m.p.g.
However, if plugs are removed, always carefully clean the area where the plug seats in the cylinder head as grit and dirt can sometimes cause gas leakage.
Also wipe any traces of oil or grease from plug leads as this may lead to arcing.

CHAMPION

DOUBLE CC COPPER

1

This photographic sequence shows the steps taken to repair the dent and paintwork damage shown above. In general, the procedure for repairing a hole will be similar; where there are substantial differences, the procedure is clearly described and shown in a separate photograph.

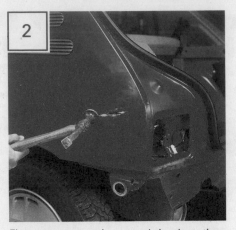

2

First remove any trim around the dent, then hammer out the dent where access is possible. This will minimise filling. Here, after the large dent has been hammered out, the damaged area is being made slightly concave.

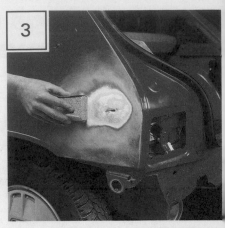

3

Next, remove all paint from the damaged area by rubbing with coarse abrasive paper or using a power drill fitted with a wire brush or abrasive pad. 'Feather' the edge of the boundary with good paintwork using a finer grade of abrasive paper.

4

Where there are holes or other damage, the sheet metal should be cut away before proceeding further. The damaged area and any signs of rust should be treated with Turtle Wax Hi-Tech Rust Eater, which will also inhibit further rust formation.

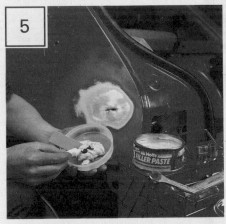

5

For a large dent or hole mix Holts Body Plus Resin and Hardener according to the manufacturer's instructions and apply around the edge of the repair. Press Glass Fibre Matting over the repair area and leave for 20-30 minutes to harden. Then ...

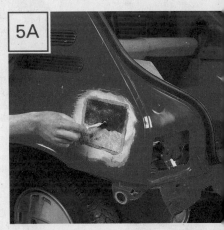

5A

... brush more Holts Body Plus Resin and Hardener onto the matting and leave to harden. Repeat the sequence with two or three layers of matting, checking that the final layer is lower than the surrounding area. Apply Holts Body Plus Filler Paste as shown in Step 5B.

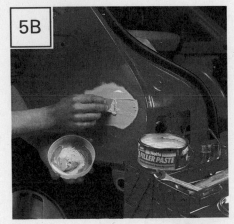

5B

For a medium dent, mix Holts Body Plus Filler Paste and Hardener according to the manufacturer's instructions and apply it with a flexible applicator. Apply thin layers of filler at 20-minute intervals, until the filler surface is slightly proud of the surrounding bodywork.

5C

For small dents and scratches use Holts No Mix Filler Paste straight from the tube. Apply it according to the instructions in thin layers, using the spatula provided. It will harden in minutes if applied outdoors and may then be used as its own knifing putty.

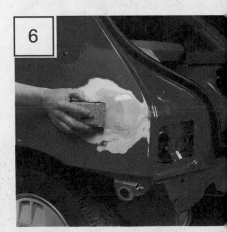

6

Use a plane or file for initial shaping. Then using progressively finer grades of wet-and dry paper, wrapped round a sanding block and copious amounts of clean water, rub down the filler until glass smooth. 'Feather' the edges of adjoining paintwork.

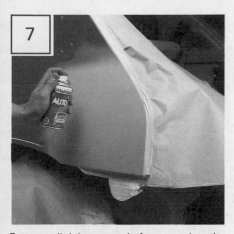

7 Protect adjoining areas before spraying the whole repair area and at least one inch of the surrounding sound paintwork with Holts Dupli-Color primer.

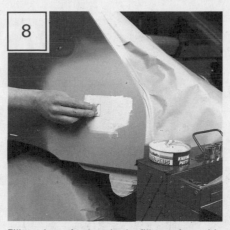

8 Fill any imperfections in the filler surface with a small amount of Holts Body Plus Knifing Putty. Using plenty of clean water, rub down the surface with a fine grade wet-and-dry paper – 400 grade is recommended – until it is really smooth.

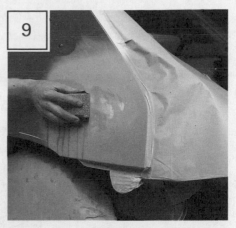

9 Carefully fill any remaining imperfections with knifing putty before applying the last coat of primer. Then rub down the surface with Holts Body Plus Rubbing Compound to ensure a really smooth surface.

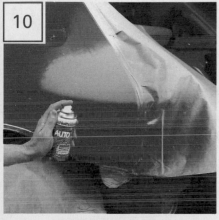

10 Protect surrounding areas from overspray before applying the topcoat in several thin layers. Agitate Holts Dupli-Color aerosol thoroughly. Start at the repair centre, spraying outwards with a side-to-side motion.

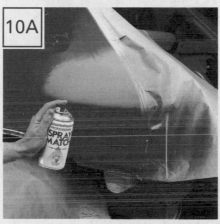

10A If the exact colour is not available off the shelf, local Holts Professional Spraymatch Centres will custom fill an aerosol to match perfectly.

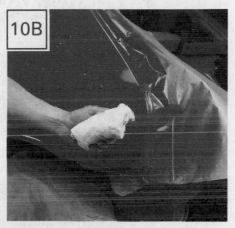

10B To identify whether a lacquer finish is required, rub a painted unrepaired part of the body with wax and a clean cloth.

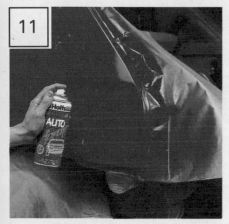

11 If *no* traces of paint appear on the cloth, spray Holts Dupli-Color clear lacquer over the repaired area to achieve the correct gloss level.

12 **13** The paint will take about two weeks to harden fully. After this time it can be 'cut' with a mild cutting compound such as Turtle Wax Minute Cut prior to polishing with a final coating of Turtle Wax Extra.

14 When carrying out bodywork repairs, remember that the quality of the finished job is proportional to the time and effort expended.

HAYNES No1 for DIY

Haynes publish a wide variety of books besides the world famous range of *Haynes Owners Workshop Manuals*. They cover all sorts of DIY jobs. Specialist books such as the *Improve and Modify* series and the *Purchase and DIY Restoration Guides* give you all the information you require to carry out everything from minor modifications to complete restoration on a number of popular cars. In addition there are the publications dealing with specific tasks, such as the *Car Bodywork Repair Manual* and the *In-Car Entertainment Manual*. The *Household DIY* series gives clear step-by-step instructions on how to repair everyday household objects ranging from toasters to washing machines.

Whether it is under the bonnet or around the home there is a Haynes Manual that can help you save money. Available from motor accessory stores and bookshops or direct from the publisher.

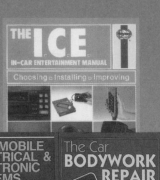

connecting rod caps to the correct rods (photo). Keep the shells with their correct rods and caps if they are to be re-used.
8   Withdraw each piston/connecting rod assembly from the liner. Do not allow the piston rings to jump out sharply or they may break. Instead restrain them with the fingers.
9   Remove the top piston ring by gently springing the ring open far enough to insert a couple of old feeler blades or similar beneath it, then 'walking' the ring off the top of the piston. Take care not to scratch the piston – it is made of softer metal than the ring – or to break the ring, which is brittle. Also mind your fingers – piston rings are sharp!
10 Repeat the process with the second and third piston rings, using the feeler blades to stop the rings falling into the grooves above. Keep the rings with the appropriate piston if they are to be re-used.

12 Crankshaft and main bearings, type 847 engine – removal

1   The engine must be out of the vehicle to permit this work to be carried out.
2   Remove the clutch and flywheel (see Section 13).
3   Remove the sump and oil pump (see Section 8).
4   Remove the timing chain and sprockets (see Section 9).
5   Mark the connecting rods and big-end caps, number one at the clutch end on the side away from the camshaft, and so on, if this has not already been done. Remove the nuts, followed by the big-end caps.
6   Mark the main bearing caps in relation to the cylinder block, remove the nuts, and take off the caps (photo).
7   Lift the crankshaft out. Recover all main and big-end bearings shells. Note the thrust washers on either side of the centre (No 3) main bearing (photos).

13 Flywheel, oil seal and spigot bearing, type 847 engine – removal and refitting

Flywheel – removal
1   The flywheel may be removed after removing either the engine or the gearbox
2   Remove the clutch assembly as described in Chapter 5.
3   Lock the flywheel, preferably by making up a suitable bracket bolted to the cylinder block and engaging the flywheel teeth.
4   Loosen all the bolts, firmly grip the flywheel to prevent it dropping off, remove the bolts and take off the flywheel (photo). Be careful, it is heavy!

Crankshaft oil seal
5   Prise out the old oil seal, taking great care that damage is not caused to the sealing surface on the shaft.
6   Offer up the oil seal after applying lubricant both to the lip and to

the outside diameter. Tap it very gently into place using either the old seal or a suitable piece of tubing. Ensure that all items are thoroughly clean, that the seal is tapped in square, and that it is flush with the rear face of the crankshaft block (photo).

Spigot bearing
7   It is not necessary to remove the flywheel.
8   Screw a tap, M14 x 2 mm pitch, into the spigot bush. Grip the tap and pull it out squarely, bringing the bush with it.
9   Use mandrel Emb 319, or a close-fitting home manufactured substitute with a shoulder of suitable size, and gently tap the new bush into position until it reaches the edge of the chamfer (photo).

Flywheel – refitting
10 Refit all parts in reverse order. Use thread locking compound on the threads of the flywheel bolts, before tightening them to the specified torque figure.

14 Inlet and exhaust manifold assembly, type 847 engine – removal and refitting

1   Disconnect the battery.
2   Remove the air filter assembly, see Chapter 3.
3   Remove the warm air pipe.
4   Remove the carburettor with its heating hoses (as applicable).
5   Remove the accelerator cable swivel.
6   Disconnect the exhaust pipe clamp.
7   Remove the warm air shroud.
8   Disconnect the oil fume hoses.
9   Disconnect the servo vacuum pipe.
10 Remove the fixings and take off the manifold.
11 If a new manifold is to be fitted, remove the vacuum union, and fit to the new part
12 Clean the manifold mating surfaces, and refit using a new gasket.
13 Refit all parts in reverse order.

15 Oil pump, engine type 847 – dismantling and reassembly

1   Holding the cover in position, take out the retaining bolts.
2   Carefully release the cover, and remove the ball seat, ball and pressure spring (photos).
3   Take out the idler gear, driving gear and shaft (photos).
4   Clean all the parts and examine them for visually evident defects.
5   Check the clearance between the gears and the pump body. If this exceeds 0.008 in (0.2 mm), preferably renew the pump, but at least renew the gears (photo).
6   To reassemble, fit the idler gear, followed by the driving gear.

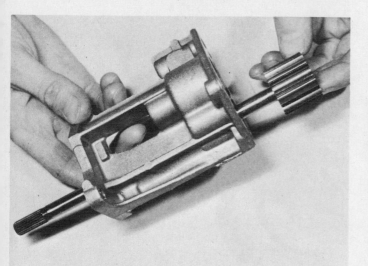

15.3b Lifting out the driving gear and shaft

15.5 Checking the gear-to-body clearance

Lubricate all parts before fitting.

7 Fit the spring guide, the spring, ball and seat, and position the cover to retain the parts.

8 Refit the cover bolts, and tighten.

16 Cylinder head, rockers and valves, type 847 engine – dismantling

1 Remove the fan by removing the fixing screws.

2 If care is exercised, the water pump, temperature transmitter and alternator bracket may be left in place if no work is required upon them.

3 Remove the warm air intake channel, rear lifting hook, and diagnostic socket bracket.

4 Remove the inlet and exhaust manifold assembly (see Section 14).

5 Remove the core plate and air filter bracket.

6 Remove the bolts and nuts securing the rocker shaft assembly, a little at a time, to avoid distortion of the shaft. Remove the shaft assembly.

7 Compress the valve springs, using a suitable compressor, and extract the collets (2 per valve) (photo).

8 Release and remove the spring compressor, and take out the top cups, springs, base washers and valves. Keep them identified for refitting purposes. Remove each valve similarly (photos).

9 To dismantle the rocker shaft, pull off the circlips at each end and slip off the component parts, noting their positions for refitting.

10 Note that the plugs in the ends of the shaft must not be removed.

17 Examination and renovation – general

The engine has presumably been dismantled to remedy some specific defect. Before committing yourself to extensive overhauling and the purchase of new parts, compare the cost of overhauling with the cost of an exchange Renault engine.

Truth of cylinder head gasket face

1 Check the gasket mating surface for flatness, by first ensuring that it is completely clean. Either place it face down on a true surface, or invert it and place a known straight-edge such as a good quality steel rule across the gasket mating surface, from end to end.

2 Using a feeler gauge, check for distortion of the face. If the maximum bow exceeds 0.002 in (0.05 mm), have the head face resurfaced (photo).

3 The cylinder head height must not, after resurfacing, be less than a specified minimum dimension. See the Specifications for details.

16.7 Removing the valve collets

16.8a Lifting off the top cup and spring

16.8b Lifting out a valve

17.2 Checking for distortion of the cylinder head surface

Cylinder head – general

4   Examine the cylinder head carefully for any obvious defects such as cracks or defective threads.

5   Any such defects will mean that a specialist engineer or Renault agent should be consulted.

Valves and valve seats

6   Examine the heads of the valves for pitting and burning, especially the exhaust valves. The valve seatings should be examined at the same time. If pitting is slight, the marks can be removed by grinding the seats and valves together with coarse, and then fine, valve grinding paste. Where bad pitting has occurred to the seats, recut them and fit new valves. If the valve seats are so worn that they cannot be recut, then it will be necessary to fit new valve seat inserts. These latter two jobs should be entrusted to the local Renault agent or engineering works. In practice it is seldom that the seats are so worn that they require renewal. Normally, it is the exhaust valve that is too badly worn for refitting, and the owner can easily purchase a new set of valves and match them to the seats by valve grinding.

7   To grind a valve, smear a trace of coarse carborundum paste on the seat face and apply a suction grinder tool to the valve head. With a semi-rotary motion, grind the head to its seat, lifting the valve occasionally to redistribute the paste. When a dull matt even surface finish is produced on both the valve seat and valve, wipe off the paste and repeat the process with fine carborundum paste, lifting and turning the valve to redistribute the paste as before. A light spring placed under the valve head will greatly ease this operation. When a smooth unbroken ring of light grey matt finish is produced on both valve and valve seat faces, the grinding operation is completed.

8   Scrape all carbon from the valve head and stem. Carefully clean away every trace of grinding compound, taking care to leave none in the ports or valve guides. Clean the valves and seats with a paraffin soaked rag, then with a clean rag, and finally, if an air line is available, blow the valves, valve guides and valve ports clean.

Valve guides

9   Examine the valve guides internally for wear. If the valves are a very loose fit in the guides and there is the slightest suspicion of lateral rocking using a new valve, then new guides will have to be fitted. Valve guide renewal should be left to a Renault agent who will have the required press and mandrel. Work of this kind in a light alloy head without the correct tools can be disastrous.

Valve springs

10   Compare the old valve springs with new ones. Renew the springs if there is any significant difference in length, or if distortion is evident.

Crankshaft and bearings

11   Wear in the main or big-end bearings will have been noticed before dismantling, evidence being a fall in oil pressure and a rumbling (main bearings) or knocking (big-end bearings) from the engine. If wear is known to exist, take the crankshaft to a motor factor or Renault agent, who will measure the journals and decide whether regrinding is necessary. Serious wear will be evident to the naked eye.

12   The main and big-end bearing shells should be renewed as a matter of course unless they are known to have covered only a nominal mileage. Take the old shells with you when buying new ones and check that you get the same size. Oversize shells (fitted after regrinding) should have the oversize dimension marked on the back.

13   The thrust washers on either side of the centre main bearing are also subject to wear and may need renewal. It may be best to delay purchase until the crankshaft can be temporarily refitted, since several thicknesses are available to give the required endfloat – see Specifications.

Camshaft, timing gears and chain

14   Inspect the camshaft lobes for wear and the bearing surfaces for tracks or other signs of wear. A new camshaft is probably the only answer if much wear has taken place. If there is suspicion that the camshaft bearings in the block are badly worn, consult your Renault dealer. A new block will probably be required.

15   Examine the timing chain and sprockets. Wear in the sprocket teeth will show as hook-shaped teeth, possibly bright and sharp on one side. Renew the sprockets as a pair if necessary. Wear in the chain will show up as side slackness; on a badly worn chain the actual rollers may be grooved or broken. It is sensible to renew the chain anyway if it has covered a high mileage.

16   Do not overlook the timing chain tensioner. This should be renewed if it is badly grooved or seems to have lost its tensioning ability.

Pistons and liners

17   The cylinder bores must be examined for taper, ovality, scoring and scratches. Start by carefully examining the top of the cylinder bores. If they are at all worn a very slight ridge will be found on the thrust side. This marks the top of the piston ring travel. The owner will have a good indication of the bore wear prior to dismantling the engine, or removing the cylinder head. Excessive oil consumption accompanied by blue smoke from the exhaust is a sure sign of worn cylinder bores and piston rings.

18   Measure the bore diameter just under the ridge with an internal micrometer and compare it with the diameter at the bottom of the bore, which is not subject to wear. If the difference between the two measurements is more than 0.006 in (0.1524 mm) then it will be necessary to fit new pistons and liner assemblies. If no micrometer is available remove the rings from a piston and place the piston in each bore in turn about ¾ in (18 mm) below the top of the bore. If an 0.010 in (0.254 mm) feeler gauge can be slid between the piston and the cylinder wall on the thrust side of the bore then remedial action must be taken.

19   Should the liners have been disturbed they **must** be completely removed from the cylinder block and new seals fitted, otherwise once the seals have been disturbed the chances are that water will leak into the sump.

20   If the old pistons are to be refitted, carefully remove the piston rings and then thoroughly clean them. Take particular care to clean out the piston ring grooves. At the same time do not scratch the aluminium in any way. If new rings are to be fitted to the old pistons then the top ring should be stepped so as to clear the ridge left above the previous top ring. If a normal but oversize new ring is fitted it will hit the ridge and break because the new ring will not have worn in the same way as the old, which will have worn in union with the ridge.

General

21   If in doubt as to whether a particular component is serviceable, consider the time and effort which will be required to renew it if it fails prematurely. In a borderline case it is probably best to decide in favour of renewal, but it must be admitted that some degree of compromise is usually inevitable unless funds are unlimited. Obviously, the owner doing a quick 'shell and ring' job will have different standards from the person undertaking a thorough overhaul.

---

18 Cylinder block, bare, engine type 847 – preparing for use

1   If a new cylinder block is being used, the points mentioned in this Section should be noted and carried out, as applicable.

2   Ensure that the aluminium plugs at the ends of the main oil gallery are in place.

3   Screw the camshaft bearing plugs into place, if necessary, and peen over.

4   Check that the oilway plug behind the timing chain tensioner is in place.

5   Ensure that the following are in place:

   (a) Two timing cover studs
   (b) The engine/gearbox studs and dowels
   (c) Distributor stud
   (d) Fuel pump studs (smear threads with jointing compound before fitting)
   (e) Oil filter threaded mounting
   (f) Coolant drain plug
   (g) Dipstick tube (smear thread locking compound on it before fitting)
   (h) Insert the plug in the camshaft locating bore (hammer the centre to spread it)
   (j) The oil recovery pipe (opposite the dipstick)
   (k) The oil pump locating dowel
   (m) The timing chain tensioner locating peg

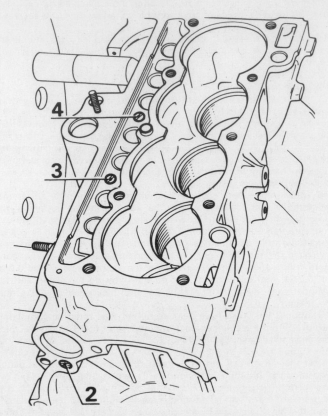

Fig. 1.6 Cylinder block plugs (Sec 18)

2   Oilway plug (one at each   3   Camshaft bearing plug
    end)   4   Camshaft bearing plug

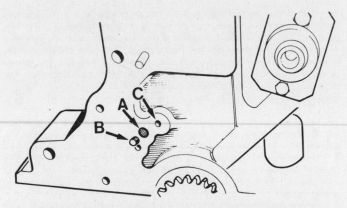

Fig. 1.7 Cylinder block – timing chain end (Sec 18)

A   Oilway plug   C   Tensioner spring hole
B   Tensioner locating dowel

19 Cylinder head, rockers and valves, 847 engine – reassembly

1   Ensure that all parts are completely clean.
2   Refit each valve in turn into the correct guide, and place over it the base washer, spring ((ensuring that the close coils are downwards towards the base washer) and the top cup.
3   Fit the spring compressor and compress the spring, fit the collets around the collet grooves, and gradually release the compressor, ensuring that the collets remain in their groove as the springs are decompressed. Remove the compressor.
4   Note that, if new pillars are required for the rocker assembly, a modified type of pillar will be supplied by the parts department of

Fig. 1.8 Rocker pillar, as supplied for service (Sec 19)

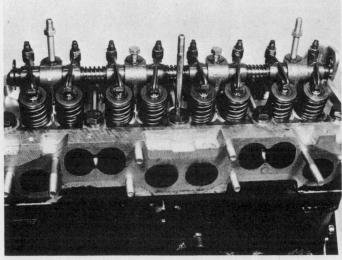

19.7 The rocker shaft assembly, repositioned

Renault agents. This pillar, possessing an extra oil hole and possibly a modified drilling for the main clamping bolt also, may be fitted in any position.
5   Refit the component parts of the rocker assembly on the shaft in the proper order, ensuring that the oilways in the rocker shaft face the pushrods, and that those in the rocker shaft pillars are in line with those in the shaft.
6   Refit the circlips to the end of the shaft.
7   Refit the rocker shaft assembly by tightening the holding-down bolts and nuts progressively, a little at a time to avoid distortlon, to the correct torque figure (photo).
8   If removed, refit parts in reverse of the order described in Section 16, paragraphs 1 to 5.

20 Pistons, cylinder liners, connecting rods and big-end bearings, type 847 engine – refitting

1   Ensure that complete cleanliness is observed, and refer to Section 25 paragraphs 3 and 4, before commencing work. Thoroughly clean the inside of the cylinder block, particularly the seal locations for the liners. Ensure that the oilways are completely clear in the crankshaft, cylinder block and cylinder head.
2   Dissolve the anti-rust film which will be found on the piston/liner new parts. Do not scrape them. Use methylated spirit or cellulose thinners as a solvent.
3   Check the oil pump condition, and the cylinder head gasket surface for truth.
4   Mark up the parts in the piston/liner kit, so that they can be kept in four separate sets from this stage onwards.
5   If old pistons and liners are being re-used, ensure that each item is refitted to its former position, that the piston rings are the right way up, and that they do not seize in the grooves when compressed. If they do, it is likely that some carbon still remains in the ring land, and that further cleaning is required.

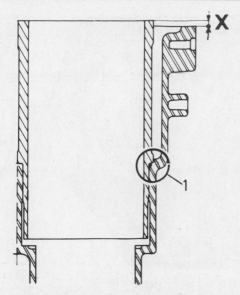

Fig. 1.9 Liner protrusion detail (Sec 20)

1   Liner seal position   X   Liner protrusion

20.11 The piston rings in place

Cylinder liner protrusion
6   The liners have a designed amount of protrusion above the cylinder block, and this should be checked by fitting the liner into the block without the O-ring seal, placing a straight-edge across the liner top, and checking the gap with a feeler gauge (see Fig. 1.9). The protrusion should be within the specified limits.
7   Note that the difference in protrusion between any two adjacent liners must not exceed 0.0016 in (0.04 mm).
8   Note that with a new set of liners, the protrusions should be stepped down from one end or the other. Number the liners from one to four to confirm their positions. If incorrect protrusion is found with a new liner set, the block is probably at fault.
9   With an old block and liners, incorrect protrusion means that either may be faulty. Eliminate one possible fault by retesting using a new liner set.

Gudgeon pin fitting
10   Fitting of pistons to connecting rods must be carried out by a Renault agent, owing to the special tooling and heating needs involved. Check when this work is complete that the pistons have been fitted with the arrows pointing in the direction of the flywheel when the big-end identification marks (see Section 11, paragraph 4) are facing away from the camshaft.

Piston rings
11   Fit the oil scraper ring, followed by the tapered compression ring (with the mark upwards), and lastly the top compression ring. Use the method described in Section 11 to assist ring fitting (photo).
12   Do not attempt to adjust the piston ring gaps, which are preset.

Piston fitting to liner
13   Space the rings round the piston so that the gaps are at 120° to one another.
14   Oil the pistons thoroughly and, with the aid of a suitable piston ring clamp, fit each piston to its respective liner, ensuring that the machined sides of the big-end are parallel with the flat on top of the liner (photo).

Fitting the liner and piston assembly to the cylinder block
15   Fit an O-ring to each liner, ensuring that it is not twisted.
16   Fit a half shell bearing to the connecting rod. Insert the piston and liner assembly into its correct bore, and the right way round, ie:

  (a) No 1 liner assembly at the clutch end
  (b) The numbers on the big-ends on the side away from the camshaft
  (c) The arrows on the pistons to be facing the flywheel

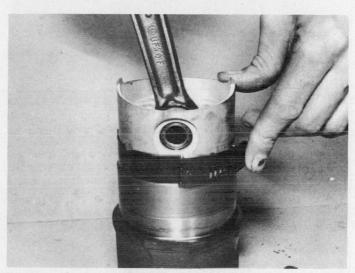

20.14 Fitting a piston assembly to a liner

17   Fit the liner clamping arrangement (see Section 7, paragraph 7). If the crankshaft is fitted, reassemble the big-end bearings as described in Section 21.

---

21 Crankshaft and main bearings, type 847 engine – refitting

1   Clean all the gasket surfaces on the cylinder block and associated components. Ensure that all parts are clean and all oilways clear.
2   Fit the bearing half shells to the connecting rods, and the main bearing half shells to their recesses in the crankcase. Note the oil hole positions.
3   Lubricate the shells, and lower the crankshaft into place.
4   Slip the thrust washers into place, with the white-metalled sides towards the crankshaft. Squirt oil onto them.
5   Place the main bearing shells in the caps (note that these have no oil feed holes), lubricate the shells, and fit the caps into their proper places. Tighten the bolts to the specified torque. Check for freedom of crankshaft rotation. Some stiffness is to be expected with new components.
6   Measure the endfloat of the crankshaft, using either a dial gauge operating on the flywheel flange, or by testing with feeler gauges between a thrust washer and the adjacent crankshaft abutment (photo). If the endfloat is outside the specified limits, change the thrust washers as necessary for a pair of suitable thickness.

21.6 Checking the crankshaft endfloat

21.8 Tightening a big-end bearing cap

7   Fit the rear oil seal as instructed in Section 13.
8   Lubricate the crankpins, pull the connecting rod big-ends on to them, and fit the big-end caps. Ensure that the caps are fitted to the correct rods, and refit and tighten the nuts to the specified torque (photo). Check again for freedom of crankshaft rotation.
9   Proceed in reverse of the dismantling procedure (Section 12, paragraphs 1 to 4).

22 Camshaft type 847 engine – refitting

1   If necessary, fit a new key, fit a new flange, and tap a new distance piece on using a piece of tube, until the distance piece touches the shoulder.
2   Fit the camshaft sprocket and tighten the nut to the specified torque. Check dimension (J), see Fig. 1.10.
3   Remove the sprocket. Lubricate the shaft, and refit it. Fit and tighten the camshaft flange bolts.
4   Refit the tensioner, timing chain and sprockets, and the timing cover (see Section 23).
5   Refit the remaining items in reverse of the removal procedure (see Section 10), noting the following paragraphs.
6   Lubricate the tappets before refitting.

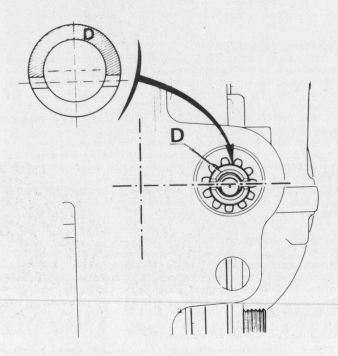

Fig. 1.11 Positioning the distributor drive pinion (Sec 22)

D   Large offset

7   To refit the distributor driving pinion, first turn the crankshaft to place No 1 piston at TDC on the compression stroke. (Note that the valves on No 4 cylinder will both be open at this point).
8   With a 12 mm dia x 1.75 mm pitch bolt inserted in the distributor drive pinion, insert the pinion so that the slot is at right angles to the centre-line of the engine, and with the large offset facing the clutch (see Fig. 1.11).

23 Timing cover and seal, timing chain, tensioner and sprockets, type 847 engine – refitting

Timing chain, tensioner and sprockets
1   Ensure that all gasket faces on the sump, cylinder block, timing cover and sump are completely clean.
2   Temporarily fit the camshaft sprocket, refit the crankshaft sprocket, and align the marks are indicated in Fig. 1.12. Remove the

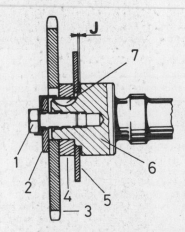

Fig. 1.10 Camshaft flange clearance (Sec 22)

1	Bolt	5	Flange
2	Washer	6	Camshaft
3	Sprocket	7	Key
4	Distance piece		

J = 0.003 to 0.004 in (0.06 to 0.11 mm)

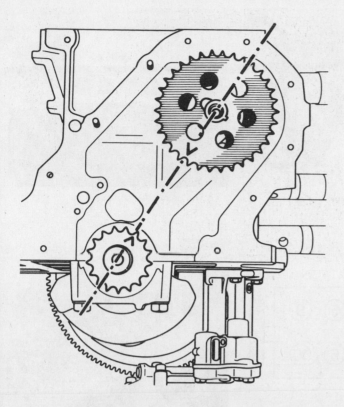

Fig. 1.12 Timing sprocket alignment (Sec 23)

camshaft sprocket without disturbing the setting.

3   Fit the timing chain over the crankshaft sprocket, and slip the camshaft sprocket into mesh with the chain. Refit the sprocket to the camshaft, ensuring that the marks remain in line.

4   Refit the timing chain tensioner assembly, with the spindle slot over the pin and the spring tag hooked into the hole in the cylinder block (photo).

Timing cover and seal

5   Support the timing cover adequately and place a little gasket cement around the periphery of the oil seal. Tap the seal carefully into place using a block of wood or other suitable item until the inner edge of the seal is flush with the inner edge of the cover (photo).

6   Use a new gasket, and loosely refit the timing cover, leaving the bolts finger tight.

7   Lubricate the oil seal lips, and refit the crankshaft pulley, or the pulley sleeve if fitted, to permit the timing cover to centralise itself about the pulley. Nip up two or three of the cover bolts, and remove the pulley or sleeve (photo).

8   Tighten down all the cover bolts, working progressively and evenly.

9   Refit the crankshaft pulley, jamming the flywheel if necessary to allow the bolt to be tightened.

10   Refit the remaining items in reverse of the removal procedures (see Section 9).

24 Sump and oil pump, type 847 engine – refitting

1   Ensure that the faces of the cylinder block and sump tray are thoroughly clean.

2   Refit the oil pump and tighten the securing bolts (photo).

3   Place the front and rear main bearing rubber seals in position.

4   Smear a suitable jointing compound on the sump side gaskets and stick them in place on the sump. Ensure that there is adequate jointing compound at the ends (photos).

5   With the pistons of Nos 1 and 4 cylinders at BDC, offer up the

23.4 The timing chain and associated components, refitted. Note that gearwheel marks are in alignment

23.5 The timing cover oil seal, fitted flush with the inside

23.7 The timing cover in place, with the pulley sleeve in position

24.2 Refitting the oil pump

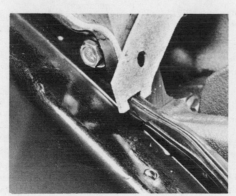

24.4a A sump side gasket, and the front rubber seal

24.4b A sump side gasket, and the rear rubber seal

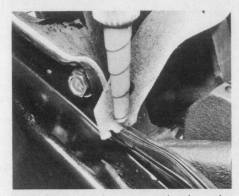
24.4c Placing jointing compound at the gasket connections

24.5 Fitting the sump (engine on bench)

25.6 The head gasket in position

25.7a Positioning the cylinder head

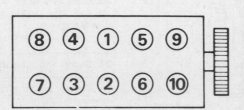
Fig. 1.13 Tightening and slackening sequence for cylinder head bolts on all engines (Sec 25)

25.7b Tightening the cylinder head bolts

sump carefully and fit the bolts (photo). Tighten progressively and evenly.
6   Refit the undertray.
7   Refill the sump with oil.

25 Cylinder head, engine type 847 – refitting

1   Before refitting, carry out an examination and any necessary renovation work, as described in Section 17.
2   It is strongly advised that no scraping of any kind should be carried out on aluminium parts, and instead a cleaning solution (available from Renault dealers) should be employed to clean the cylinder head and other mating surfaces. This liquid should be used according to the instructions supplied, and must not be permitted to contact bare skin or any paintwork.
3   Do not allow foreign matter to enter the oilways upon the cylinder block and head surfaces, or serious damage due to oil shortage may result.
4   Ensure before refitting the cylinder head that all oil in the cylinder head bolt holes on the block surface is removed, thus ensuring that the precise head tightening torques necessary are obtainable.
5   Remove the cylinder liner clamping arrangement, and check that cylinder liner protrusion is satisfactory (see Section 20).
6   Place the new cylinder head gasket in position with the marking 'HAUT-TOP' uppermost. Ensure that the correct gasket has been supplied (photo).
7   Carefully position the cylinder head on the gasket, screw the head bolts in by hand, and tighten them in stages to the specified torque in the order shown in Fig. 1.13 (photos).
8   Fit the pushrods, and check the valve clearances (see Section 4).
9   If the engine is still in the vehicle, reverse the procedure given in Section 7, paragraphs 1 and 3, as applicable. Leave off the air filter assembly.
10 Fill the sump with oil.

11 Refill the cooling system (see Chapter 2), bleed, and then refit the air cleaner assembly.
12 Check the ignition timing and idle speed.
13 Run the engine for twenty minutes, switch it off, and leave it for two and a half hours before retightening the cylinder head. First loosen bolt No 1 (see Fig. 1.13) and tighten again to the specified torque figure. Repeat this procedure for all the other bolts in the specified sequence. No further retightening should be necessary.
14 Adjust the valve clearances again (see Section 4).

26 Engine, type 847 – refitting

1   To ease fitting the clutch shield, use two studs threaded 8 mm dia x 1.25 pitch x 20 mm long inserted in the holes through which the clutch shield bolts pass, thus providing a means of temporarily locating the shield.
2   Sparingly lubricate the splines on the gearbox shaft, using Molykote BR2 grease, or equivalent.
3   Raise the engine on the lifting tackle, and lower carefully into place, whilst locating the gearbox shaft carefully in the clutch centre.
4   Refit the top bolts, securing the engine and gearbox together.
5   Refit the nuts and washers securing the engine mountings to the body mountings.
6   Remove the two temporary studs (see paragraph 1) and fit the lower engine-to-gearbox bolts.
7   Refit all the remaining parts in reverse of the removal procedure, noting the following points:

   (a) Adjust the accelerator cable (see Chapter 3)
   (b) Remove the plugs or clamps from the fuel pipes
   (c) Adjust the clutch cable (see Chapter 5)
   (d) Ensure that the sump plug is in position and refill with engine oil
   (e) Fill the cooling system and bleed, leaving off the air cleaner assembly until this has been completed

27 Engine type 847 and gearbox – refitting

1   Attach the lifting tackle, and lower the combined unit into the vehicle until the engine mountings are located, and the unit is resting on the crossmember. Remove the lifting tackle.
2   Raise the gearbox and refit the gearbox mountings.
3   Reverse the procedure given in Section 5, paragraphs 1 to 5, ensuring that the splines of the driveshafts are lubricated with Molykote BR2 grease or equivalent before refitting.
4   Observe the points mentioned in Section 26, paragraph 7.

PART C : TYPE 841 ENGINE (1647 cc)

28 Engine type 841 and gearbox – removal

1   Proceed as described in Section 29, paragraphs 1 to 12.
2   Detach the clutch cable at the bellhousing lever, and take the cable from the stop.
3   Remove the distributor (see Chapter 4).
4   Disconnect the driveshafts (see Chapter 8).
5   Disconnect the speedometer cable.
6   Disconnect the gear selector mechanism (manual gearbox) at the

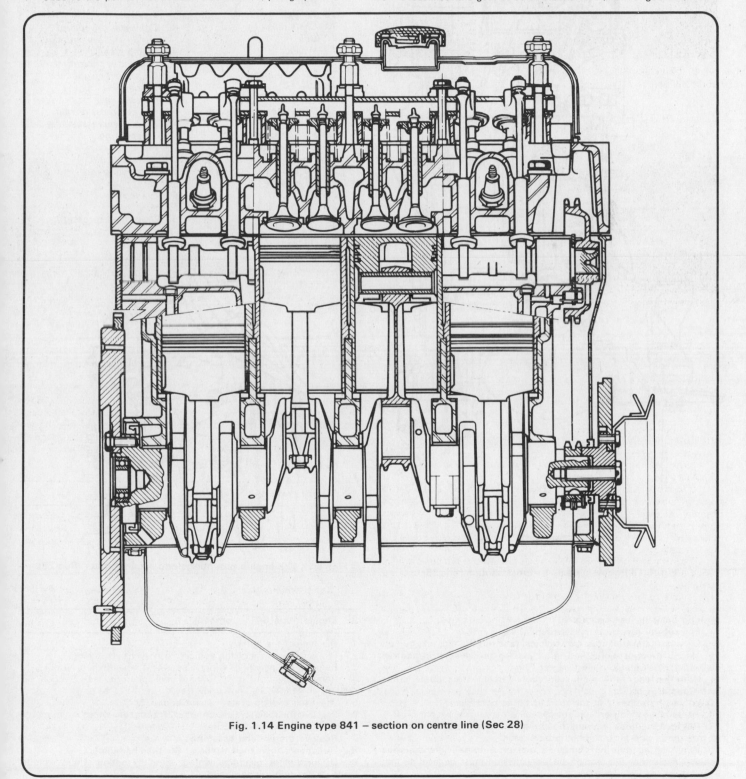

Fig. 1.14 Engine type 841 – section on centre line (Sec 28)

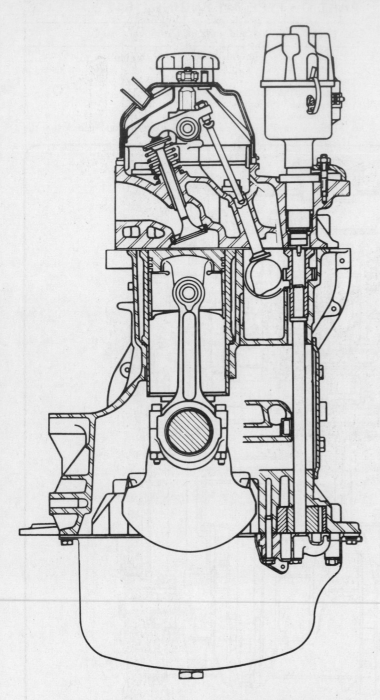

Fig. 1.15 Engine type 841 – cross section (Sec 28)

Fig. 1.16 Engine type 841 (less cylinder head) – exploded view
(Sec 28)

1	Camshaft sprocket	30	Bolt
2	Timing chain	31	Washer
3	Chain guide	32	Gasket
4	Chain guide	33	Core plug
5	Washer	34	Oilway blanking plug
6	Bolt	35	Washer
7	Bolt	36	Flywheel bolt
8	Washer	37	Locking plate
9	Bolt	38	Dowel
10	Washer	39	Flywheel
11	Chain tensioner	40	Converter driveplate
12	Bolt		(automatic transmission)
13	Washer	41	Clamp plate
14	Dowel		(automatic transmission)
15	Locating plate	42	Crankshaft spigot
16	Key		bearing
17	Camshaft	43	Cylinder liner
18	Distributor drive pinion	44	Top compression ring
19	Endplate	45	Taper compression ring
20	Endplate gasket	46	Scraper ring
21	Cylinder block	47	Piston
22	Dowel	48	Big-end bolt
23	Gasket	49	Connecting rod
24	Cover plate	50	Nut
25	Bolt	51	Crankshaft
26	Washer	52	Key
27	Plug	53	Crankshaft sprocket
28	Washer		
29	Oil pump driveshaft blanking plate		

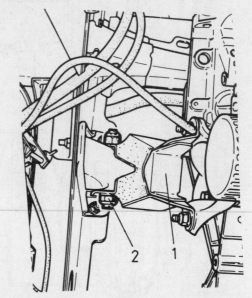

Fig. 1.17 The engine mountings and securing nuts (Sec 28)

1   Mounting   2   Securing nut

pivoting balljoint (see Chapter 6).
7   Pull off the reversing light switch wires.
8   On automatic transmission vehicles, take out the dipstick, pull off the electrical connections, and (after placing the selector in neutral) disconnect the selector (see Chapter 7).
9   With the rear of the gearbox supported, detach the rubber mounts (see Chapter 6 or 7).
10   Lower the gearbox at the rear, as far as possible.
11   Fit suitable lifting equipment and take the engine weight on it.
12   Remove the nuts securing the engine mountings, raise the engine to free them, and lift out the combined unit.
13   When lifting out the combined unit on automatic transmission vehicles, note that the rear of the transmission must be level so that it will clear the crossmember.

29 Engine, type 841 – removal

1   Disconnect the battery.
2   Protect the wings of the vehicle, to prevent scratching.
3   Remove the headlight wiper blades and pipes (where fitted).
4   Remove the front grille, bonnet lock platform (see Section 6), radiator screen, and engine undertray.
5   Drain the cooling system (see Chapter 2).
6   Remove the radiator (see Chapter 2), complete with the fan motor assembly and expansion bottle.
7   Remove the air filter assembly.
8   Disconnect the exhaust pipe at the manifold joint.
9   Disconnect all electrical wiring, carefully noting where various leads go, for refitting purposes.

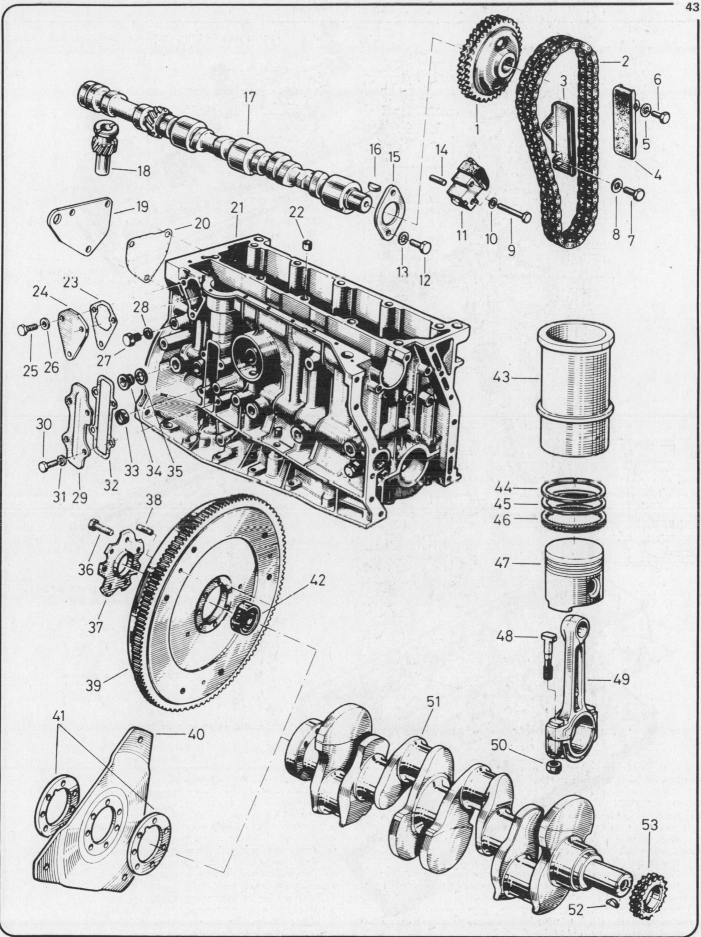

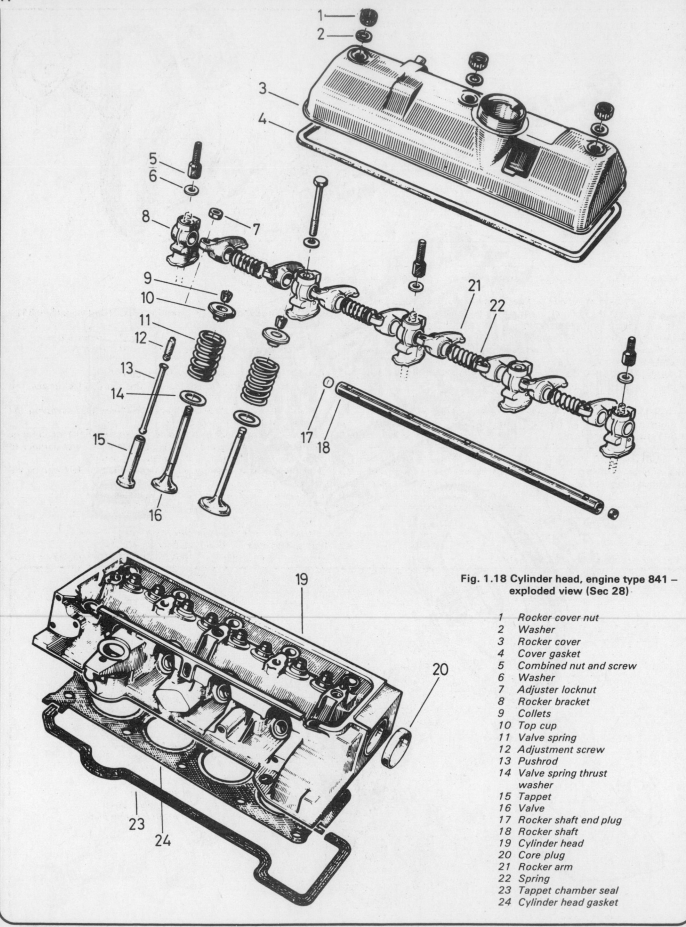

Fig. 1.18 Cylinder head, engine type 841 –
exploded view (Sec 28)

1   Rocker cover nut
2   Washer
3   Rocker cover
4   Cover gasket
5   Combined nut and screw
6   Washer
7   Adjuster locknut
8   Rocker bracket
9   Collets
10 Top cup
11 Valve spring
12 Adjustment screw
13 Pushrod
14 Valve spring thrust
    washer
15 Tappet
16 Valve
17 Rocker shaft end plug
18 Rocker shaft
19 Cylinder head
20 Core plug
21 Rocker arm
22 Spring
23 Tappet chamber seal
24 Cylinder head gasket

10 Disconnect the heater hoses at the water pump, and the vacuum hoses to the servo (and automatic transmission capsule, where applicable).

11 Remove the accelerator cable swivel and (where applicable) the governor cable, to avoid upsetting the cable adjustment.

12 Clamp the fuel pipes from the tank, and disconnect them at the pump.

13 Remove the lower bolts securing the engine to the gearbox.

14 On automatic transmission vehicles only, remove the converter shield, and take out the converter-to-driveplate bolts.

15 Take the upper engine-to-gearbox bolts out.

16 Take the engine weight on suitable lifting gear, disconnect the engine side mountings, slightly raise the engine and pull it forwards. Lift the engine out.

17 On automatic transmission models only, fit a retaining plate as soon as separation takes place to prevent the converter from moving. Refer to Chapter 7 if necessary.

30 Cylinder head, engine type 841 – removal

1 Proceed largely as described in Section 7, with the following differences.

2 The air filter assembly need not be removed, but instead remove the elbow between the carburettor and air filter.

3 The alternator need not be removed.

4 After tapping the cylinder head to unstick it, raise the head slightly and either push the tappets back into the cylinder head, or withdraw them and note their positions.

5 Take the same precautions over the cylinder liners as described in Section 7, paragraphs 5 to 7.

31 Timing cover and seal, timing chain, tensioner and sprockets, type 841 engine – removal

1 It is possible for the timing cover and seal to be removed and refitted without removing the cylinder head. However, the complexity of the procedure combined with the need for special tools leads us to advise the home mechanic not to undertake the work in this way. The procedure is described as it may be undertaken when the cylinder head is removed.

Timing cover
2 Remove the cylinder head (see Section 30).

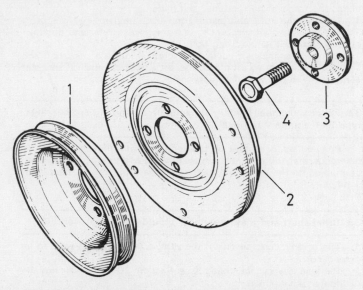

Fig. 1.19 Crankshaft pulley assembly – exploded view (Sec 31)

1	Pulley	3	Pulley boss
2	Damper	4	Boss bolt

3 Remove the sump as described in Section 34.

4 Remove the radiator as described in Chapter 2. Also remove the expansion bottle.

5 Remove the headlight wiper blades (if fitted), disconnecting the water tubing at the ends of the arm.

6 Remove the front grille by taking out the screws in the top corners.

7 Remove the bolts securing the crankshaft pulley and damper to the pulley boss, and withdraw the pulley and damper.

8 Lock the flywheel, and remove the pulley hub bolt, followed by the hub.

9 Remove the bolts and withdraw the timing cover.

10 Carefully tap out the oil seal in the cover.

Timing chain, tensioner and sprockets
11 Proceed as described in paragraphs 1 to 10.

12 Secure the tensioner shoe with a length of soft wire, take out the

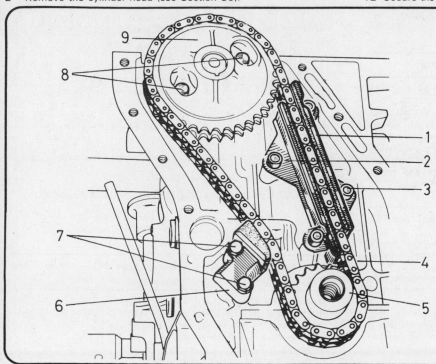

Fig. 1.20 Timing gear (Sec 31)

1   Guide
2   Guide
3   Guide bolt
4   Chain
5   Crankshaft sprocket
6   Chain tensioner
7   Tensioner bolts
8   Locating plate bolts
9   Camshaft locating plate

bolts, and remove the tensioner and plate.
13 Remove the nuts, and take off the chain guides.
14 Set the camshaft sprocket to give the necessary access, and take out the camshaft flange fixing bolts.
15 Using the crankshaft pulley bolt screwed into the crankshaft and a suitable puller, draw off the crankshaft sprocket, camshaft sprocket complete with the camshaft, and chain.

32 Pistons, cylinder liners, connecting rods and big-end bearings, type 841 engine – removal

Proceed as described in Section 11, with the one important difference that the connecting rods and big-end bearing caps should be marked on the *same* side as the camshaft (Section 11, paragraph 4 refers).

33 Crankshaft and main bearings, type 841 engine – removal

1 The engine must be out of the vehicle to permit this work to be carried out.
2 Remove the cylinder head (see Section 30), and take out the distributor drive pinion.
3 Remove the clutch assembly (see Chapter 5).
4 Remove the timing cover, timing chain and sprockets (see Section 31).
5 Mark the connecting rods and big-end caps, number one at the clutch end, on the same side as the camshaft. Remove the bolts, and take off the big-end caps and bearings.
6 Remove the flywheel (see Section 36).
7 Mark the main bearing caps relative to the cylinder block (No 1 at the flywheel end), take out the bolts, and remove the bearing caps and bearings, tapping bearing cap number one lightly underneath to remove it if necessary.
8 Remove the oil seal and side seals, and lift out the crankshaft, bearings and thrust washers.

34 Sump and oil pump, type 841 engine – removal and refitting

1 Proceed as described in Sections 8 and 24, noting the following paragraphs.
2 Tilting of the sump is unnecessary.
3 Remove the bolts shown in Fig. 1.21. Remove the pump and both pinions together.
4 When refitting the sump, note the bolt positions shown in Fig. 1.22.

35 Camshaft, type 841 engine – removal and refitting

1 These operations are described in Sections 31 and 44, but note the points which follow.
2 If the camshaft sprocket has to be removed from the shaft, it must not be re-used. A new sprocket must always be fitted, together with a new flange and key. When pressing on the sprocket, take the load by supporting the camshaft behind the first bearing. Press the sprocket fully home.
3 Check the clearance at the flange (see Fig. 1.23) and check against the camshaft endfloat as given in the Specifications.

36 Flywheel (or driveplate), oil seal and spigot bearing, type 841 engine – removal and refitting

1 Removal and refitting is largely as described in Section 13. However, the following should be noted.
2 On manual gearbox vehicles, fit a new lockplate under the flywheel retaining bolts, and after tightening to the specified torque bend the lockplate over a flat on each bolt.
3 A suitable tool will be necessary to extract the spigot bearing, which is in fact a ball race. Tap the new race into position using a suitable diameter piece of tube bearing only upon the outer bearing track. Do not clean the bearing before fitting, as it is grease-packed ready for use.

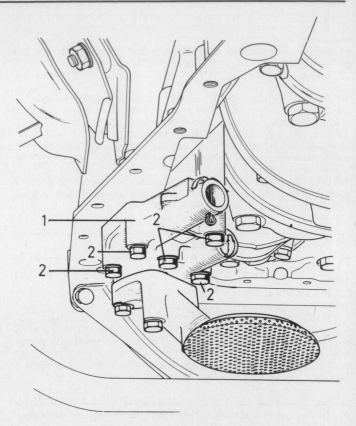

Fig. 1.21 Oil pump in position (Sec 34)

1   Pump   2   Pump locating bolts

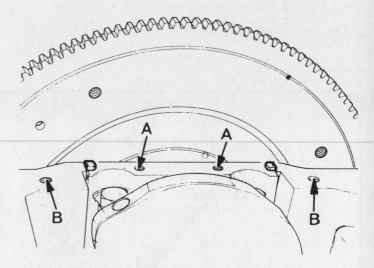

Fig. 1.22 Bolt positions when refitting the sump (Sec 34)

A   Use bolts with deep slotted heads
B   Use bolts with normal slotted heads

37 Inlet and exhaust manifold, engine type 841 – removal and refitting

1   Disconnect the battery.
2   Remove the air filter elbow and the warm air intake pipe.
3   Disconnect the fuel pipes.
4   Disconnect the oil fume hose.
5   Disconnect all vacuum hoses.
6   Remove the accelerator cable swivel.

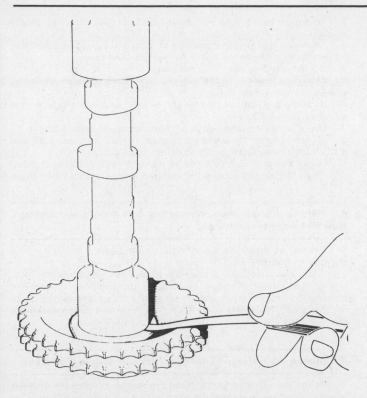

Fig. 1.23 Checking the clearance at the camshaft sprocket flange (Sec 35)

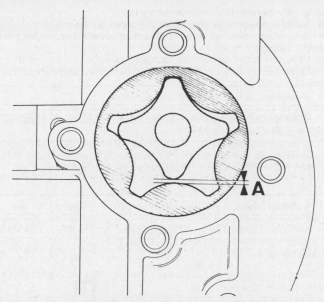

Fig. 1.25 Rotor clearance, position 1 (Sec 38)

A = 0.002 to 0.012 in (0.04 to 0.29 mm)

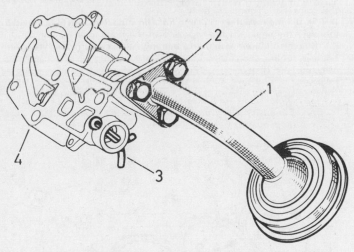

Fig. 1.24 Oil pump (Sec 38)

| 1 | Suction pipe | 3 | Split pin |
| 2 | Suction pipe bolts | 4 | Pump body |

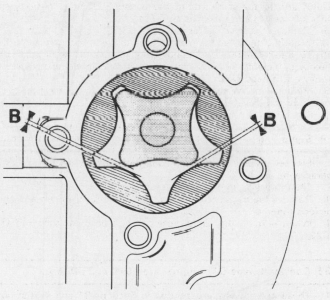

Fig. 1.26 Rotor clearance, position 2 (Sec 38)

B = 0.001 to 0.006 in (0.02 to 0.14 mm)

7 Clamp the coolant hoses to the carburettor, and disconnect them.
8 Remove the exhaust pipe clamp.
9 Remove the warm air pipe and hose clip.
10 If the manifold is being renewed, remove the vacuum unions and the carburettor.
11 Unbolt and remove the manifold.
12 To refit, reverse the removal procedure. Ensure that the manifold mating surfaces are clean, and that a new gasket is used.

38 Oil pump, engine type 841 – dismantling and reassembly

1 Remove the bolts, followed by the suction pipe.
2 Take out the split pin from the pressure relief valve, and then the cup, spring and piston.
3 Clean and check all parts.

4 Place the pinions in the cylinder block and check the clearances shown in Figs. 1.25 and 1.26. If either clearance is excessive, discard both pinions complete with the driving spindle.
5 Refit all parts in reverse order, tighten the suction pipe bolts and bend over the locking tabs.
6 Insert the pinions in the block, and refit the pump.

39 Cylinder head, rockers and valves, type 841 engine – dismantling and reassembly

1 With the cylinder head removed from the engine, remove the water pump pulley and pump, and the manifold assembly complete with carburettor.
2 Tie string, or stretch a rubber band, round the rocker shaft assembly, remove the bolts and nuts, and lift the assembly off.
3 Remove the valve spring assemblies and dismantle the rocker shaft assembly (see Section 16, paragraphs 7 to 10), noting that

circlips are not fitted to the rocker shaft.
4   To refit, proceed as described in Section 19, paragraphs 1 to 3, and 5 to 7, again noting that circlips are not fitted to the rocker shaft. Ensure that the end pillars are located on the dowels.
5   Refit the items listed in paragraph 1 above.

40 Examination and renovation – general

1   Refer to Section 17. The remarks made concerning the 847 type engine are broadly applicable to the 841 type also.
2   Note additionally that if the tappets (cam followers) and/or their bores are worn, the bores can be reamed and oversize tappets fitted. Consult a Renault dealer or motor engineer if this is necessary.

41 Cylinder block, bare, engine type 841 – preparing for use

1   No scraping of aluminium parts is permissible, and a cleaner as available from Renault agents should be employed, used as directed on the container.

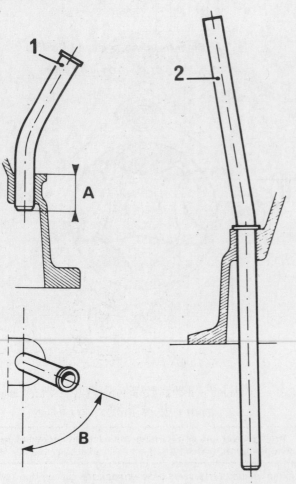

Fig. 1.27 Dipstick tube position (Sec 41)

A = $\frac{29}{32}$ in (23 mm)
B = 70° (tube pointing towards front)
1   Dipstick tube

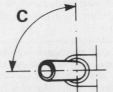

Fig. 1.28 Oil recovery pipe (Sec 41)

2   Pipe   C   90°

2   Take great care that foreign matter does not cause blockage of any oilways.
3   To ensure that proper tightening of bolts is achieved, suck or blow all remaining oil from the bolt holes and oil feed holes.
4   Do not interfere with the core plugs.
5   Check for freedom of the cylinder head bolts in their threads by screwing them in and out.
6   If necessary, tighten the main oilway screwed plugs to the specified torque figures.
7   Use a suitable sealant on the fuel pump studs, before fitting.
8   Fit the dipstick tube and oil recovery pipe (see Figs. 1.27 and 1.28). Use thread locking compound on them before fitting.
9   Using a new gasket, refit the oil pump driveshaft blanking plate.
10   Clean (but do not scrape) the surfaces for the cylinder liner seals.

42 Pistons, cylinder liners, connecting rods and big-end bearings, type 841 engine – refitting

1   Proceed as described in Section 20, paragraphs 1 to 17, but note the following difference.
2   When the pistons have been assembled to the connecting rods, ensure that with the arrows pointing to the flywheel, the big-end identification marks made upon dismantling are on the same side as the camshaft. Therefore, when fitting the liner and piston assembly, this will also apply.

43 Crankshaft and main bearings, type 841 engine – refitting

1   Fit the main bearing shells in the cylinder block, with the oil holes aligned, lubricate the shells and crankshaft journals, and lower the shaft into the block.
2   Slip the thrust washers into place with the white-metalled sides to the crankshaft. Lubricate them.
3   Fit the main bearing shells to bearing caps numbers 2, 3, 4 and 5, lubricate the shells, fit the caps and fit the bolts loosely.
4   Select suitable number one bearing cap side seals, by fitting the bearing to the cap, fitting the cap, and nipping up the cap bolts. Measure the dimension C (Fig. 1.31), and select suitable seals as follows:

(a) *Dimension C = 5 mm or less, use 5.10 mm thick seals*
(b) *Dimension C = more than 5 mm, use 5.4 mm thick seals (coded white)*

Fig 1.29 Direction of the piston crown arrows, when the piston/liner assemblies are fitted (Sec 42)

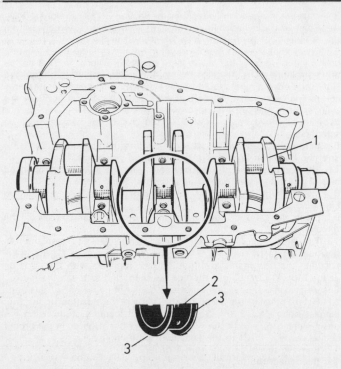

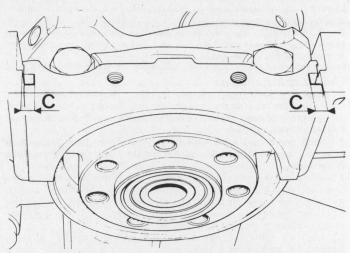

Fig. 1.31 Dimension C to be measured when selecting side seals for No 1 main bearing cap (Sec 43)

Fig. 1.30 Crankshaft, half shells and thrust washers in position (Sec 43)

1 Crankshaft   3 Thrust washers
2 Half shell bearing

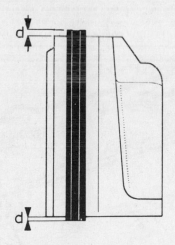

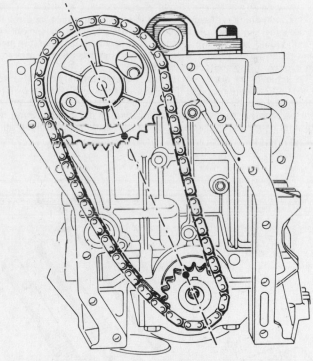

Fig. 1.32 Main bearing cap side seal fitted (Sec 43)

d = 0.008 in (0.2 mm) approximately

Fig. 1.33 Timing marks and crankshaft key, correctly aligned (Sec 44)

44 Timing cover and seal, timing chain, tensioner and sprockets, type 841 engine – refitting

Timing chain, tensioner and sprockets

1   Turn the crankshaft to place the key at the top.
2   Fit the chain to the sprockets, with the timing marks and keyway aligned as in Fig. 1.33.
3   Start entering the camshaft, register the crankshaft sprocket with the crankshaft and key at the appropriate point, and tap the crankshaft sprocket on gently as necessary.
4   Fit and lock the camshaft flange retaining bolts.
5   If re-using the old chain tensioner, take the shoe from the tensioner body and lock the piston in the shoe using a 3 mm hexagon key. Refit the shoe in the body with a shim about 0.080 in (2 mm) thick between body and shoe to prevent premature setting.
6   If using a new chain tensioner, leave the plastic keep plate in position.
7   Fit the tensioner and tighten the securing bolts. Set the tensioner,

Remove the bearing cap.
5   Fit both seals to the bearing cap with the seal groove facing outwards. The seal should be approximately 0.008 in (0.2 mm) proud of the joint cap surface.
6   Fit the bearing shell, and lubricate the shell and side seals.
7   Obtain two studs of suitable length, 10 mm diameter by 1.5 mm pitch, screw them into the bearing bolt holes, offer up the bearing cap to the studs, and with two foil strips between the block and seals to prevent damage to the seals, persuade the cap down. As the cap is nearly home, check that the side seals are still projecting by checking with a rule. Take out the foil strips and the studs, fit the cap bolts, and tighten all bolts to the specified torque. Cut off the protruding tips of the seal at the cap and sump surface.
8   Proceed as described in Section 21, paragraphs 6 to 8.
9   Fit the remaining items in reverse of the removal procedure.

by removing the plastic keep or shim, pressing the shoe to the bottom, and releasing it.

8   Fit the chain guides with the bolts left loose. Place a $\frac{1}{32}$ in (0.8 mm) spacer strip between the chain and each guide, tighten the guide bolts, and remove the spacer strips.

Timing cover and seal

9   Carefully fit a new oil seal to the crankshaft aperture, by supporting the cover at the rear and tapping the seal in from the front, using a piece of tube of suitable diameter. The seal should be flush at the rear of the aperture.

10   Stick the timing cover gaskets to the cylinder block face, using grease or a little jointing compound. Fit the cover, leaving the bolts finger tight.

11   Lubricate the timing cover seal lip, and fit the crankshaft pulley boss, allowing it to centralise the cover (and thus the oil seal) around the boss. Check that the timing cover top face is flush with the cylinder block, and tighten the cover bolts. Trim the edges of the gasket flush, at the surface of the timing cover and sump.

12   Refit the remaining items in reverse of the procedure given in Section 31, paragraphs 2 to 7.

45 Cylinder head, engine type 841 – refitting

1   The positioning of the cylinder head is very important, determining as it does the alignment of the distributor drive spindle and pinion. This in itself would present no problem, but Renault say that the head gasket must not be moved once fitted, nor may the head be moved to align it once it has contacted the gasket.

2   To achieve the above conditions, Renault special tools Mot 451 and Mot 412.01 are required (Fig. 1.34). If these tools cannot be hired or borrowed locally – it is unlikely to be economic to buy them – it might be possible to improvise the locating studs at least, using two long bolts of the same diameter and pitch as the cylinder head bolts.

3   Observe the cautions described in Section 25, paragraphs 1 to 4.

4   Remove the cylinder liner clamping arrangement, and check that the cylinder liner protrusion is satisfactory (see Section 42).

5   Check that the distributor drive pinion and cylinder head locating dowel are in place in the block. With the No 1 cylinder at TDC on the firing stroke (ie with the cams for No 4 cylinder both in operation) the slot in the drive pinion should be parallel with the engine longtitudinal centre-line, with the smaller offset adjacent to the camshaft. Place the cylinder head gasket into position on the block surface, and *do not allow it to move again*. If this should occur, the gasket must be scrapped.

6   Screw the studs from tool Mot 451 into the holes (see Fig. 1.35) until the balls just hold the gasket down to the block.

7   Fit the tappet chamber rubber seal, with the ends properly linked into the cylinder head gasket as shown in Fig. 1.35.

8   Insert gauge Mot 412.01 in the cylinder block hole indicated in Fig. 1.35.

9   Clean the cylinder head bolts, and put a little engine oil on the threads.

10   Tap each tappet into place in the cylinder head to keep them in position. Smear them with a little grease if necessary. Fit the head, ensuring that the manifold mounting face registers with the arms on gauge Mot 412.01 before the head is permitted to touch the gasket. Ensure that the tappet chamber seal does not come out of position.

11   Fit six bolts in the vacant holes, noting the position of the four longest bolts (Fig. 1.36). Remove the two locating studs using the unscrewing tool, and fit the remaining two bolts. Tighten in the proper sequence (Fig. 1.13) to the specified torque figure.

12   Remove the alignment tool Mot 412.01.

13   Reconnect the fuel pipes.

14   Fit the pushrods.

15   Refit the distributor (see Chapter 4) and connect the wires. Set the points gap and the static ignition timing if necessary.

16   Use a suitable spanner on the crankshaft pulley bolt to turn the engine, and set the valve clearances.

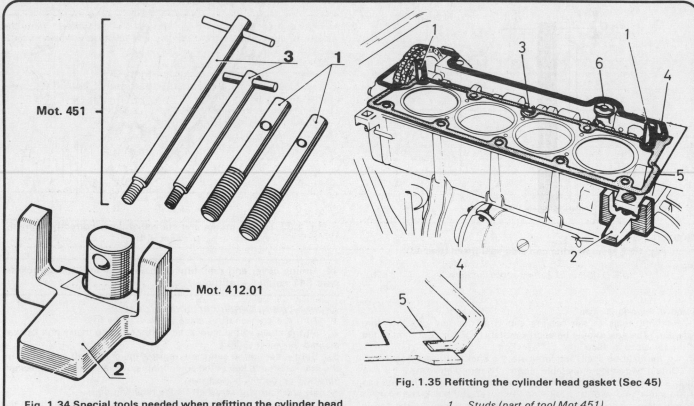

Fig. 1.34 Special tools needed when refitting the cylinder head (Sec 45)

1   Gasket locating studs   3   Stud removers (short one only required)
2   Head positioning gauge

Fig. 1.35 Refitting the cylinder head gasket (Sec 45)

1   Studs (part of tool Mot 451)
2   Gauge (tool no. Mot 412-01)
3   Head locating dowel
4   Tappet chamber rubber seal
5   Head gasket
6   Distributor drivegear

17 Refit the rocker cover. Make sure the seal is in good condition.
18 Connect all wires, hoses and cables, securing as appropriate.
19 Fit and tension the drivebelt.
20 Fill and bleed the cooling system (see Chapter 2).
21 Adjust the ignition and carburettor settings as necessary (refer to Chapters 3 and 4).
22 Run the engine, switch off when the cooling fan starts to operate, and allow it to cool for two and a half hours.
23 Retighten the cylinder head, and reset the valve clearances. There is no need for any further tightening.

46 Engine – type 841 – refitting

1   Lightly lubricate the clutch shaft splies (manual gearboxes) or converter location (automatic gearboxes) using Molykote BR2 grease or equivalent.
2   Refit the cylinder block coolant drain plug.
3   Lower the engine into the vehicle and mate it with the transmission.
4   On automatic transmission vehicles, the driveplate blade with sharp corners must be lined up with the hole in the converter marked with a paint spot (see Chapter 7). Remove the converter retaining plate fitted during removal of the unit, at the appropriate point.
5   Refit the upper engine-to-gearbox bolts.
6   Reconnect and tighten the engine side mounting fixings.
7   Refit the lower engine-to-gearbox bolts.
8   Refit the flywheel shield.
9   On automatic transmission vehicles, refit the converter driving plate bolts and the converter shield.
10   Reverse the procedure given in Section 29, paragraphs 1 to 12.
11   Refill the engine with oil, and bleed the cooling system (see Chapter 2).

47 Engine type 841 and gearbox – refitting

1   Attach the lifting tackle to the combined unit and lower it into the vehicle. Where automatic transmission is fitted, this must be horizontal to allow it to clear the crossmember.
2   Raise the rear of the gearbox.
3   Refit the gearbox and engine mountings.
4   Proceed in reverse of the procedure given in Section 28, paragraphs 1 to 8.
5   Refill the sump and/or gearbox, as necessary.
6   Bleed the cooling system.

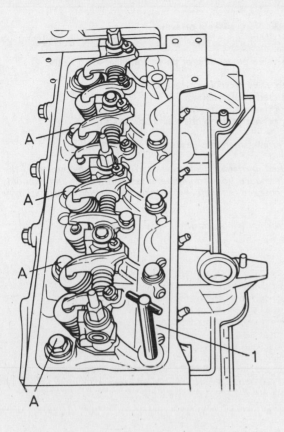

Fig. 1.36 Bolting down the head (Sec 45)

1   *Extracting tool Mot 451, for removing tooling studs*
A   *Positions of the four long cylinder head bolts*

PART D: ALL ENGINES

48 Fault diagnosis – engine

Symptom	Reason(s)
Engine will not turn when starter switch is operated	Flat battery Bad battery connections Bad connections at solenoid switch and/or starter motor Defective solenoid Starter motor defective
Engine turns normally but fails to start	No spark at plugs No fuel reaching engine Too much fuel reaching the engine (flooding)
Engine starts but runs unevenly and misfires	Ignition and/or fuel system faults Incorrect valve clearances Burnt out valves Worn out piston rings
Lack of power	Ignition and/or fuel system faults Incorrect valve clearances Burnt out valves Worn out piston rings

Symptom	Reason(s)
Excessive oil comsumption	Oil leaks from gaskets or seals Worn piston rings or cylinder bores resulting in oil being burnt by engine Worn valve guides and/or defective valve stem seals
Excessive mechanical noise from engine	Wrong valve to rocker clearances Worn crankshaft bearings Worn cylinders (piston slap) Slack or worn timing chain and sprockets
Poor idling	Leak in inlet manifold gasket

Note: *When investigating starting and uneven running faults, do not be tempted into snap diagnosis. Start from the beginning of the check procedure and follow it through. It will take less time in the long run. Poor performance from an engine in terms of power and economy is not normally diagnosed quickly. In any event, the ignition and fuel systems must be checked first before assuming any further investigation needs to be made.*

Chapter 2 Cooling system

For modifications, and information applicable to later models, see Supplement at end of manual

Contents

Specifications

System type ....
Sealed and pressurised. Centrifugal pump and thermostat. Electric cooling fan on some models

Coolant
Type ....
Ethylene glycol based antifreeze (Duckhams Universal Antifreeze and Summer Coolant)

Capacity, including heater:
R1340 ....
10.5 Imp pts (6.3 US qts, 6.0 litres)
R1341 ....
11.0 Imp pts (6.7 US qts, 6.3 litres)

Drivebelt adjustment, deflection
Engine type 841 ....
$\frac{3}{32}$ to $\frac{9}{64}$ in (2.5 to 3.5 mm)
Engine type 847 ....
$\frac{9}{32}$ to $\frac{5}{16}$ in (7 to 8 mm)

Thermostat
Location ....
Top hose, between water pump and radiator
Type ....
Wax
Opening temperature ....
86°C (187°F)
Fully open at ....
92° to 96°C (198° to 205°F)

1 General description and maintenance

The system is of pressurised type and sealed, but with the inclusion of an expansion bottle to accept coolant displaced from the system when hot and to return it when the system cools.

Coolant is circulated by thermosyphon action and is assisted by means of the impeller in the belt-driven water pump.

A thermostat is fitted in the outlet of the water pump. When the engine is cold, the thermostat valve remains closed so that the coolant flow which occurs at normal operating temperatures through the radiator matrix is interrupted.

As the coolant warms up, the thermostat valve starts to open and allows the coolant flow through the radiator to resume.

The engine temperature will always be maintained at a constant level (according to the thermostat rating) whatever the ambient air temperature.

The coolant circulates around the engine block and cylinder head and absorbs heat as it flows, then travels in an upward direction and out into the radiator to pass across the matrix. As the coolant flows across the radiator matrix, air flow created by the forward motion of the car cools it and it returns via the bottom tank of the radiator to the cylinder block. This is a continuous process, assisted by the water pump impeller.

Some models are fitted with an electric cooling fan which is actuated by the thermostat switch according to coolant temperature.

The car interior heater operates by means of water from the cooling system.

The carburettor is fitted with water connections to permit coolant from the cooling system to circulate to the carburettor base, and (on some models) to the automatic choke operating mechanism.

Maintenance

Apart from renewing the coolant at the prescribed intervals, maintenance is confined to checking the coolant level in the expansion bottle. The level should be between the MAX and MIN lines. The need for persistent topping up should be investigated.

The hoses and their clamps should also be inspected regularly for security and good condition, and the drivebelt checked and adjusted or renewed as necessary.

2 Antifreeze mixture – general

1   It is essential that an approved type of antifreeze is employed, in order that the necessary antifreeze and anticorrosion proportions are maintained.

2   Whilst the life of the coolant originally used in the vehicle is stated to be 3 years or 30 000 miles (45 000 km), owners are recommended to consider removing the coolant yearly to ensure that all the essential properties of the solution are fully maintained.

3   Make up the solution, ideally using distilled water or rain water, in the proportions necessary to give protection in the prevailing climate. Percentages of antifreeze necessary are usually found on the container. Do not use too low a percentage of antifreeze, or the anticorrosion properties will not be sufficiently effective.

4   If it is suspected that the coolant strength is unsatisfactory, a check may be made using a hydrometer. Employ the instrument as instructed by the manufacturer, and using the correction tables normally supplied. If the protection is found to be insufficient, drain off some of the coolant and replace it with pure anti-freeze. Recheck the coolant with the hydrometer.

3.3 Expansion bottle, with cap fitted

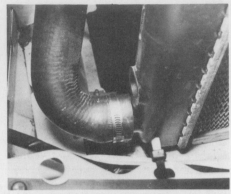
3.5a Radiator bottom hose

3.5b Radiator filler cap

3.6 Thermostat in place in the hose

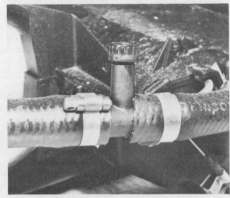
4.4a The heater hose bleed screw

4.4b The top hose bleed screw

5.2 Top crossmember bolts

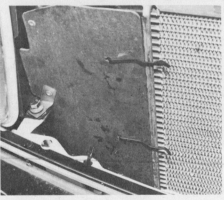
5.3 Screen clips

5.5 Electrical connection to the temperature sender

5.6 Lifting out the radiator

6.7a The water pump securing bolts (847 engine)

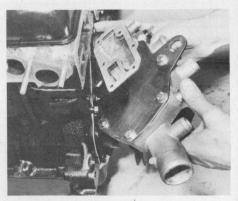
6.7b Lifting away the water pump (847 engine)

3   Cooling system – draining and flushing

Draining

1   If the coolant is known to be in acceptable condition for further use, have suitable containers available in which to catch it.
2   Set the heater facia control to 'Hot'.
3   Remove the expansion bottle cap (photo). Take care to avoid scalding if the system is hot.
4   Remove the drain plug from the cylinder block (where fitted).
5   Open the drain tap at the base of the radiator, where one is fitted. Alternatively, remove the bottom hose at the radiator stub. When the expansion bottle is empty, remove the filler plug from the radiator. Open all bleed screws (photos).

Flushing

6   Remove the thermostat from the hose, and reconnect the hose (photo).
7   Set the heater facia control to 'Hot'.
8   With all drain plugs open, insert a hose at the radiator filler and keep it running until the water emerges clean.
9   If, after a reasonable period, the water still does not run clear, remove the radiator (see Section 5) and reverse-flush it. In severe cases of neglect, the radiator may be flushed with a good proprietary cleaning agent, such as Holts Radflush or Holts Speedflush. It is important that the manufacturer's instructions are followed carefully.

4   Cooling system – refilling and bleeding

Refilling

1   Prepare a sufficient quantity of coolant.
2   Ensure that the drain plugs and holes are tight, and in good condition.
3   Place the heater control lever to 'Hot'.
4   Open the bleed screws, removing the air filter if necessary to give access to the bleed screw on the carburettor automatic choke (photos).
5   Temporarily raise the expansion bottle as high as possible by any convenient means, leaving the tube attached.
6   Fill the radiator, and fit the cap.
7   Complete the filling, by pouring coolant into the expansion bottle, and tighten the bleed screw as coolant starts to come from them. Top up the bottle to $2\frac{3}{4}$ in (40 mm) above the 'maximum' mark.

Bleeding

8   With the bottle cap in place, run the engine until it is warmed up, so that the thermostat is open.
9   Open the bleed screws, retightening them once a steady flow of bubble-free coolant appears.
10 Refit and secure the expansion bottle.
11 When the engine has cooled, top up the coolant level in the bottle if necessary.

5   Radiator – removal and refitting

Note: *If the reason for removing the radiator is concern over coolant loss, note that minor leaks may be repaired by using a radiator sealant, such as Holts Radweld, with the radiator in situ. Extensive damage should be repaired by a specialist or the unit exchanged for a new or reconditioned radiator*

Type 841 engine

1   Disconnect the battery.
2   Remove the front grille and top crossmember, leaving the bonnet lock in position on it, by removing the bolts (photo).
3   Remove the screen retaining clips from the radiator (photo).
4   To prevent loss of coolant from the water pump, radiator and expansion bottle, clamp them off before disconnecting them from the radiator. Alternatively, completely drain the coolant, and keep it for re-use.
5   Take the electrical connections from the temperature sender and cooling fan motor (photo).

6   With the radiator pressed towards the engine, lift it out (photo).
7   As necessary, remove from the radiator the cooling fan motor, the thermostat, and the expansion bottle pipe shield.
8   To refit, reverse the dismantling procedure.
9   Fill and bleed the cooling system as described in Section 4.

Type 847 engine

10 Proceed as for the 841 engine, but ignore details relating to the cooling fan motor.

6   Water pump – removal and refitting

1   The water pump is a sealed unit and cannot be repaired. If defective, a complete new unit must be obtained.

Type 841 engine

2   Disconnect the battery, and drain the cooling system.
3   Remove the radiator (see Section 5).
4   Remove the hoses at the water pump.
5   Slacken the pump drivebelt (see Section 7).
6   Remove the water pump pulley.
7   Remove the pump securing bolts, gently tap it to free it, and lift the pump away (photos).
8   To refit, reverse the removal procedure. Ensure that the mounting face is completely clean, and that a new gasket is used without jointing compound.
9   Refill the cooling system (see Section 4), bleeding as necessary.

Type 847 engine

10 Proceed as for the 841 engine, but take particular care when handling the plastic fan blades, as these are easily damaged.

7   Drivebelt – adjustment, removal and refitting

Adjustment

1   Correct drivebelt adjustment is very important. A tight belt will shorten both belt and bearing life, whilst one that is loose will also reduce belt life and provide an inefficient drive to the units involved.
2   To adjust a belt, loosen the alternator mounting and adjuster link bolts, to the point where the alternator can just be moved (photo).
3   Move the alternator to give the required deflection to the belt (see Fig. 2.1 or 2.2).
4   Tighten the alternator mounting and adjuster link bolts.
5   Recheck the adjustment.

7.2 Alternator mounting and adjustment arrangements (847 engine)

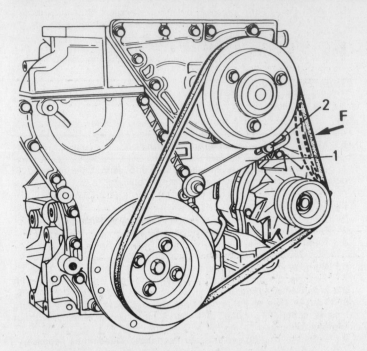

Fig. 2.1 Drivebelt tensioning – 841 engine (Sec 7)

1   Alternator adjustment bracket
2   Alternator adjustment bolt

$F = \frac{3}{32}$ to $\frac{9}{64}$ in (2.5 to 3.5 mm)

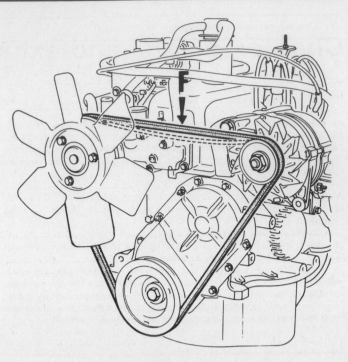

Fig 2.2 Drivebelt tensioning – 847 engine (Sec 7)

$F = \frac{9}{32}$ to $\frac{5}{16}$ in (7 to 8 mm)

Removal and refitting

6   Proceed as in the foregoing paragraphs, slackening the alternator adjustment off completely to permit the old belt to be removed.
7   A new belt will stretch after a short period of use, and should be adjusted once or twice as necessary, quite early in its life.
8   Note that rapid belt wear will result if the pulleys are bent, damaged or out of alignment. Oil or grease contamination is also to be avoided.

8   Thermostat – removal, testing and refitting

1   The thermostat is located in the hose attached to the water pump outlet pipe.
2   To remove the thermostat, drain sufficient coolant to allow the level to fall below the pump outlet. Remove the hose, and extract the thermostat.
3   Test the thermostat by suspending it in a pan of water, together with a thermometer. Commence warming the water, watch when the thermostat begins to open, and check the temperature. Compare the opening temperature with the temperature stamped on the thermostat.
4   Transfer the thermostat to cold water, and check that it closes promptly.
5   If the thermostat does not operate as outlined, it should be replaced.
6   Refit in the reverse order to dismantling, top up the coolant, and bleed the system.
7   In an emergency it is permissible to run the car without a thermostat fitted, but this will lead to prolonged warm-up time, poor heater output and (possibly) poor fuel economy. All these drawbacks are to be preferred to overheating, however!

9   Fault diagnosis – cooling system

Symptom	Reason(s)
Overheating	Coolant loss due to leakage Electric cooling fan malfunction (if applicable) Drivebelt slack or broken Thermostat jammed shut Radiator matrix clogged internally or externally Brakes binding New engine not yet run-in
Overcooling	Thermostat missing or jammed open
Coolant loss	External leakage (hose joints etc) Overheating (see above) Internal leakage (head gasket or cylinder liner) – look for water in oil and/or oil in coolant

Chapter 3 Fuel and exhaust systems

For modifications, and information applicable to later models, see Supplement at end of manual

Contents

Specifications

General

System type	Rear mounted fuel tank, mechanical fuel pump, single barrel Solex or Zenith carburettor

Air cleaner element:
R1340	Champion W146
R1341	Champion W109
Fuel filter	Champion L101

Fuel pump

Drive	From camshaft

Fuel pressure (pump not delivering):
Minimum	2.5 lbf/in^2 (0.17 bar)
Maximum	4.0 lbf/in^2 (0.27 bar)

Carburettor type

Engine.
847	Solex 32 EITA 690, 32 SEIA or Zenith 32 If 7
841.25	Solex 35 EITA 691
841.26	Solex 35 EITA 709

Idle speed

R1340	775 ± 25 rpm
R1341 (manual)	800 ± 50 rpm
R1341 (auto)	650 ± 50 rpm (in D)

Carburettor specifications

	32 EITA 690	35 EITA 691	35 EITA 709
Choke (venturi) size	24 mm	26 mm	26 mm
Main jet	127.5	130	130
Air compensating jet	160	205	205
Idling jet	45	43	43
Enrichment device	70	80	85
Needle valve	1.5	1.7	1.7
Econostat	–	60	–
Accelerator pump jet	35	45	45

Carburettor adjustment data

	32 EITA 690	32 EITA 691	35 EITA 709
Float level (all models)	0.5 ± 0.04 in (12.5 ± 1 mm)		
Butterfly angle (all models)	0.154 in (3.91 mm)		
Initial throttle opening:			
Medium cold	0.030 in (0.75 mm)	0.037 in (0.95 mm)	0.041 in (1.05 mm)
Extreme cold	0.037 in (0.95 mm)	0.041 in (1.05 mm)	0.051 in (1.30 mm)
Pneumatic part-open setting	0.142 in (3.6 mm)	0.142 in (3.6 mm)	0.118 in (3.0 mm)
Choke flap opening in deflooded position	0.394 in (10 mm)	0.433 in (11 mm)	0.433 in (11 mm)
Idle mixture CO% (maximum)	2	2	2

Emission control

System fitted	Crankcase emission control
Fume jet diameter	0.059 in (1.5 mm)

1 General description

A conventional fuel system is employed, with fuel from a rear-mounted tank being drawn by an engine-mounted, mechanically operated diaphragm pump. A filter is fitted to assist with the elimination of sediment.

A Solex or Zenith downdraught carburettor is fitted, with manual or automatic choke according to model. When an automatic choke is fitted, it is operated by heat from the engine coolant.

A crankcase emission control system is fitted, its purpose being to duct oil fumes and blow-by gases from the engine to the intake manifold, whence they are burnt in the combustion chambers.

2 Fuel pump – routine servicing

1 On the 841 engine the pump is bolted to the RH side of the cylinder block, whilst on the 847 engine it is on the LH side.
2 Periodically check that the cover screw(s) have not come loose. However, do not overtighten.
3 On the pump illustrated in Fig. 3.2, also check the security of screws (E).
4 At the specified service interval, clamp off the hoses from the tank, remove the cover screw(s) and take off the cover and gasket (photo).
5 Remove the filter, clean it, and clean out the pump chamber, using clean petrol.
6 Check the serviceability of the gaskets, and refit all parts in the reverse order. Do not overtighten the cover screw(s). Remove the clamps.

3 Fuel pump – removal and refitting

1 Clamp the fuel pipes from the tank.
2 Loosen all the pipe clips adjacent to the pump, and pull off the pipes.
3 Remove the pump securing nuts or bolts, and withdraw the pump (photo).
4 To refit, place the pump in position on the engine block, preferably using new gaskets, and refit the nuts or bolts. Tighten progressively to avoid distortion.
5 Reconnect the fuel pipes, and tighten the clips.
6 Remove the pipe clamps.

4 Fuel pump – testing

1 The recommended fuel pump test is made with the engine running, using a Renault test gauge. However, useful tests can be made as described in the paragraphs which follow.
2 Using a container to catch ejected fuel, and with the fuel delivery pipe removed from the carburettor, spin the engine on the starter. (Disable the ignition first by disconnecting an LT lead from the coil). If

2.4 Remove the fuel pump cover

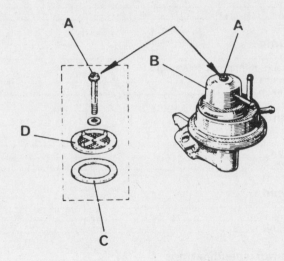

Fig. 3.1 Fuel pump – vehicle type R1341 (Sec 2)

A   Cover screw
B   Cover
C   Sealing ring

D   Filter (note that designs may vary)

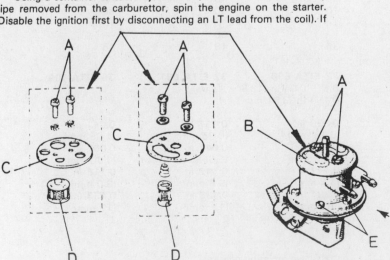

Fig. 3.2 Fuel pump – vehicle type R1340
(Sec 2)

A   Cover screws
B   Cover
C   Sealing ring
D   Filter (note that designs may vary)
E   Diaphragm flange screws

3.3 Pump with hoses removed and securing bolts slackened (engine removed for clarity)

5.4 Air filter cover removed, and element exposed

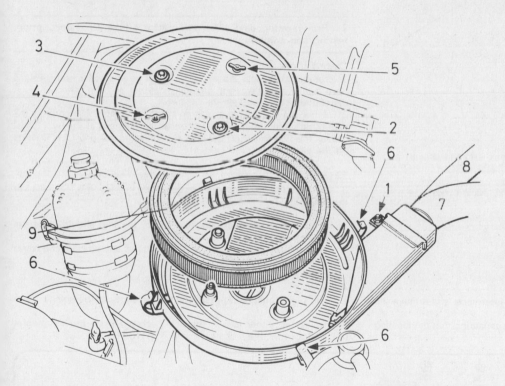

Fig. 3.3 Air filter case and element –
R1340 (Sec 5)

1 Nut
2 Bolt and washers
3 Bolt and washers
4 Wing nut
5 Wing nut
6 Clips
7 Cold air inlet
8 Warm air inlet
9 Filter element

a good spurt of petrol comes from the hose, the pump is functioning correctly.

3 If the pump is removed from the vehicle, place a finger over the inlet port and actuate the operating lever. If the pump is serviceable, suction should be felt.

4 If the pump is defective it must be renewed, no repair being possible.

5 Air filter element – removal and refitting

Model R1340

1 Refer to Fig. 3.3. Spring off the four clips, item (6).
2 Remove the bolts, items (2) and (3).
3 Remove the wing nuts, items (4) and (5).
4 Lift off the cover, and take out the element (photo). Wipe the inside of the housing clean.
5 To fit a new element, insert it, and refit all parts in reverse order,

noting that the cover holes are not symmetrically spaced.

Model R1341

6 Referring to Fig. 3.4, remove the wing nut, item (1), and the washer (2).
7 Take off the cover (15) and remove the element (3). Wipe the inside of the housing clean.
8 To fit a new element, reverse the removal procedure.

6 Air filter assembly – removal and refitting

Model R1340

1 Referring to Fig. 3.3, remove the nut, item (1), and the bolts, items (2) and (3).
2 Disconnect the warm and cold air inlet hoses, items (7) and (8).
3 Lift away the air filter assembly (photo).
4 To refit, reverse the removal procedure.

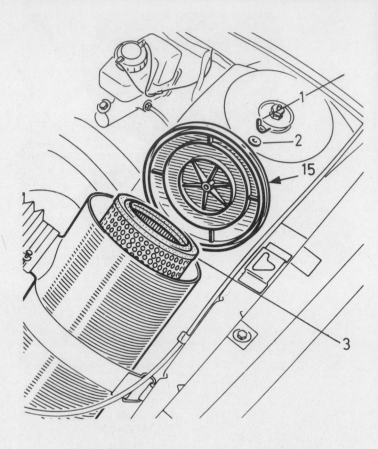

Fig. 3.4 Air filter case and element – R1341 (Sec 5)

1 Wing nut	3 Element
2 Washer	15 Cover

Model R1341

5   Slacken the carburettor air inlet elbow clip.

6   Unclip the air filter band.

7   Disconnect the warm air hose at the filter case, and remove the filter assembly.

8   To refit, reverse the removal procedure, ensuring that the locating pegs are correctly positioned.

6.3 Lifting off the air filter assembly

7 Pre-warming device for carburettor intake air – description, testing and adjustment

Description

1   Referring to Figs. 3.5 and 3.6, it will be seen that both designs have a dual air intake, one providing cold air and the other warm. A distribution flap regulates the mixture.

2   A wax-type thermostat is fitted in the air stream, and controls the distribution flap setting. Contraction of the element at below 17.5°C causes the flap to cut off the cold air, whilst expansion at 26°C shuts off the warm air.

Testing

3   Remove the filter assembly as described in Section 6, and take out the element.

4   Place the filter body in water to element height.

5   With a water temperature of 17.5°C, the flap should close the cold air intake after 5 minutes.

6   With a water temperature of 26°C, the flap should close the warm air intake after 5 minutes.

Adjusting

7   The only adjustment possible is to the length of the spindle, item 14.

8   To adjust, either loosen the fixing screw (R) on the thermostat (R1340), or turn the adjuster screw (R) (R1341).

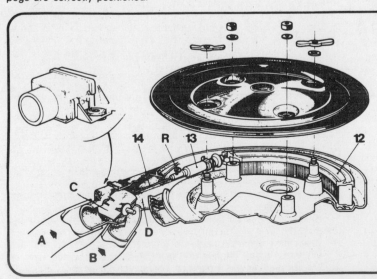

Fig. 3.5 Pre-warming device for carburettor air intake – R1340 (Sec 7)

A Cold air intake	
B Warm air intake	
C Flap	
D Mixed air stream	
R Spindle adjustment screw	
12 Filter	
13 Thermostat	
14 Spindle	

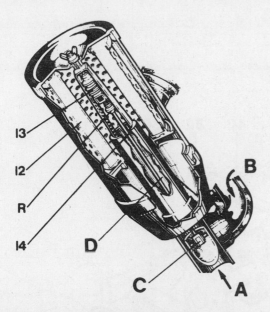

8.1a Carburettor – general view with air filter assembly removed

Fig. 3.6 Pre-warming device for carburettor air intake – R1341.
See Fig. 3.5 for key (Sec 7)

8 Carburettor – general description

1 The vehicle is fitted with a Solex or Zenith carburettor of single barrel downdraught design, incorporating either a manual or semi-automatic choke (photos).

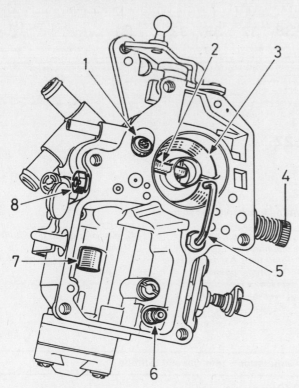

Fig. 3.7 Carburettor – internal detail (Sec 8)

1	Air compensating jet	6	Accelerator pump
2	Diffuser		defuming valve
3	Choke tube	7	Main jet
4	Idle speed screw	8	Idling jet
5	Accelerator pump jet		

8.1b Carburettor – rear view

2 The semi-automatic choke operates as the result of heat from the engine coolant acting on a bi-metallic spring.
3 Engine coolant flows through the carburettor base, thereby providing improved carburettor efficiency during the warm-up period.
4 When the engine is running with the automatic choke in use, the fast idle speed may be reduced by pressing the accelerator and then releasing it.
5 When starting the engine from cold, the automatic choke must be 'cocked' by depressing the throttle pedal once and then releasing it.

9 Carburettor – idling speed adjustment

1 Idle speed may be adjusted if necessary using the screw on the left-hand side of the carburettor (photo).
2 The idle mixture screw has a 'tamper-proof' cap fitted (photo). The object of the manufacturers in fitting such a cap is to prevent adjustment by unqualified operators. If adjustment is necessary, ideally a tachometer and exhaust gas analyser should be used; failing such equipment, proceed as follows, but satisfy yourself that you are not breaking any local or national emission control regulations by so doing.
3 Break off the tamperproof cap.
4 With the engine at normal operating temperature and with all other adjustments correct (ignition timing, valve clearances etc), turn

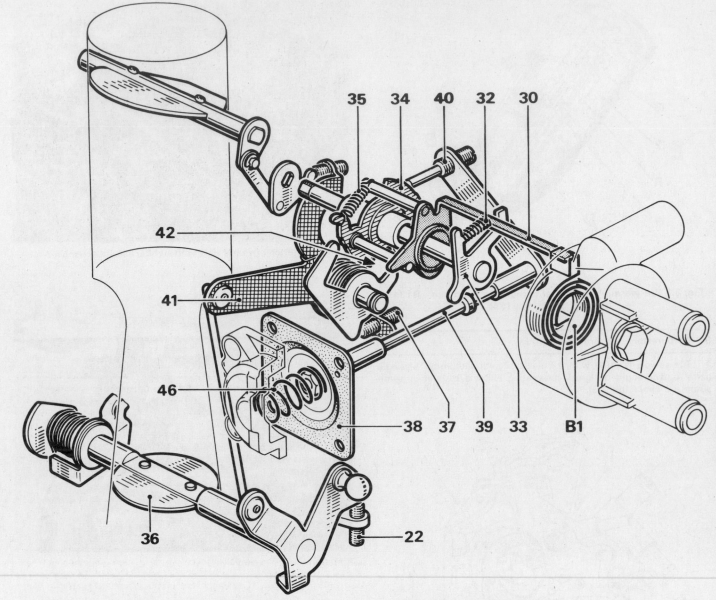

Fig. 3.8 Semi-automatic choke operating mechanism (Sec 8)

B1 Bi-metal spring
22 Butterfly angle setting
    screw
30 Bi-metal spring movement
    relaying lever
32 Spring

33 Lever integral with
    choke flap
34 Double cam
35 Double cam spring
36 Butterfly
37 Initial throttle opening
    screw

38 Pneumatic opening
    diaphragm
39 Pneumatic opening spindle
40 Pneumatic opening
    setscrew
41 Lever initial throttle
    opening cam

42 Locking notch on double
    cam
46 Pneumatic opening return
    spring

the idle speed screw to achieve an engine speed close to that specified.
5   If a CO meter (exhaust gas analyser) is available, turn the mixture screw to achieve a CO level within the specified limits.
6   If a CO meter is not available, turn the mixture screw to achieve the highest engine speed consistent with even running, then screw it in one quarter of a turn.
7   Readjust the idle speed if necessary. Fit a new tamperproof cap if required.
8   It is emphasised that haphazard carburettor 'adjustment' is unlikely to give satisfactory results. The modern carburettor is a reliable instrument which rarely gives trouble.

10 Carburettor – removal and refitting

R1340 vehicles
1   Remove the air filter assembly (see Section 6).
2   Clamp off the coolant hoses from the carburettor base and (if applicable) from the automatic choke. Disconnect the hoses.
3   Disconnect the fuel pipe and the vacuum advance hose.
4   Disconnect the crankcase breather pipe.
5   Disconnect the accelerator linkage at the balljoint. If applicable, disconnect the choke operating cable.
6   Remove the carburettor fixing nuts and washers, and lift the

9.1 Idle speed adjustment screw (arrowed)

9.2 Idle mixture adjustment screw (arrowed). Note tamper-proof cap

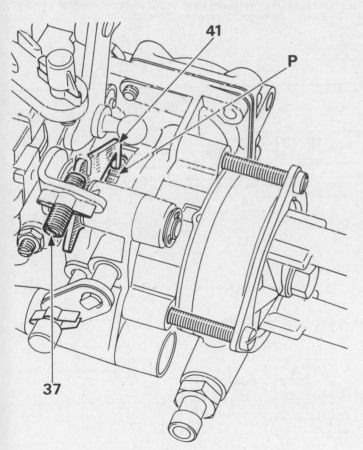

Fig. 3.9 Initial opening – setting details (Sec 11)

P   Cam   41 Lever
37 Adjusting screw

R1341 vehicles

9   Proceed as described in paragraphs 1 to 6 above, but first remove the air intake elbow.

10 To refit, proceed as described in paragraphs 7 and 8 above, but in addition refit the air intake elbow.

11 Carburettor – general adjustments (after rebuild or overhaul)

The information below relates to carburettors equipped with automatic choke. No specific information relating to carburettors equipped with manual choke was available at the time of writing, but those adjustments below which do not relate directly to the automatic choke mechanism will apply in principle to all carburettors.

Throttle butterfly angle

1   Special equipment is necessary for correct setting of this angle, and reference must be made to a Renault agent.

Initial throttle opening

2   Refer to Fig. 3.9. Ensure that the choke mechanism is set with the large radius of the initial throttle opening cam (P) towards the carburettor base flange. Ensure that the lever (41) is on the cam apex.

3   Refer to Fig. 3.10. Using a rod of the diameter quoted in the

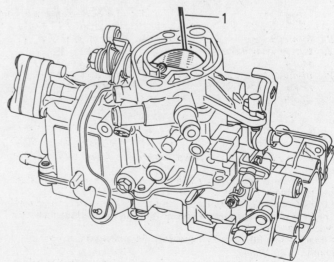

Fig. 3.10 Initial throttle opening – test rod (1) in position (Sec 11)

instrument off.

7   To refit, preferably renew the gaskets, and refit the carburettors in reverse of the removal procedure. Ensure that the flange nuts are evenly tightened to avoid distortion.

8   Adjust the idling speed, as described in Section 9.

Specifications, check the initial throttle opening. Adjust as necessary, using screw (37) (Fig. 3.9).

Choke flap opening
4 Set the choke as in paragraph 2 above.
5 With spindle (39) pressed, see Fig. 3.11, turn screw (40) until it contacts cam (R).
6 Using a gauge rod (1) as quoted in the Specifications, measure the choke flap opening on the higher side, and adjust if necessary using screw (40).
7 Remove the choke body, correctly position the double cam, and measure the initial throttle opening and choke flap pneumatic part-open setting.

Accelerator pump
8 Refer to Fig. 3.13. With the butterfly in the idling position, bring roller (25) to touch cam (15), turn screw (26) until it just touches plunger (27), and screw in another ¾ turn.

Float level
9 Invert the float chamber top.
10 Keeping the top level, measure dimension (A) (Fig. 3.14). Note that this is from the 0.866 (2.2 mm) diameter on the float, to the gasket face (gasket not fitted).
11 Bend the float stem (3) if necessary, to reset the level.
12 Check that the float pivots easily at the pivot pin, and that it clears the econostat immersed pipe (when fitted).

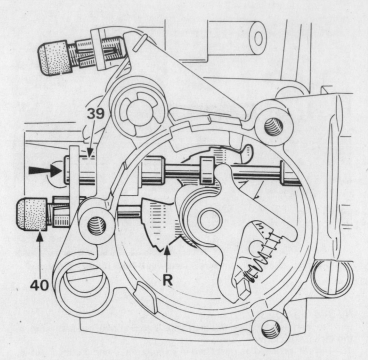

Fig. 3.11 Choke flap opening – setting details (Sec 11)

39 Spindle   R Pneumatic opening cam
40 Adjusting screw

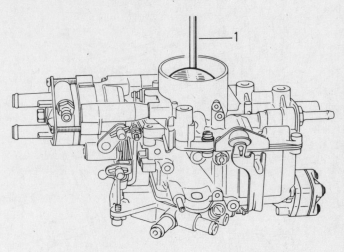

Fig. 3.12 Choke flap opening – test rod (1) in position (Sec 11)

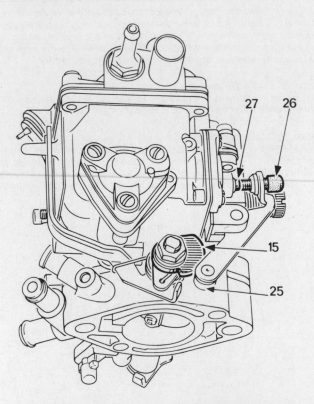

Fig. 3.13 Accelerator pump – setting details (Sec 11)

15 Cam   26 Screw
25 Roller   27 Plunger

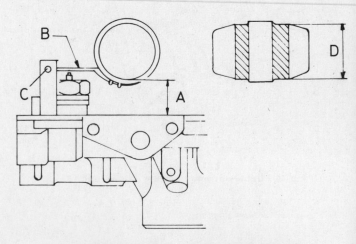

Fig. 3.14 Float level – setting details (Sec 11)

A = 12.5 mm (0.5 in)   D = 22 mm (0.866 in)
B   Stem   diameter
C   Pivot

12 Fuel filter – removal and refitting

1 Clamp the fuel lines on either side of the filter (see Fig. 3.15).
2 Loosen the filter clips, and remove the filter.
3 To fit a new filter, connect it to the fuel pipes with the arrow on the body pointing in the direction of the carburettor (ie in the direction of flow).
4 Tighten the clips, and remove the clamps.

13 Fuel tank – removal and refitting

1 Disconnect the battery.
2 Preferably choose a time when the fuel level in the tank is low, and syphon off the remaining contents of the tank via the filler cap.
3 Remove the tank shield.
4 Disconnect the luggage compartment light wires.
5 Disconnect the wire at the fuel gauge sender.
6 Take off the fuel pipes, and clamp them to prevent leakage.
7 Take the fuel pipe clips from the tank.
8 Remove the bolt at each side of the tank, and the two nuts beneath it.
9 With the cap removed from the filler, take the weight to free the tank from the lower studs, move it to the left to bring the filler pipe clear of the body, and take the tank out.
10 Remove the fuel gauge sender unit if necessary.
11 Refit the fuel gauge sender unit using a new washer, and refit the wiring harness sleeve.
12 Reverse the removal procedure to refit the tank.
13 Never attempt to solder or weld a leaking fuel tank unless it has been thoroughly steamed or boiled first. Fuel tank repair is a specialist field. Temporary repairs with the tank in situ can sometimes be made using proprietary plugging compounds, but these are unlikely to prove satisfactory in the long run.

14 Throttle and choke cables – removal, refitting and adjustment

Throttle cable

1 Disconnect the battery.
2 At the carburettor end, free the locknut and the cable clamping screw (photo).
3 At the panel end, remove the split pin, extract the clevis pin, and take the scuttle fitting from its location.
4 Pass the clevis pin through into the engine compartment, and withdraw the cable.
5 To refit, reverse the removal procedure. Check that the scuttle fitting is properly in place, and ensure that no sharp bends are present in the cable run. Then adjust as follows.
6 Have an assistant hold the throttle pedal down to the floor, whilst you hold the throttle butterfly on the carburettor wide open.
7 Tension the cable against the compensator spring, so that the spring is compressed about $\frac{5}{64}$ in (2 mm). Tighten the cable securing screw at the carburettor trunnion, and lock with the nut (see Fig. 4.16).
8 Where automatic transmission is fitted, check for correct operation of the kickdown switch (see Chapter 7).
9 Ensure that where the cable leaves the sleeve, at the stop adjacent to the carburettor, it makes good alignment with the carburettor lever. The stop may be bent if necessary to this end. Poor alignment causes excessive friction, producing heavy operation and wear. Make sure also that, when the throttle pedal is released, the cable has enough tension to prevent the sleeve from jumping out of the stop.

Choke cable (where applicable)

10 Disconnect the battery earth lead.
11 Slacken the bolt at the carburettor which retains the outer component of the cable (photo). Withdraw the cable from the carburettor.
12 The choke cable and control can be withdrawn from the steering column lower surround after releasing the surround. Note the wire connector for the choke warning light (photo).
13 Refitting is the reverse of the removal procedure. Adjust the cable so that with the choke control fully home, there is a small amount of slack in the inner cable. Check for correct operation on completion.

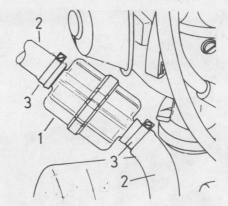

Fig. 3.15 Fuel filter (typical) (Sec 12)

1 *Filter*
2 *Fuel pipes*
3 *Pipe clips*

14.2 Throttle cable, carburettor end, showing clamping screw and compensator spring

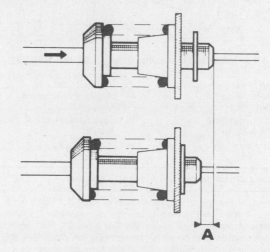

Fig. 3.16 Compensator setting (Sec 14)

A = 0.078 in (2 mm)

14.11 Choke cable attachment at carburettor

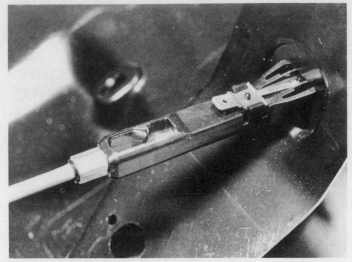

14.12 Choke cable attachment at control

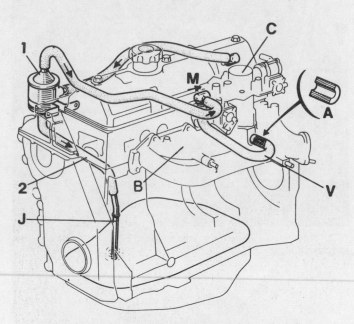

Fig. 3.17 Crankcase emission details – R1340 (Sec 15)

1   Oil separator   J   Dipstick tube
2   Sump pipe   M   Upstream circuit for
A   Jet   fumes
B   Inlet manifold   V   Downstream circuit for
C   Carburettor   fumes

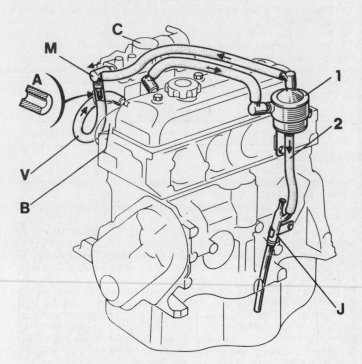

Fig. 3.18 Crankcase emission details – R1341. See Fig. 3.17 for key (Sec 15)

15 Crankcase emission control system – general

1   The layout of the system is as shown in Figs. 3.17 and 3.18.
2   Crankcase fumes are drawn through the oil separator (1), and any oil separated out goes to the sump via the pipe (2). The remaining fumes are recirculated, and burnt in the combustion chambers, the jet (A) assisting with correct distribution of the flow of fumes.
3   Check that the hoses are in good condition, and not blocked, when inspecting the system. Clean out the restrictor jet from time to time. The crankcase emission control system should be the first suspect if there is a build-up of 'mayonnaise' in the valve cover and oil filler cap, or if symptoms of crankcase compression are evident.

16 Exhaust system – description, removal and refitting

1   The exhaust system consists of four sections, namely the front pipe, the expansion box, the intermediate pipe, and the rear silencer/pipe. Clamps are employed to join the sections together (photo).
2   To check the condition of the exhaust system, raise the car as necessary, then start the engine and examine the entire length of the exhaust, while an assistant temporarily places a wad of cloth over the tailpipe. If a leak is evident, stop the engine, and use a good proprietary repair kit to seal it. Holts Flexiwrap and Holts Gun Gum exhaust repair systems can be used for effective repairs to exhaust

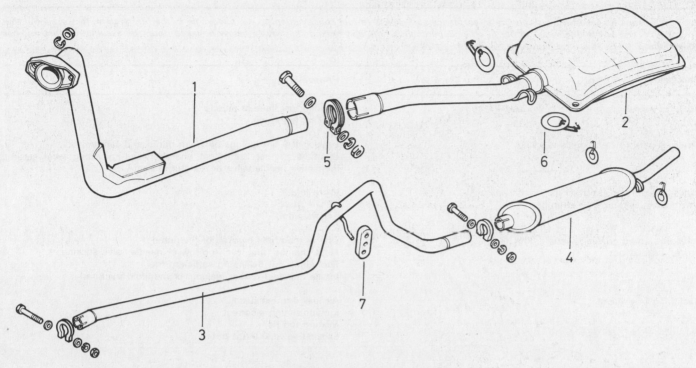

Fig. 3.19 Typical exhaust system (Sec 16)

1 Front pipe	3 Intermediate pipe	5 Clamp (typical)	7 Hanger bracket (typical)
2 Expansion box	4 Rear silencer/pipe	6 Rubber ring (typical)	

16.1 The exhaust front pipe, clamped to the exhaust manifold

16.4 The exhaust hanger arrangement, in front of the rear pipe/silencer

pipes and silencer boxes, including ends and bends. Holts Flexiwrap is an MOT-approved permanent exhaust repair. If the leak is large, or if serious damage is evident, it may be better to renew the relevant exhaust section.

3 Each section can be removed, thus providing for renewal of individual sections as these become worn out. Corrosion at the joints can sometimes make separation of the sections difficult, and this problem can sometimes be eased by using a wire brush and penetrating oil. Corroded bolts are probably best cut off with a hacksaw, and replaced with new parts.

4 Rubber rings and a hanger bracket are used to suspend the system from the vehicle (photo).

5 When refitting the system, check the condition of all hanger rings and renew if necessary. Smear an exhaust sealant such as Holts Firegum on all exhaust joints before assembly. Align the system carefully before finally tightening the joint clips, to make sure that it does not chafe or knock against any part of the vehicle.

6 Tighten all joints and clips, and check for leaks. Tighten the joints again when the engine has been run for a little while and then switched off.

17 Fault diagnosis — fuel and exhaust systems

The reasons for unsatisfactory engine performance and excessive fuel consumption do not always lie in the fuel system. It is essential that the general tune of the engine is correct in other departments. The fault finding chart given below is based upon the assumption that this has been done.

Symptom	Reason(s)
Smell of petrol when engine is stopped	Leaking fuel lines or unions Leaking fuel tank
Smell of petrol when engine is idling	Leaking fuel line unions between pump and carburettor Overflow of fuel from float chamber due to wrong level setting, ineffective needle valve or punctured float
Excessive fuel consumption for reason not covered by leaks or float chamber	Worn jets Choke stuck Engine worn
Difficult starting, uneven running, lack of power, cutting out	One or more jets blocked or restricted Float chamber fuel level too low or needle valve sticking Fuel pump not delivering sufficient fuel Intake manifold gasket leaking, or manifold fractured
Backfiring in exhaust	Air leak into exhaust Ignition timing incorrect Mixture too rich Exhaust valve(s) burnt out

Chapter 4 Ignition system

For modifications, and information applicable to later models, see Supplement at end of manual

Contents

Specifications

General

System type	12 volt battery, coil and distributor with contact breaker
Firing order	1-3-4-2
Location of No 1 cylinder	Flywheel end of engine

Spark plugs

Type:

European models	Champion RN9YCC or RN9YC
US and Canadian models	Champion RN12YC

Electrode gap:

Champion RN9YCC and RN12YC	0.8 mm (0.031 in)
Champion RN9YC	0.6 mm (0.024 in)

HT leads

HT leads	Champion CLS 1, boxed set

Distributor

Direction of rotation	Clockwise viewed from above
Contact breaker points gap	0.016 in (0.4 mm) approx
Dwell angle	57° ± 3 (63% ± 3)

Centrifugal/vacuum curve numbers:

R1340	R310 C33
R1341 (manual)	R308 D61
R1341 (auto)	R309 D61

Ignition timing*

R1340	10° ± 1
R1341 (manual)	6° ± 1
R1341 (auto)	10° ± 1

* Initial advance (static, or idling with vacuum pipe disconnected). Refer also to clip on HT lead

Torque wrench setting

	lbf ft	Nm
Spark plugs	23	31

1 General description

In order that the engine may run correctly it is necessary for an electrical spark to ignite the fuel/air mixture in the combustion chambers at exactly the right moment in relation to engine speed and loading. The ignition system is based on feeding low tension voltage from the battery to the coil where it is converted to high tension voltage. The high tension voltage is powerful enough to jump the gap between the electrodes of the spark plugs in the cylinders many times a second under high compression, providing that the system is in good condition and all the adjustments are correct.

The system is divided into two circuits; the low tension and high tension.

The low tension (sometimes called primary) circuit consists of the battery, lead wire to the starter solenoid, lead from the starter solenoid to the ignition switch, lead from ignition switch to the coil low tension windings (SW or + terminal) and from the coil low tension windings (CB or – terminal) to the contact breaker points and condenser in the distributor.

The high tension circuit comprises the high tension or secondary windings in the coil, the heavily insulated lead from the coil to the distributor cap centre contact, the rotor arm, and the leads from the four distributor cap outer contacts (in turn) to the spark plugs.

Low tension voltage is stepped up by the coil windings to high tension voltage intermittently by the operation of the contact points and the condenser in the low tension circuit. High tension voltage is then fed via the centre contact in the distributor cap to the rotor arm.

The rotor arm rotates clockwise at half engine revolutions inside the distributor. Each time it comes in line with one of the outer contacts in the cap, the contact points open and the high voltage is discharged, jumping the gap from rotor arm to contact and thence along the plug lead to the centre electrode of the plug. Here it jumps the other gap – sparking in the process – to the outer plug electrode and hence to earth.

The static timing of the spark is adjusted by moving the outer body of the distributor in relation to the distributor shaft. This alters the position at which the points open in relation to the position of the crankshaft (and thus the pistons).

The timing is also altered automatically by a centrifugal device, which further alters the position of the complete points mounting assembly in relation to the shaft when engine speed increases, and by a vacuum control working from the inlet manifold which varies the timing according to the position of the throttle and consequently load on the engine. Both of these automatic alterations advance the timing of the spark at light loads and high speeds. The mechanical advance mechanism consists of two weights, which move out from the distributor shaft as engine speed rises due to centrifugal force. As they move out, so the cam rotates relative to the shaft and the contact breaker opening position is altered.

The degree to which the weights move out is controlled by springs, the tension of which significantly controls the extent of the advance to the timing.

The vacuum advance device is a diaphragm and connecting rod attached to the cam plate. When the diaphragm moves in either direction the cam plate is moved, thus altering the timing.

The diaphragm is acutated by depression (vacuum) in the inlet manifold and is connected by a small bore pipe to the carburettor body.

On both engine variants the firing order is 1-3-4-2 and the No. 1 cylinder is at the flywheel end of the engine as shown in Figs. 4.1 and 4.2.

A diagnostic socket is fitted to enable Renault mechanics to quickly pinpoint any problem within the ignition system and also to time it accurately. Unfortunately, without the necessary associated equipment, the diagnostic socket is of little use to the home mechanic.

2 Routine maintenance

Spark plugs
1 Remove the plugs. Examine the porcelain insulation round the central electrodes inside the plug, and if damaged discard the plug. Reset the gap between the electrodes. Do not use a set of plugs for more the 10 000 miles. It is false economy.
2 Note that, as the cylinder head is of aluminium alloy, it is recommended that a little anti-seize compound (such as Copaslip) is applied to the plug threads before they are fitted. Refitting the spark plugs is a reversal of the removal procedure. Screw in the plugs by hand until the sealing washer just seats, then tighten by no more than a quarter-turn using a spark plug spanner. If a torque wrench is available, tighten the plugs to the specified torque.
3 At the same time check the plug caps. Always use the straight tubular ones normally fitted. Good replacements are normally available at your Renault agency.

Distributor
4 Every 10 000 miles remove the cap and rotor arm and put one or two drops of engine oil into the centre of the cam recess. Smear the surfaces of the cam itself with petroleum jelly. Do not over lubricate as any excess could get onto the contact breaker points surfaces and cause ignition difficulties. At the same time examine the contact breaker point surfaces. If there is a build-up of deposits on one face and a pit in the other it will be impossible to set the gap correctly and they should be refaced or renewed. Set the gap when the contact surfaces are in order.

General
5 Examine all leads and terminals for signs or broken or cracked insulation. Also check all terminal connections for slackness or signs of fracturing of some strands of wire. Partly broken wire should be renewed. The HT leads are particularly important as any insulation faults will cause the high voltage to 'jump' to the nearest earth and this will prevent a spark at the plug. Check that no HT leads are loose or in a position where the insulation could wear due to rubbing against part of the engine.

3 Spark plugs and leads – general

1 The correct functioning of the spark plugs is vital for the correct running and efficiency of the engine. It is essential that the plugs fitted are appropriate for the engine (the correct type is specified at the beginning of this Chapter). If this type is used, and the engine is in good condition, the spark plugs should not need attention between scheduled service renewal intervals. Spark plug cleaning is rarely necessary, and should not be attempted unless specialised equipment is available, as damage can easily be caused to the firing ends. At the intervals specified in 'Routine maintenance' at the beginning of

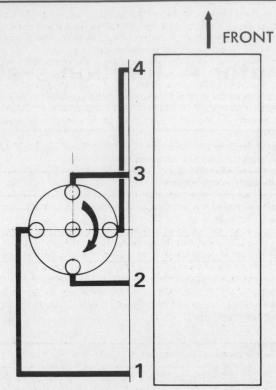

Fig. 4.1 847 engine – HT lead layout (Sec 1)

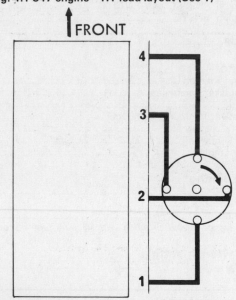

Fig. 4.2 841 engine – HT lead layout (Sec 1)

this manual, the plugs should be removed and renewed.
2 Spark plugs can also be used as good indications of engine condition, particularly as regards the fuel mixture being used and the state of the pistons and cylinder bores. Check each plug as it is possible that one cylinder condition is different from the rest. Plugs come in different types to suit the particular type of engine. A 'hot' plug is for engines which run at lower temperatures than normal and a 'cold' plug is for the hotter running engines. If plugs of the wrong rating are fitted they can either damage the engine or fail to operate properly. Under normal running conditions a correctly rated plug in a properly tuned engine will have a light deposit of brownish colour on the electrodes. A dry black, sooty deposit indicates an over-rich fuel mixture. An oily, blackish deposit indicates worn bores or valve guides. A dry, hard, whitish deposit indicates too weak a fuel mixture. If plugs of the wrong heat ranges are fitted they will have similar symptoms to a weak mixture together with burnt electrodes (plug too hot) or to an

over-rich mixture caked somewhat thicker (plug too cold). Do not try to economise by using plugs beyond 10 000 miles. Unless the engine remains in exceptionally good tune, reductions in performance and fuel economy will outweigh the cost of a new set.

3　The HT leads and their connections at both ends should always be clean and dry and, as far as possible, neatly arranged away from each other and nearby metallic parts which could cause shorting. The end connectors should be a firm fit and free from deposits. If any lead shows signs of cracking or chafing of the insulation it should be renewed. Remember that radio interference suppression is required when renewing any leads.

4　The spark plug lead connections are as shown in Figs. 4.1 and 4.2.

4　Distributor – identification

1　The initial ignition advance value is stamped on a clip on one of the HT leads (see Fig. 4.3).

2　The centrifugal and vacuum advance curves are identified by the numbers stamped on the distributor body.

3　Whilst the initial advance is a function of the engine, and can be adjusted when a new distributor is fitted, the advance curves are particular to the distributor concerned and cannot easily be adjusted. Make sure therefore when a replacement distributor is obtained, be it from the scrapyard or from a Renault dealer, that it carries the same identification number.

5　Distributor points – adjustment

1　The contact points should ideally be adjusted to the dwell angle given in the Specifications, whilst the engine is running. Access for adjustment is provided to allow this to be carried out without switching the engine off. Alternatively the contacts may be adjusted

using a feeler gauge, but the adjustment should be checked by means of an instrument capable of measuring the dwell angle, as soon as practicable.

2　Dwell angle is the angle in degrees through which the distributor cam turns between the instant of closure and opening of the contact breaker points. It is directly proportional to the contact breaker gap. Increasing the points gap reduces the dwell angle, and vice versa. Dwell angle measurement is superseding contact breaker gap measurement because the former is (potentially!) quicker and more accurate. Various proprietary instruments are available for measuring dwell angle; most of them operate with the engine idling, though some only require the engine to be cranked on the starter motor.

Conventional contact breaker arrangement

3　Remove the distributor cap, the rotor arm and the dust shield.

4　By engaging a gear and pushing the car, turn the engine so that the points are fully opened by one of the peaks on the cam spindle.

5　Place a feeler gauge 0.016 in (0.4 mm) thick between the contact point faces, and adjust as necessary using the adjuster nut (A) indicated in Fig. 4.5. The feeler gauge should be a firm sliding fit.

6　Refit all parts.

7　To adjust the dwell angle, connect the instrument to the engine as recommended by the instrument manufacturer.

8　Start the engine, and run it at idling speed, or crank it on the starter motor (according to the type of motor used).

9　Adjust as necessary using screw (A) indicated in Fig. 4.4 until the correct dwell angle is achieved. Stop the engine, and disconnect the test instrument.

Cassette type contact breaker arrangement

10　Remove the contact breaker capsule (see Section 6).

11　Insert a length of ground bar 0.6677 in (16.96 mm) diameter through the cassette, to take the place of the drive spindle. (This diameter corresponds to the maximum diameter across the cams).

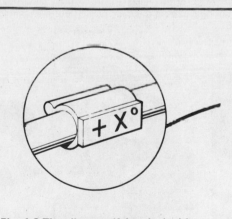

Fig. 4.3 The clip, specifying the ignition advance (Sec 4)

Fig. 4.4 Conventional contact breaker points – details (Sec 5)

A　Adjuster
B　Adjuster bracket screws
C　Retaining lug
D　Adjuster spring and rod
E　Fixed contact screw
F　Vacuum capsule
G　LT terminal
H　LT lead
J　Clip
K　Spring blade
L　Serrated cam
M　Vacuum capsule retaining screws

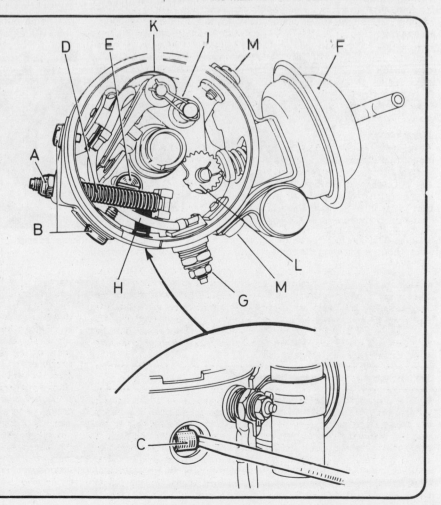

12 Using a 3 mm Allen key in the adjusting screw, adjust the points to give a gap of 0.016 in (0.40 mm) (photo).
13 Refit the capsule.
14 To adjust the dwell angle, proceed as described in paragraphs 7 to 9, using the 3 mm Allen key in the head of the adjusting screw.

6 Distributor – removing and refitting contact points

Conventional contact breaker arrangement

1 Remove the cap, and then the rotor arm and dust shield.
2 Remove the adjuster nut (A), see Fig. 4.4.
3 Unscrew screws (2).
4 Take out the small plug covering retaining lug (C), and use a suitable small tool to prise out the lug.
5 Remove adjustment rod and spring (D).
6 Take out screw (E) and remove the fixed contact.
7 Loosen the LT terminal (G), and detach the LT lead (H).
8 Free clip (J) from the pivot post and take off the insulating washer. Remove the moving contact whilst keeping the spring blade (K) pressed inwards, thereby releasing it from its mounting.
9 Examine the point faces for a pip on one, for a corresponding pit on the other, and for generally poor condition. Where necessary, contacts may be resurfaced using an oil stone, with care being taken to maintain the original contours. However, where the overall condition is poor, or where considerable burning has occurred, they should be renewed.
10 Clean the contact surfaces, whether old or new, with methylated spirit, to ensure complete freedom from greasy deposits.
11 Refit in the reverse order to removal.
12 Adjust the contact breaker gap as described in Section 5. Check the ignition timing and adjust if necessary.

Cassette type contact breaker arrangement

13 Remove the distributor cap and rotor arm.
14 Pull off the electrical connections to the condenser support assembly.

15 Lift the condenser support assembly and cassette upwards and out of the body (photo).
16 Disconnect the cassette wiring connector, and remove the cassette (photos).
17 Fit a new cassette in the reverse order to the removal procedure, ensuring when placing the cassette over the cam that the contact point heel is not in line with the high point of a cam, as this might cause damage to the heel.
18 Assuming that a new cassette assembly has been fitted, there will be no need to set the contact breaker gap as this should have been done at manufacture. Start the engine and check the dwell angle as described in Section 5. Check the ignition timing and adjust if necessary.

7 Distributor – removal, overhaul and refitting

Removal

1 Unless complete dismantling of the engine is being undertaken, turn the crankshaft to bring No 1 piston to TDC on the firing stroke.
2 Remove the distributor cap. Make a mark on the rim of the distributor body to show the position of the tip of the rotor arm. Make further alignment marks on the distributor body and on the engine so that the distributor can be refitted in the same position (photo).
3 Disconnect the LT lead from the distributor.
4 Undo and remove the distributor clamp nut. Remove the clamp and lift out the distributor (photo). Access to the clamp nut may be limited if the engine is in the car; Renault specify a cranked spanner (tool Elé 556), but something similar could no doubt be made up locally.

Overhaul – conventional contact breaker type

5 Remove the contact breaker points as described in Section 6.
6 Remove the condenser (see Section 8).
7 Note the relationship of the serrated cam to the vacuum capsule rod. Remove the spring clip from the cam, and lift out the points pivot unit (10) (Fig. 4.5).

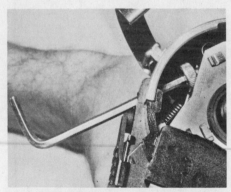

5.12 An Allen key inserted in the adjustment screw, prior to adjusting the contact points

6.15 Lifting out the condenser support assembly and cassette

6.16a Disconnecting the wiring connector

6.16b Disconnecting the cassette (underneath view)

6.16c The distributor with condenser and cassette assembly removed

7.2 Alignment marks (arrowed) on distributor flange and engine block

7.4 Removing the distributor (engine removed for clarity)

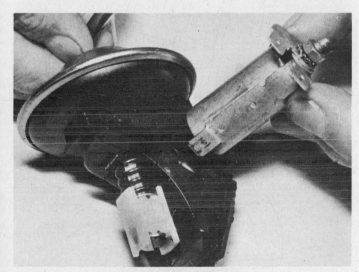

8.5 Removing the condenser from the support

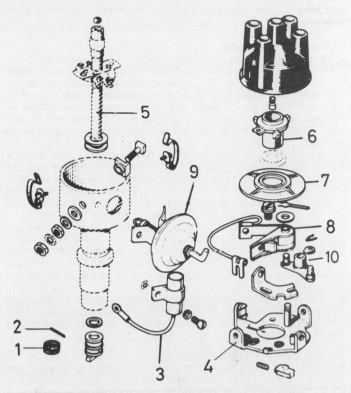

Fig. 4.5 Distributor with conventional contact breaker
arrangement – exploded view (Sec 7)

1	Spring	6	Rotor
2	Retaining pin	7	Dust shield
3	Condenser	8	Moving contact
4	Baseplate	9	Advance/retard unit
5	Rotor spindle	10	Points pivot unit

8 Remove the vacuum capsule retaining screw, followed by the capsule.
9 Remove the spring from the end of the driveshaft.
10 Mark the drive dog in relation to the cam slot in the top of the driving spindle. Tap out the pin, and remove the washers and drive dog.
11 Remove the screws which secure the baseplate, and lift the baseplate off.
12 Lift out the shaft and counterweight assembly.
13 Defects in the bearing bushes, counterweight assembly or springs will probably mean that the most economical course is to obtain a factory exchange unit. Care should be taken that the correct distributor is obtained, as outlined in the Specifications. Individual spare parts are unlikely to be available, but it may be possible to have items such as bushes locally made and fitted.
14 Clean all parts, lightly lubricate as necessary, and reassemble in the reverse order to the dismantling procedure.

Overhaul – cassette contact breaker type
15 No specific information is available concerning dismantling at the time of compilation of this manual.
16 It must be assumed that, if items of servicing other than those given elsewhere in this Chapter are necessary, then the purchase of an exchange distributor will be the best course for the home mechanic to take.

Refitting
17 If the crankshaft has been turned since the distributor was removed, turn it to bring No 1 piston to TDC on the firing stroke.
18 Align the marks made in paragraph 2 and refit the distributor to the engine. Fit the clamp and nut, but only do the nut up finger tight. Reconnect the LT lead.
19 Set the points gap or dwell angle if necessary (Section 5), then check and adjust the ignition timing (Section 9).
20 Tighten the clamp bolt.

8 Condenser – removal and refitting

1 Failure to start, misfiring, or excessive burning or pitting of the contact breaker point faces, can be caused by a failed condenser.

Conventional contact breaker
2 To remove the condenser, disconnect the wire at the terminal post, and remove the condenser mounting screw.
3 Remove the condenser, and fit a new part.

Cassette type contact breaker
4 Remove the condenser support/cassette assembly (see Section 6), and disconnect the cassette.
5 Push out the condenser from the support (photo).
6 Fit a new condenser, and reassemble in the reverse order to the dismantling procedure.

9 Ignition timing – checking and adjusting

1 Whenever the contact breaker gap (or dwell angle) has been altered, when new points have been fitted, or after distributor removal

and refitting, the ignition timing must be checked and if necessary adjusted.

Timing from scratch

2   If the timing has been lost completely, it will be necessary to turn the engine so that the timing marks are aligned, with No 1 cylinder on the compression stroke. (Turn the engine with a spanner on the crankshaft pulley bolt, or on manual transmission models by pushing the car with top gear engaged. Remove the spark plugs and feel for compression being generated in No 1 cylinder.) Cylinder and HT lead identification is shown in Figs. 4.1 and 4.2. Typical timing marks are shown in Fig. 4.6.

3   Look for the clip on one of the HT leads giving the initial advance figure (Fig. 4.3). With the flywheel timing mark aligned with the specified initial advance point on the scale, and the engine set up as described in paragraph 2, the distributor rotor arm should be pointing at No 1 HT lead segment in the distributor cap, and the points should just be opening. Slacken the clamp bolt and adjust if necessary.

4   If difficulty is experienced in telling precisely when the points are open, connect a 12 volt test lamp across the points and switch on the ignition. When the points are open, the lamp will light. With the timing marks aligned, turn the distributor if necessary to the point where the test lamp is just coming on. Tighten the clamp bolt when adjustment is complete.

5   The procedure above is accurate enough to enable the engine to be started. For the best results however, it is advisable to check the timing with the engine running, using a stroboscopic lamp as described below.

Timing using a strobe light

6   Connect a timing light (strobe) to the engine in accordance with the maker's instructions – most lights will require some connection to No 1 HT lead, and some will need to be connected to the car battery or to mains electricity as well.

7   Disconnect and plug the vacuum advance pipe at the distributor.

8   Start the engine and shine the light onto the timing marks (Fig. 4.6). (Depending on the strength of the timing light and the brightness of the ambient light, it may be necessary to highlight the timing marks with quick-drying white paint. Typist's correcting fluid is ideal for this purpose). If the ignition timing is correct, the specified marks will appear stationary and in alignment.

9   If the timing marks are not aligned, slacken the distributor clamp bolt and rotate the distributor slightly in the direction necessary to correct the misalignment. Tighten the clamp bolt and recheck.

10 If the timing marks appear unsteady and will not stay in alignment, this may be due to wear in the distributor, or to wear in the timing gear generally.

11 Increasing the engine speed above idle should make the timing marks appear to drift apart as the centrifugal advance mechanism comes into operation.

12 Applying suction to the distributor vacuum unit should similarly cause the timing marks to move in the direction showing advance.

13 When the timing is correct, disconnect the timing light, remake the original connections and reconnect the vacuum pipe.

---

10 Diagnostic socket – removal and refitting

1   Disconnect the battery earth lead.

2   Remove the diagnostic socket, leaving the supporting plate in place.

3   Remove the socket cover, and disconnect the earth wire, contact breaker wire, and ignition coil wire.

4   Working under the vehicle, remove the pick-up retaining screw and withdraw the pick-up (see Chapter 6).

5   Refit in the reverse order to the removal procedure. Adjust the pick-up as described in Chapter 6, Section 7.

---

11 Fault diagnosis – ignition system

Engine troubles normally associated with, and usually caused by faults in, the ignition system are:

(a) Failure to start when the engine is turned
(b) Uneven running caused by misfiring or mistiming

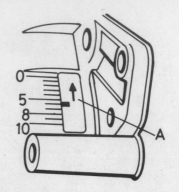

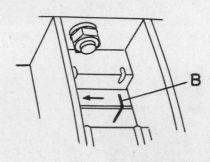

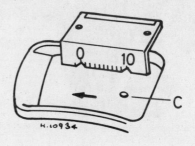

Fig. 4.6 Typical timing marks (Sec 9)

A   R1341 manual   C   R1341 automatic
B   R1340 (at TDC)

(c) Even running at low engine speed and misfiring when engine speeds up or is under the load of acceleration or hill climbing
(d) Even running at higher engine speeds and misfiring or stoppage at low speed

For (a), first check that all wires are properly connected and dry. If the engine fails to catch when turned on the starter do not continue turning or the battery will be flattened and the problem made worse. Remove one spark plug lead from a plug and turn the engine again and see if a spark will jump from the end of the lead to the top of the plug or the engine block. (Hold the lead with an insulated tool!). It should jump a gap of $\frac{1}{4}$ inch with ease. If the spark is there, ensure that the static ignition timing is correct and then check the fuel system.

If there is no spark at the plug lead, proceed further and remove the HT lead from the centre of the distributor which comes from the coil. Try again by turning the engine to see if a spark can be obtained from the end. If there is a spark the fault lies between the contact in the distributor cap and the plug. Check that the rotor arm is in good condition and making proper contact in the centre of the distributor cap and that the plug leads are properly attached to the cap. The four terminals inside the cap should be intact, clean and free from corrosion.

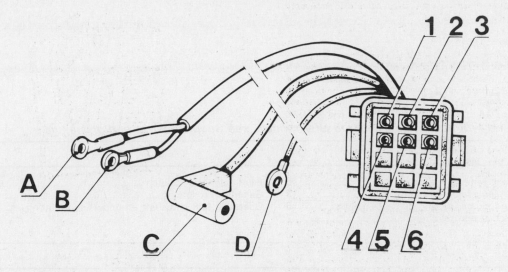

Fig. 4.7 Leads from the diagnostic socket – details (Sec 10)

A	Black/red sleeve, to ignition coil negative	D Yellow, to earth
B	Grey/blue sleeve, to ignition coil positive	1 Red, TDC terminal pickup
C	TDC pickup, to engine	2 Yellow, distributor earth

A   Black/red sleeve, to
    ignition coil negative
B   Grey/blue sleeve, to
    ignition coil positive
C   TDC pickup, to engine

D   Yellow, to earth
1   Red, TDC terminal pickup
2   Yellow, distributor earth
3   Black, contact breaker
4   White, TDC signal pick-up

5   TDC signal pick-up
    screening
6   Grey, ignition coil
    positive

If no spark comes from the coil HT lead, check next that the contact breaker points are clean and the gap is correct. If there is still no spark obtainable it may be assumed that the low tension circuit is at fault. To check the low tension circuit properly it is best to have a voltmeter handy or a 12V bulb in a hold with two wander leads attached. The procedure now given is arranged so that the interruption in the circuit – if any – can be found.

Starting at the distributor, put one of the two leads from the tester (be it lamp or voltmeter) to the moving contact terminal and the other to earth. A reading (or light) with the ignition on indicates that there is no break in the circuit between the ignition switch and the contact point (which should be open). Check the condenser by substitution. If this is satisfactory it means that the coil is not delivering HT to the distributor and must therefore be renewed.

If there is no LT reading on the first check point, repeat the test between the CB (–) terminal of the coil and earth. If a reading is now obtained there must be a break in the wire between the CB (–) terminal and the contact points.

If there is no reading at this second check point, repeat the test between the SW (+) terminal of the coil and earth. If this produces a reading then the low tension part of the coil windings must be open-circuited and the coil must be renewed.

If there is no reading at this third check point there must be a break between the ignition switch and the coil. If this is the case, a temporary lead between the + terminals of both coil and battery will provide the means to start the engine until the fault is traced.

For (b), uneven running and misfiring should be checked first by ensuring that all HT leads are dry and properly connected. Ensure also that the leads are not short-circuiting to earth against metal pipework or the engine itself. If this is happening an audible click can usually be heard from the place where the unwanted spark is being made.

For (c), misfiring occurs at high speeds, the points gap is too small or the spark plugs need renewal due to failure under more severe operating pressures.

For (d), if misfiring is occurring at low engine speeds and the engine runs satisfactorily at high speeds, the points gap is probably the cause – being too great. If not, check the slow running adjustment of the carburettor (see Chapter 3).

Chapter 5 Clutch

For modifications, and information applicable to later models, see Supplement at end of manual

Contents

Specifications

Type	Single dry plate, with diaphragm spring pressure plate. Cable operated

Identification	R1340	R1341
Clutch cover, part number	180 DBR 335	200 DBR 350
Release bearing, outside diameter of the thrust face	1.280 in (32.5 mm)	1.50 in (38 mm)
Disc (driven plate):		
Number of splines	20	21
Spring colour	Green, White, Dark grey	Blue
Outside diameter	7.145 in (181.5 mm)	7.874 in (200 mm)
Friction face thickness	0.303 in (7.7 mm)	0.303 in (7.7 mm)

Operating clearance	$\frac{3}{32}$ in (2.5 mm) measured at the end of the operating lever

Torque wrench setting	lbf ft	Nm
Clutch pressure plate-to-flywheel	37	50

1 General description

A cable operated clutch is employed, with an adjustable cable which permits wear to be taken up.

The clutch pedal pivots on the same shaft as the brake pedal. The release arm activates a thrust bearing (clutch release bearing) which bears on the diaphragm spring of the pressure plate. The diaphragm then releases or engages the clutch driven plate which is splined onto the gearbox primary shaft. The clutch driven plate (disc) spins in between the clutch pressure plate and the flywheel face when it is released, and is held there when engaged, to connect the drive from the engine to the transmission unit.

Different items are employed, depending upon whether the vehicle type is R1340 or R1341. Details are given of the differences in the Specifications. Care should be taken also when renewing the clutch shaft or flywheel, as these items also differ between models. The servicing instructions are, however, identical.

2 Clutch – removal, inspection and refitting

Removal

1 Remove the gearbox as described in Chapters 1 or 6, as applicable.
2 Mark the position of the clutch cover in relation to the flywheel. Remove the cover bolts and lift the cover away (photo).
3 As the cover is lifted away collect the friction disc.

Inspection

4 Examine the driven plate friction linings for wear and loose rivets. Check the disc for distortion, cracks, broken hub springs or worn splines in its hub. The surface of the linings may be highly glazed provided the woven pattern of the friction material can be clearly seen,

then the plate is serviceable. Any signs of oil staining will necessitate renewal of the driven plate and investigation and rectification of the oil leak (probably crankshaft rear main bearing and seals) being required.
5 Check the amount of wear which has taken place on the friction linings and, if they are worn level with or within 1.6 mm ($\frac{1}{16}$ in) of the heads of the securing rivets, the driven plate should be renewed as an assembly. Do not attempt to re-line it yourself as it is rarely successful.
6 Examine the machined faces of the flywheel and the pressure plate and if scored or grooved, renew both components on a factory exchange basis.
7 Check the segments of the pressure plate diaphragm spring for cracks and renew the assembly if apparent.
8 Where clutch engagement has been fierce, or clutch slip has occurred in spite of the driven plate being in good condition, renew the pressure plate assembly complete.
9 Examine the release bearing and mechanism (see Sections 3 and 4). Renew the bearing as a matter of course unless it is known to be virtually new.
10 Ensure that all frictional surfaces are clean and free from greasy deposits. Clean the contact faces of the flywheel and clutch cover using petrol.

Refitting

11 Offer up the friction plate to the flywheel, with the damper plate and springs facing towards the gearbox (photo).
12 Offer up the cover plate assembly, lining up the marks made when dismantling. Refit the cover plate bolts, but leave them just finger tight.
13 Align the centre of the friction plate with the bearing in the end of the crankshaft. Use either a Renault or proprietary alignment tool, an old gearbox primary shaft, or failing this align the friction plate by eye by moving it about using a bar placed through the centre (photo).
14 Tighten the cover plate bolts progressively, and in a diametrically opposite sequence, to avoid distortion of the cover.
15 **Very lightly** lubricate the release bearing surface on the clutch cover spring.

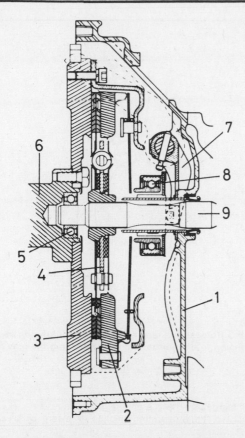

Fig. 5.1 Clutch — sectional view (Sec 1)

1	Clutch housing	5	Spigot bearing
2	Clutch cover (pressure plate)	6	Crankshaft
3	Flywheel	7	Release fork
4	Friction plate (driven plate)	8	Release bearing
		9	Gearbox clutch shaft (primary shaft)

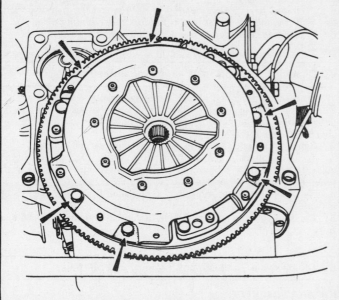

Fig. 5.2 Clutch cover, showing retaining bolts (arrowed) (Sec 2)

2.2 Removing the clutch cover bolts

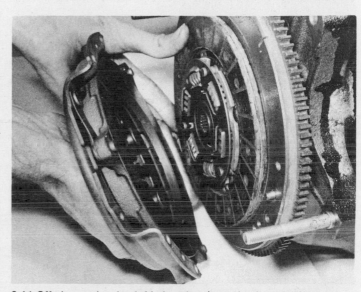

2.11 Offering up the clutch friction plate (note the damper plate and springs), followed by the clutch cover

2.13 Aligning the clutch plate, using the gearbox primary shaft

3.2 Freeing the ends of the spring

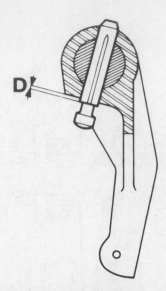

Fig. 5.3 Release fork and retaining pins, showing essential dimension to be maintained when refitting the pins (Sec 4)

$$D = \tfrac{1}{32} in (1 mm)$$

16 Refit the gearbox to the engine.
17 Adjust the clutch operating mechanism clearance, as described in Section 5.

---

3 Clutch release bearing – removal and refitting

1 Remove the gearbox as described in Chapter 1 or 6, as applicable.
2 Free the ends of the spring, thereby releasing the operating fork (photo).
3 Remove the clutch release bearing (photos).
4 To refit, lubricate the sleeve on the gearbox shaft and the fingers of the operating fork, using Molykote BR2 grease, or equivalent.
5 Place the new release bearing in position on the gearbox shaft, and refit the ends of the spring in the fork and bearing holder.
6 Proceed as described in Section 2, paragraphs 15 to 17.

3.3a Removing the clutch release bearing

---

4 Operating fork and shaft – removal and refitting

1 Proceed as described in Section 3, paragraphs 1 to 3.
2 Pull out the pins which retain the fork to the shaft. This can be difficult without the correct tool, and if necessary a Renault agent should be consulted.
3 Pull out the fork shaft, and withdraw the fork and spring.
4 To refit, lubricate the shaft with Molykote BR2 grease or an equivalent. Place the fork and spring in position, and insert the shaft together with the rubber seal.
5 Refit the pins, ensuring that dimension D in Fig. 5.3 is as indicated.
6 Continue refitting as described in Section 3, and paragraphs 5 and 6.

---

5 Clutch cable – adjustment, removal and refitting

Cable adjustment
1 Refer to Fig. 5.4. Maintain hand pressure on lever L in the direction of arrow A, thereby ensuring that the release bearing is in the contact with the diaphragm spring.
2 Release locknut G, and screw nut E in or out as necessary until clearance J is $\tfrac{3}{32}$ in (2.5 mm), when the cable is pulled by hand pressure in direction C.
3 Lock nut G against nut E.

3.3b A view of the operating mechanism, with the release bearing removed

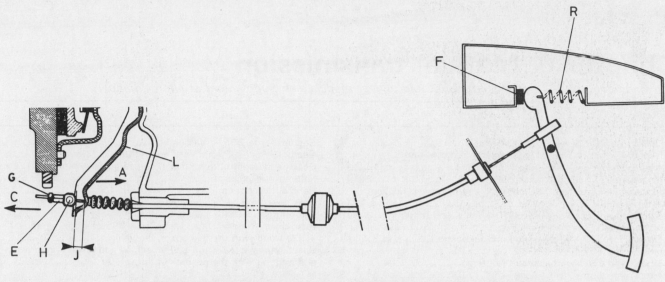

Fig. 5.4 Clutch cable and release mechanism (diagrammatic representation) (Sec 5)

L Clutch housing operating lever	G Adjuster nut locknut	R Return spring	H Trunnion
	E Adjuster nut	F Stop pad	For A, C and J, see text

Cable removal and refitting

4   Remove the nuts and trunnion from the cable at the gearbox end.
5   Disconnect the cable at the pedal end.
6   Remove the cable.
7   To refit, reverse the removal procedure.
8   Adjust the cable as described in paragraphs 1 to 3.

6   Clutch spigot bearing and clutch housing oil seal — general

1   See Chapter 1 for operations which may be performed on the clutch spigot bearing.
2   See Chapter 6, Section 5, for details of operations on the clutch housing oil seal.

7   Fault diagnosis — clutch

Symptom	Reason(s)
Judder when taking up drive	Loose engine/gearbox mountings or over-flexible mountings Badly worn friction surfaces or friction plate contamination with oil carbon deposit Worn splines in the friction plate hub or on the gearbox input shaft
Clutch spin (or failure to disengage) so that gears cannot be meshed	Clutch actuating cable clearance too great Clutch friction disc sticking because of rust on splines (usually apparent after standing idle for some length of time) Damaged or misaligned pressure plate assembly Incorrect release bearing fitted
Clutch slip — (increase in engine speed does not result in increase in car speed — especially on hills)	Clutch actuating cable clearance from fork too small resulting in partially disengaged clutch at all times Clutch friction surfaces worn out (beyond further adjustment of operating cable) or clutch surfaces oil soaked

Chapter 6 Manual transmission

For modifications, and information applicable to later models, see Supplement at end of manual

Contents

Specifications

Type

352	Four forward gears and one reverse. Synchromesh on all forward gears. Centre-mounted gear lever
395	As for type 352, but with five forward gears

Gear ratios

	352	395
1st gear	3.82	3.82
2nd gear	2.24	2.24
3rd gear	1.48	1.48
4th gear	0.97	1.04
5th gear	–	0.86
Reverse	3.08	3.08
Final drive ratio	9 x 34 (3.78)	9 x 34 (3.78)

Setting-up limits

Backlash crownwheel and pinion	0.0047 to 0.010 in (0.12 to 0.25 mm)
Preload, differential bearings:	
Re-used bearings	No play, but free to turn
New bearings	2 to 7 lbf (9 to 31 N)
Differential pinion protrusion	2.323 in (59 mm)

Oil type/specification

Oil type/specification	Hypoid gear oil, viscosity SAE 80EP to API GL5 (Duckhams Hypoid 80S)

Torque wrench settings

	lbf ft	Nm
Half-housing bolts, 7 mm	15 to 19	20 to 26
Half-housing bolts, 8 mm	22.5	30.5
Rear cover fixing bolts	9	12
Clutch housing fixing bolts, 8 mm	18	24.5
Clutch housing fixing bolts, 10 mm	26	35
Reverse selector fixing bolt	18	24
Speedometer worm nut	75 to 90	102 to 122
Primary shaft nut (type 395)	75 to 90	102 to 122
Differential ring nut locking bolt	18	24

1 General description

Two types of manual gearbox are fitted to the Renault 18, type 352 or 395. The first of these has four forward speeds, and the second has five.

The design of both types is similar, with an aluminium gearcase in four sections, namely the clutch housing, the right- and left-hand housings (housing the main gear assemblies and also the differential unit), and the rear cover unit (containing the selector control shaft and selector finger). On the 395 gearbox, the rear cover also houses the fifth gear and its synchromesh assembly. All forward gears are fitted with synchromesh. Drive to the gears from the flywheel and clutch unit is via the clutch shaft, which is engaged with the primary shaft splines. The clutch shaft passes through the differential compartment. The primary shaft transmits motion via the respective gears to the pinion (secondary) shaft. The reverse idler gear is located on a separate shaft in the rear of the main casing.

Motion is transmitted when a particular gear is engaged transferring the drive from the primary to the secondary shaft, which in turn drives the differential unit and consequently the driveshafts.

The gear selector forks and shafts are located in the side of the gear casing and are actuated by the selector rod mechanism located in the rear casing. The selector forks are in constant engagement with the synchro sliding hubs which move to and fro accordingly to engage the gear selected.

The speedometer drivegear is attached to the end of the secondary shaft, and this in turn drives the drive gear unit to which the drive cable is attached. Both the speedometer cable and driven gear unit can be removed with the gearbox in place.

On the type 395 gearbox, work upon the 5th gear and associated parts is feasible with the unit in place in the vehicle, by first removing the rear cover. All other gear assemblies, and the differential unit, are accessible only with the unit removed from the car.

Although the transmission unit is basically simple in operation, certain dismantling, adjustment and reassembly operations require the use of specialised tools. Therefore, if you are contemplating overhauling the gearbox it is essential that you read through the relevant sections concerning your gearbox before starting any work.

Another point to consider is the availability of parts. You will not know what you require until the casing sections are separated. At this stage an assessment should be made of the work required and parts that will be needed. In some instances, certain items are only supplied as complete assemblies, and in many cases the simplest course of action is to reassemble the casings and get an exchange unit – usually the most satisfactory and economical solution.

The gearbox is identified by a plate attached under one of the rear end cover bolts. The plate carries details of the type, and also the suffix and fabrication numbers.

---

2 Gearbox, type 352 – removal and refitting

Removal

1 Although it is stated by the manufacturer that the gearbox may be removed with the engine either in or out of the vehicle, it is our opinion that the weight and bulk of the gearbox, together with the restricted space available for manoeuvering it, make the second course the one to be recommended. Removal of the gearbox and engine as a unit is considered in Chapter 1.

2 Disconnect the battery

3 Disconnect the electrical leads to the starter motor and solenoid, and to the reversing light switch.

4 Remove the starter motor and solenoid as described in Chapter 10.

5 Disconnect the clutch cable at the lower end (see Chapter 5).

6 Remove the front brake calipers (see Chapter 9).

7 Disconnect the driveshafts at their inner ends (see Chapter 8).

8 Disconnect the speedometer cable.

9 Disconnect the bolt (1) in the balljoint (see Fig. 6.36), without splitting the joint. Remove bolt (2) (photo).

10 Remove the clutch shield.

11 Support the gearbox, and remove the mountings (see Section 15) (photo).

12 If the engine is still in the vehicle, remove the clutch housing bolts, noting their correct positions.

13 Withdraw the gearbox carefully, ensuring that an assistant is present to take some of the weight. Do not allow the weight to hang on the input shaft!

Refitting

14 Lubricate the splines of the clutch shaft lightly, using Molykote BR2 grease, or an equivalent.

15 Ensure that the clutch shaft spigot bearing has not become dislodged from the crankshaft.

16 Refit the gearbox in reverse of the removal procedure, ensuring that the splines for the driveshafts are lubricated as in paragraph 14.

17 Check the gearchange adjustment (see Section 16).

18 Check the clutch adjustment (see Chapter 5).

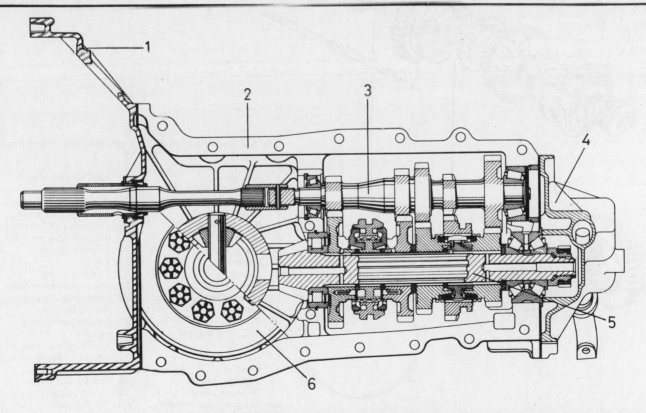

Fig. 6.1 Gearbox type 352 – sectional view (Sec 1)

1 Clutch housing	3 Primary shaft assembly	5 Secondary shaft assembly
2 Half gearbox case	4 Rear cover	6 Differential assembly

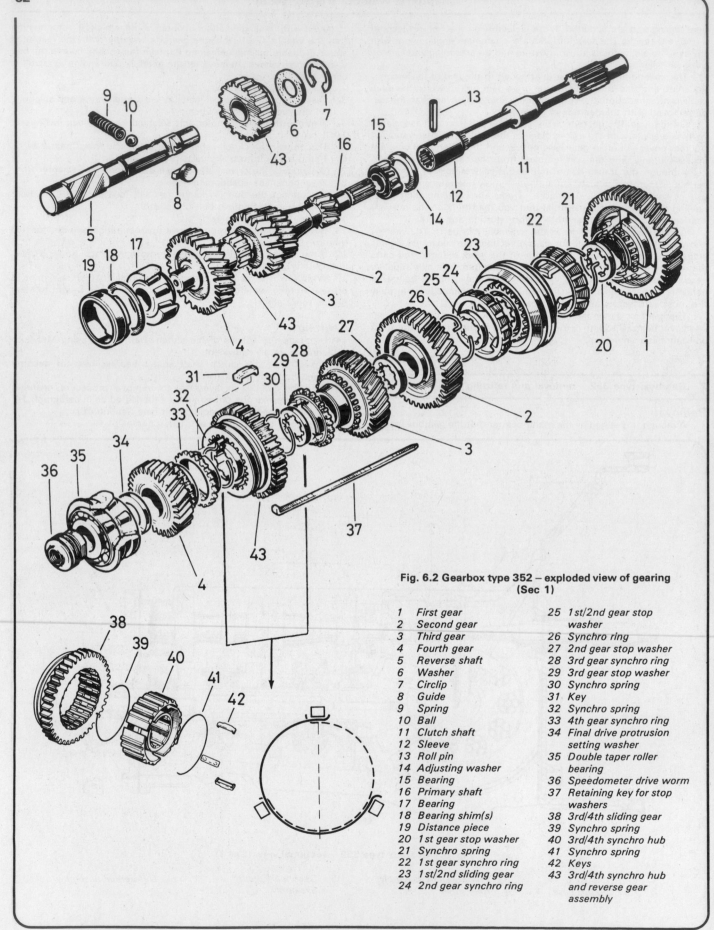

Fig. 6.2 Gearbox type 352 – exploded view of gearing (Sec 1)

1 First gear	25 1st/2nd gear stop washer
2 Second gear	26 Synchro ring
3 Third gear	27 2nd gear stop washer
4 Fourth gear	28 3rd gear synchro ring
5 Reverse shaft	29 3rd gear stop washer
6 Washer	30 Synchro spring
7 Circlip	31 Key
8 Guide	32 Synchro spring
9 Spring	33 4th gear synchro ring
10 Ball	34 Final drive protrusion setting washer
11 Clutch shaft	35 Double taper roller bearing
12 Sleeve	36 Speedometer drive worm
13 Roll pin	37 Retaining key for stop washers
14 Adjusting washer	38 3rd/4th sliding gear
15 Bearing	39 Synchro spring
16 Primary shaft	40 3rd/4th synchro hub
17 Bearing	41 Synchro spring
18 Bearing shim(s)	42 Keys
19 Distance piece	43 3rd/4th synchro hub and reverse gear assembly
20 1st gear stop washer	
21 Synchro spring	
22 1st gear synchro ring	
23 1st/2nd sliding gear	
24 2nd gear synchro ring	

2.9 Disconnecting the balljoint bolt in the gear linkage

2.11 Bolts loosened on a gearbox mounting

3.1 Removing the reversing light switch

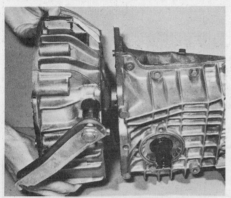

3.4 Removing the clutch housing

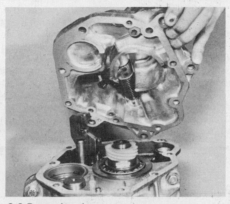

3.6 Removing the rear cover

3.7a The primary shaft distance piece

3.7b The primary shaft bearing shims

3.9 Removing a half gearcase

3.10 Lift out the differential assembly

3.11 The secondary geartrain assembly

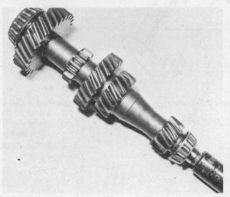

3.12 The primary gear shaft assembly

3.13 Selector forks, general view before dismantling

3   Gearbox, type 352 – dismantling

General

1   Drain the oil and remove the reversing light switch (photo).
2   Remove the clutch release bearing (see Chapter 5).
3   Remove the clutch housing bolts, washers and spring washers, noting the correct positions for the bolts of various lengths (12 bolts).
4   Remove the clutch housing, and keep the gasket (photo).
5   Remove the 8 bolts securing the rear cover, noting the identification tag under one bolt.
6   Remove the cover, and preserve the gasket (photo).
7   Remove the primary shaft distance piece and bearing shims from under the rear cover (photos).
8   Remove the half casing securing bolts, noting their correct positions.
9   Separate the half casings, if necessary by gentle use of a soft-headed mallet (photo).
10   Lift out the differential assembly (photo).
11   Lift out the secondary geartrain assembly with the stop peg from the double taper roller bearing cup (photo).
12   Lift out the primary gear shaft assembly (photo).
13   Tap the roll pin from the 3rd/4th gear selector fork, and withdraw the shaft and fork. Catch and keep the ball and spring, and remove the locking disc from between the shafts (photo).
14   With 1st speed selected, take the reverse selector shaft right back to the gear shift control end. Punch out the 1st/2nd selector fork roll pin, remove the shaft and fork, and retain the ball and spring.

15   Remove the pivot bolt from the reverse selector lever, remove the selector lever, and withdraw the reverse selector shaft.
16   Remove the circlip against the reverse gearwheel, and remove the shaft, gearwheel, washer and sleeve (photo).

Primary shaft

17   Remove the bearing track rings and associated washers.
18   Support the shaft adequately, and drive out the roll pin, thereby separating the clutch shaft from the primary shaft. Note the 'Grower' washer inside the sleeve (photo).
19   Hold the primary shaft in a soft-jawed vice, and pull the bearings from each end of the shaft.

Secondary shaft

20   Hold the shaft by the 1st gear in a soft-jawed vice. Select 1st gear, and unlock the speedometer drivegear using a suitable open-ended spanner. In the absence of such a spanner, we employed a suitable size exhaust clamp on the flats of the gear, gripped and turned by a large open-ended wrench (photo).
21   Withdraw the double taper roller bearing followed by the final drive pinion adjusting washer (photos).
22   Remove the 4th gear and synchromesh ring (photo).
23   Remove the 3rd/4th gear sliding synchromesh unit. Mark the relative position of the gear to the shaft and retain the hub keys.
24   Remove the 3rd/4th synchro hub by supporting it carefully to prevent damage, and driving the shaft through with a soft-headed mallet.
25   Remove the gearwheel stop washer retaining key (see Fig. 6.4).

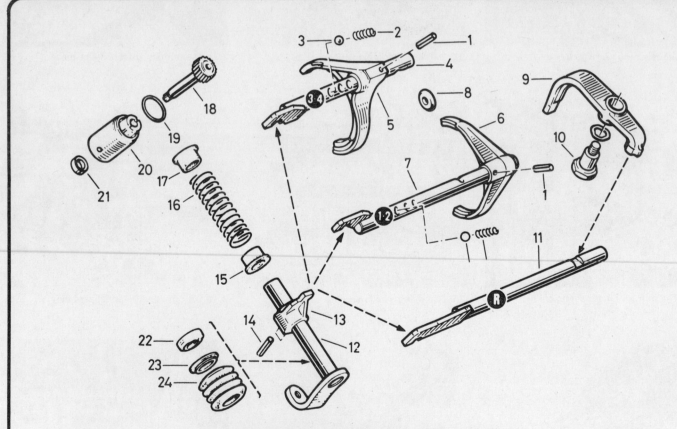

Fig. 6.3 Selector assemblies – exploded view (Sec 3)

1 Roll pin	6 1st/2nd gear selector fork	11 Reverse selector shaft	18 Speedo wheel
2 Spring		12 Control shaft	19 O-ring
3 Ball	7 1st/2nd gear selector rod	13 Selector finger	20 Bush
4 3rd/4th gear selector rod	8 Locking disc	14 Roll pin	21 Seal
	9 Reverse selector lever	15 Collar	22 Control shaft seal
5 3rd/4th gear selector fork	10 Reverse selector pivot	16 Spring	23 Bellows washer
		17 Collar	24 Bellows

3.16 The reverse gear and associated items, partially withdrawn

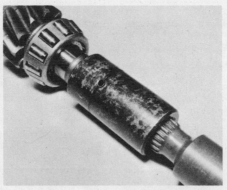

3.18 The primary shaft roll pin, before removal

3.20 Removing the speedometer drivegear

3.21a Removing the double row taper roller bearing

3.21b Removing the pinion adjusting washer

3.22 Removing the 4th gear synchro ring

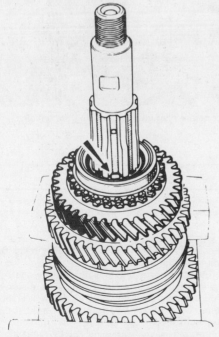

Fig. 6.4 Retaining key for gearwheel stop washers (arrowed) (Sec 3)

26 Remove the 3rd speed gear stop washer, gear and synchro ring.
27 Remove the 2nd speed gear stop washer, gear and synchro ring.
28 Mark the position of the 1st/2nd sliding synchro gear in relation to the hub. Remove the gear and stop washer (photo).
29 Remove the 1st/2nd speed synchro hub as described in paragraph 24, taking care as this is done not to disturb the front roller bearing

outer track.
30 Remove the 1st gear synchro ring, stop washer and gear (photo).
31 The front roller bearing cannot be renewed independently as the inner track is bonded to the final drive pinion. To prevent the rollers and outer track becoming dislodged, fit a suitable retaining clip or clamp over the bearing as shown in Fig. 6.5.

Rear cover
32 Tap out the roll pins which secure the control finger, and slide out the shaft. Recover the bushes, spring, finger and bellows (photo).
33 Prise out the shaft oil seal.
34 Remove the speedo drive retaining bolt, and take out the pinion unit and seal.

Differential unit
35 Major dismantling of this unit is not a task for the home mechanic, and it is recommended that the work is either entrusted to a Renault agent, or that an exchange unit be obtained.
36 Renewal of the taper roller bearings is feasible, and these should be drawn from the differential unit using a suitable puller. Hold the unit in a soft-jawed vice, and if necessary remove two diametrically opposed crownwheel holding bolts to permit the legs of the puller to be properly secured.
37 Remove the bearing outer cups from the half gearbox cases, by removing the adjuster rings, and seals (after first marking their approximate positions to aid refitting) and then pressing or driving out the cups using a suitable sized piece of tube (photo).

4   Gearbox, type 352 – examination and renovation

1   Clean all parts and examine the gears for chipping and obvious wear. Check that all bearings, when cleaned and lightly oiled, are completely smooth in operation.
2   It is advised that all oil seals and gaskets be renewed as a matter of course when the unit is dismantled.
3   Obtain replacements where possible for small items such as roll pins or clips, which may have altered dimensionally when dismantled.

3.28 Removing the 1st/2nd speed sliding gear

3.30 Removing the 1st speed gear

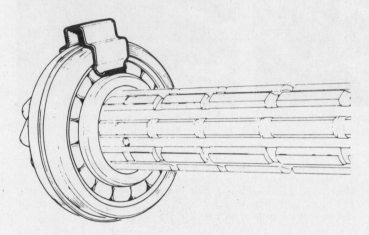

Fig. 6.5 Suitable clip in position, to prevent separation of front roller bearing items (Sec 3)

3.32 The cover before dismantling, showing the roll pins

3.37 The differential adjuster rings and seals

5 Gearbox type 352 – reassembly

1   With all items properly cleaned, spread them on a clean work-bench, using paper or cardboard 1 to provide a grit-free surface.

Secondary shaft

2   Engage the 1st gear synchro spring as shown in Fig. 6.6, ensuring that it contacts the three segments.

3   Slide the 1st speed gear into position on the shaft (against the pinion bearing) and locate the synchro ring. Fit the stop washer and rotate it to align the keyway. A suitable temporary key should now be slid into position down the keyway in the shaft to hold the washer in position during subsequent operations. This 'dummy' key can be fabricated from an old washer retaining key by removing the hooked lug. Ensure that the keyway spline chosen is one with an oil hole as shown in Fig. 6.7.

4   Detach the bearing outer race retaining clip.

5   When fitting the 1st/2nd speed synchro hub, first heat it up to an electric oven temperature of 120°C. Leave the hub in the oven at this temperature for about 15 minutes, then extract and assemble it onto the pinion shaft so that an unsplined section is aligned with the dummy key. The matching mark on the hub should be towards the second gear or, if no mark exists, orientation must be as in Fig. 6.8.

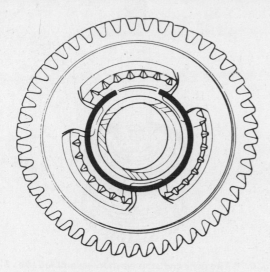

Fig. 6.6 First gear synchro spring, correctly fitted (Sec 5)

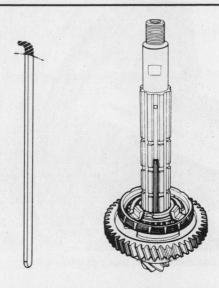

Fig. 6.7 Dummy key in position in secondary shaft (note the oil hole). Note detail, showing the end removed (Sec 5)

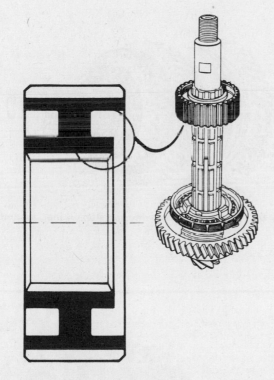

Fig. 6.8 Orientation of 1st/2nd gear synchro hub (Sec 5)

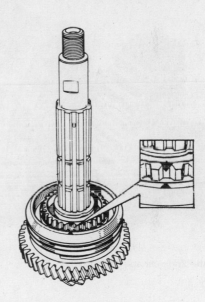

Fig. 6.9 Fitting 1st/2nd synchro sliding gear, synchro ring and stop washer (Sec 5)

Press the hub fully into position so that it just comes into contact with the stop washer. As the hub is pressed into position, centralise the synchro ring with the lugs below the stop washer level in order not to damage the spring. Withdraw the dummy key and allow the hub to cool.

6 Next, assemble the synchro hub sliding gear with the chamfered side facing the 2nd gear and the relative hub match markings in alignment.

7 Locate the stop washer with its splines aligned with those on the shaft (photo).

8 Fit the synchro spring to the 2nd gear (in a similar fashion to that for the 1st) and assemble the 2nd gear with synchro ring, as shown in Fig. 6.10. Fit the stop washer and align the splines with those of the shaft (photo).

9 Next, assemble the 3rd gear and synchro ring.

10 Slide the stop washer into position and rotate it to align the splines.

11 Slide the stop washer location key into position down a keyway in the shaft (choose a keyway with an oil hole in it) (photo).

12 Press or drive the 3rd/4th synchro hub into position, so that it is flush against the 3rd gear stop washer. Check when fitting that the notch on the hub is facing the 3rd gear and is aligned with the stop key. The three synchro ring notches must be aligned with the keys.

13 Locate the 4th gear and its synchro ring (photo).

14 Fit the pinion protrusion adjustment washer and double taper roller bearing.

15 Fit the speedometer worm drivegear.

16 Support the shaft assembly vertically in a vice with soft jaws, fastened to the 1st gear. Select the 1st gear to lock the shaft, and tighten the speedometer worm pinion to the specified torque. When tightened, do not lock the nut until the pinion shaft is adjusted on assembly.

Primary shaft

17 Check for cleanliness of the bearings and their mounting areas on the shaft. Support the shaft and fit the bearings at each end using a press, or alternatively use a piece of tube and a hammer. Refit the clutch shaft (reverse of Section 3, paragraph 18).

5.7 The 1st/2nd speed sliding gear and stop washer

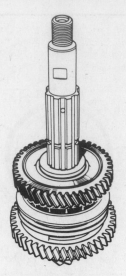

Fig. 6.10 2nd gear and stop washer, assembled (Sec 5)

5.8 Fitting the 2nd gear and synchro ring

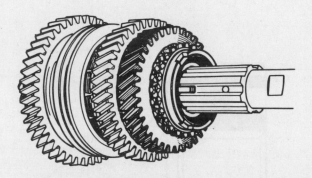

Fig. 6.11 3rd gear, synchro ring and stop washer, assembled (Sec 5)

5.11 The 3rd gear, synchro ring, stop washer and key

5.13 The 4th gear and synchro ring

Final drive pinion protrusion – adjustment

18 The correct positioning of the front face of the final drive pinion in relation to the crown wheel centre is very important (see Fig. 6.12), and must be checked and reset as necessary if the component parts of the secondary shaft have been renewed (other than the bearings which, by virtue of the close limits to which they are manufactured, should not affect the position of the pinion by an unacceptable amount).

19 A Renault agent must be requested to carry out this check, as the home mechanic will not have access to the necessary checking tools. The dimension should be corrected, if necessary, by changing the pinion protrusion adjustment washer for another of the appropriate thickness.

Differential bearings – fitting and adjustment

20 The bearing outer tracks should be carefully pressed or tapped into each half of the gearcase, so that they are slightly below the inner face of the casting.

21 After fitting new bearings, or refitting old ones, place the secondary shaft into the RH half gearcase, fit the differential assembly, and fit the LH half gearcase. Loosely fit the bolts. Fit the rear cover and gasket, so as to retain the rear secondary shaft bearing in the correct location, and tighten the half casing bolts in the correct order and to the correct torque figure.

22 Fit the differential ring nuts, but without the oil seals, as they may be damaged by the splines during setting-up. Screw the ring nuts home until they contact the bearing ring tracks.

23 When refitting the old bearings, the differential should be adjusted so that the assembly will revolve smoothly, but without any play. Effect the adjustment by screwing the differential ring nuts in or out as necessary, mark the final positions on the case halves and rings, and remove the rings.

24 When new bearings have been fitted, proceed basically as in paragraph 23. However, in this case a preload is necessary, and the ring nuts should be fitted so that the differential assembly is a little stiff to turn. Check the preload by wrapping a piece of cord round the differential housing, and checking with a spring balance the loading necessary to turn the differential, which should be within the specified limits.

25 If the bearings have not been removed, and providing that the

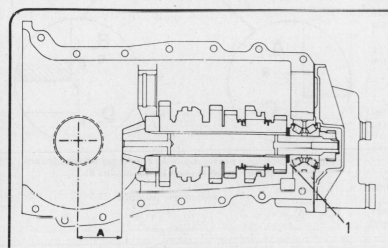

Fig. 6.12 Final drive pinion position – crownwheel centre-to-pinion front face (Sec 5)

A   2.323 in (59 mm)
1   Adjust washer to suitable thickness

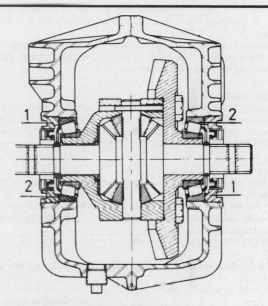

Fig. 6.13 Differential assembly – sectional view on centre-line (Sec 5)

1   Differential ring nuts
2   Differential bearings

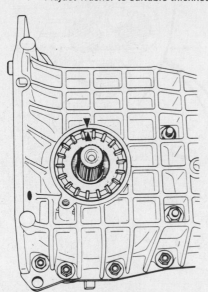

Fig. 6.14 Alignment of markings on the differential ring nut and case (Sec 5)

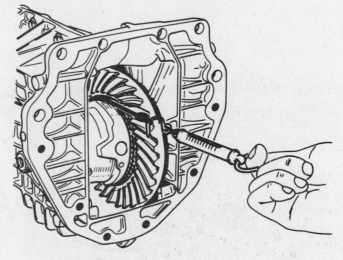

Fig. 6.15 Checking the differential preload setting (Sec 5)

differential ring nuts are marked to ensure correct refitting, then
bearing adjustment should not be necessary.

Primary shaft positioning

26 Assemble the bearing rings and adjusting washer to the primary
shaft.
27 Fit the primary and secondary shafts into the RH half casing, and
check the steps (R), which must be equal (Fig. 6.16). Correct if
necessary by changing washer (1), using a replacement of the
appropriate thickness.
28 Remove the primary and secondary shaft assemblies.

Roll pin refitting

29 It should be noted that roll pins must always be fitted with the slot
towards the rear cover.

Selector forks

30 Slide the reverse gear selector shaft home.
31 Position the reverse gear selector with the end in the slot in the
shaft, fit the pivot pin and tighten.
32 Position the 1st/2nd speed spring and ball in the appropriate hole
(Fig. 6.17), position the selector fork, slide in the shaft, and roll pin the
two together.
33 Position the 3rd/4th speed locking ball and spring and place the
locking discs between the shafts. Position the selector fork, slide in the
shaft, and roll pin the two together.

Reverse gear

34 Position the ball and spring in the LH half gearcase. Start to enter
the shaft, position the gearwheel and friction washer (bronze side to
gearwheel), fit the guide from inside the bore, and slide the shaft
home. Fit the circlip (photos).

Rear cover

35 Fit the oil seal in the cover hole, and position the collars, spring
and selector finger internally.
36 Fit the bellows over the shaft, insert the shaft through the case
and internal components, and line up the roll pin holes in the shaft and
selector finger. Fit new roll pins.

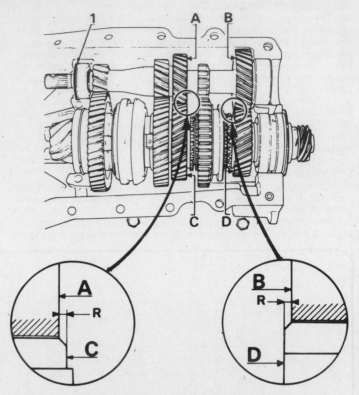

**Fig. 6.16 Checking the relationship of primary and secondary
shafts. Steps R must be equal (Sec 5)**

1   Adjustment washer

Fig. 6.17 Selector fork arrangement, type 352 gearbox (Sec 5)

1 3rd/4th selector shaft and fork	*4 Reverse selector lever*
2 1st/2nd selector shaft and fork	*5 Reverse selector lever pivot*
3 Reverse selector shaft	*6 Locking disc*

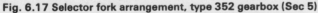
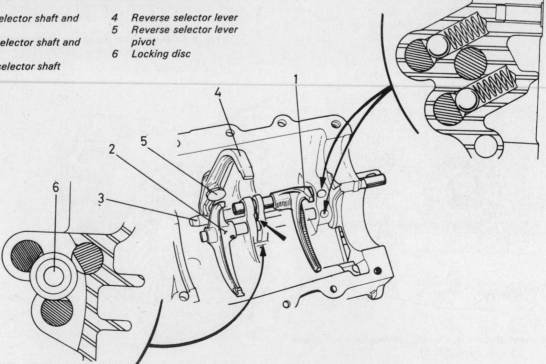

5.34a Spring and ball, reverse gear shaft

5.34b Guide, reverse selector shaft

5.34c Fitting the reverse shaft circlip

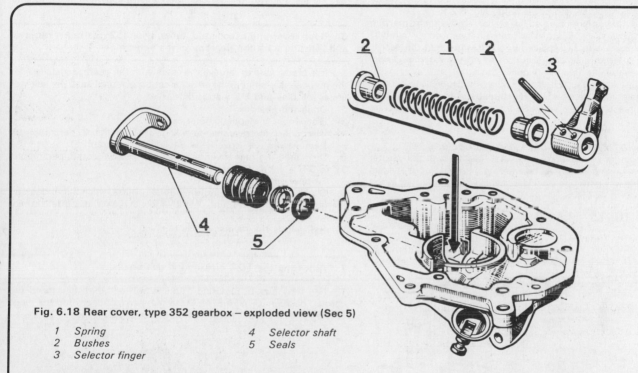

Fig. 6.18 Rear cover, type 352 gearbox – exploded view (Sec 5)

1 Spring
2 Bushes
3 Selector finger
4 Selector shaft
5 Seals

5.37 The RH case, with geartrains in place

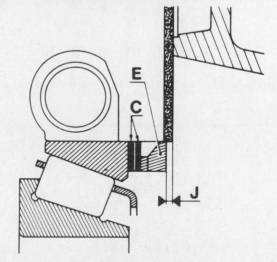

Fig. 6.19 Checking the endfloat on the primary shaft – type 352 gearbox (Sec 5)

C Shims
E Distance piece

$J = 0.0010$ to 0.0047 in
$(0.02$ to 0.12 mm$)$

Half casings

37 Fit the primary shaft, the secondary gear train and locking peg and differential assembly, into the RH half casing (photo).

38 Smear a non-hardening jointing compound on the half gearcases, offer the LH casing up to the RH casing, and fit the half casing bolts without tightening at this stage.

39 Adjust the primary shaft endplay if necessary by fitting the adjusting washers and distance washers, tapping the distance washers lightly to settle the bearings, and fitting the rear cover gasket. Place a straight-edge across the gasket, and measure the clearance between the distance washer and straight-edge (Fig. 6.19).

40 Change the adjusting washer if the clearance is outside the limits shown. Use as few washers as possible.

41 Smear jointing compound on the rear gasket and offer up the cover, engaging the selector finger in the selector shaft slots. Nip up the securing bolts, without tightening them.

42 Tighten the half casing bolts in the correct sequence, to the correct torque setting. Tighten the end cover bolts (Fig. 6.20).

43 If the differential ring nuts are still in place, remove them, marking both nuts and gearcase if this has not already been done. Carefully remove the oil seals and fit replacements, ensuring that the seal is flush with the outer surface of the nut. Prise off the external O-ring, and fit a new one using only the fingers.

44 Wind a little plastic tap round the splines to protect the oil seal, smear a little jointing compound on the threads of the ring nuts, and screw them home until the marks are correctly aligned. Remove the plastic tape.

Crownwheel and pinion – backlash rechecking

45 Fit a dial gauge to the end of the half casing with the pointer resting on a crownwheel tooth at the extreme outer edge, but still just on the tooth flank. Check the backlash, which should be within the specified limits (Fig. 6.21).

46 If backlash is excessive, loosen the ring nut on the differential side a little and screw in that on the crownwheel side by the same amount. If backlash is insufficient, reverse the procedure. Recheck and readjust as necessary.

47 Lock the rings, using the locking plates.

Clutch housing

48 Check the condition of the oil seal, and renew it if there is any doubt about its condition. Carefully drive the seal in using a suitable piece of tube.

49 Wind a little plastic tape onto the splines of the primary shaft to protect the clutch housing oil seal, smear jointing compound on the paper gasket, and fit the gasket and housing. Remove the plastic tape, and fit and tighten the housing bolts.

50 Refit the clutch release bearing (see Chapter 5) and the reversing light switch.

51 Fill the unit with oil, but preferably after installation in the vehicle.

---

6   Rear cover and associated items, type 352 gearbox – removal and refitting with the gearbox in the vehicle

---

1   It is practicable to remove the cover with the gearbox installed, for the purpose of changing the selector control shaft oil seal, the selector finger roll pins and the rear cover gasket.

2   Drain the gearbox.

3   Disconnect the speedometer cable.

4   Disconnect the gearchange linkage (see Section 2, paragraph 9).

5   Remove the rear cover (see Section 3, paragraph 5).

6   Dismantle the rear cover as necessary (see Section 3, paragraphs 32 to 34).

7   Fit new parts as necessary.

8   Check the selector control shaft for any burrs which might damage the new oil seal upon fitting. Remove as necessary, using fine emery cloth.

9   Refitting is the reverse of the removal procedure.

---

7   Adjusting the TDC pick-up (all gearboxes)

---

1   If a new pick-up is being fitted, it will automatically be in the correct position by virtue of the three pegs on it. With the pegs touching the flywheel tighten screw (1) (Fig. 6.22).

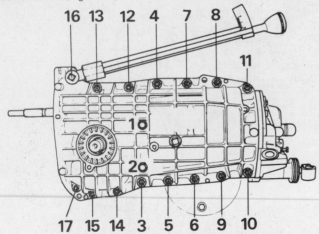

Fig. 6.20 Gearcase, type 352 gearbox – bolt tightening sequence
(Sec 5)

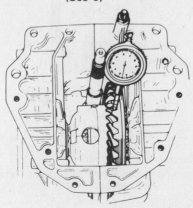

Fig. 6.21 Backlash check – crownwheel-to-pinion (Sec 5)

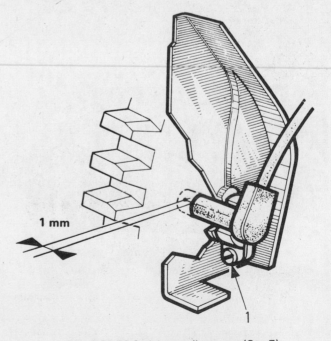

Fig. 6.22 TDC pick-up adjustment (Sec 7)

1 *Screw*

2   Where an old pick-up with worn pegs is being fitted, first set the pick-up to touch the flywheel, then withdraw it 0.04 in (1 mm). Tighten the securing screw.

8   Gearbox, type 395 – removal and refitting

Proceed as in Section 2, but note that the gearshift arrangement differs.

9   Gearbox, type 395 – dismantling

1   Remove the gearbox and drain the oil.
2   Remove the clutch thrust bearing, the clutch housing bolts, and the housing.
3   At the rear of the unit, remove the 5th speed shaft locking bolt, followed by the spring and ball beneath it (Fig. 6.24).
4   With neutral selected, remove the bolts from the rear cover. Remove the cover, whilst tilting the selector finger.
5   With 5th and reverse gears selected together, unlock the 5th speed synchro hub and speedometer drive worm nuts.
6   Select 4th gear.
7   Tap out the roll pin in the 5th gear selector fork, and remove the

5th gear synchro hub assembly with the fork, the 5th speed gears, and the needle bearing and ring (Fig. 6.25).
8   Remove the pivoting interlock (Fig. 6.26).
9   Mark the positions of the differential ring nuts, unscrew the locking tabs, and remove the ring nuts. Count and record the number of turns required to unscrew them.
10 Split the casings and remove the geartrains as described in Section 3, paragraphs 8 to 12.
11 Place the 3rd/4th gear selector shaft in the neutral position, and remove the 5th gear selector shaft. Proceed as in Section 3, paragraph 13 (note that no locking disc is fitted).
12 Remove the reverse selector pivot bolt, remove the selector and pull out the reverse selector shaft.
13 Punch out the roll pin from the 1st/2nd selector fork, remove the shaft and fork, and retain the ball and spring (Fig. 6.27).
14 Remove reverse gear and shaft as described in Section 3, paragraph 16.
15 If necessary, remove the bearing cups from the half gearbox cases (see Section 3, paragraph 37). Prise out the seals from the differential ring nuts.

Primary shaft
16 Support the shaft adequately and drive out the roll pin to separate the clutch shaft from the primary shaft.

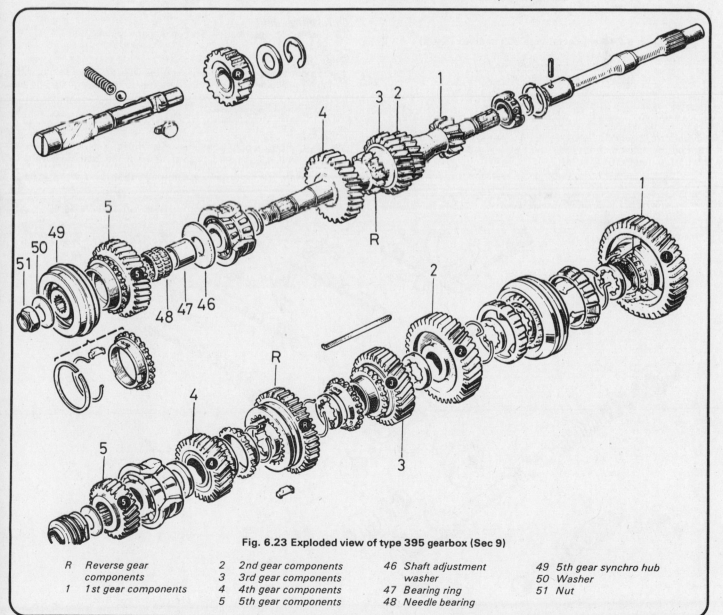

Fig. 6.23 Exploded view of type 395 gearbox (Sec 9)

R	Reverse gear components	2 2nd gear components	46 Shaft adjustment washer	49 5th gear synchro hub
1	1st gear components	3 3rd gear components	47 Bearing ring	50 Washer
		4 4th gear components	48 Needle bearing	51 Nut
		5 5th gear components		

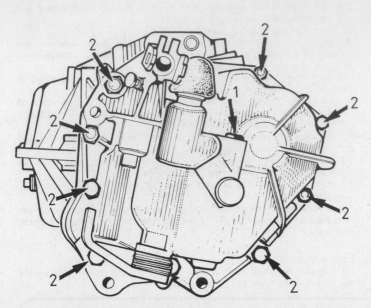

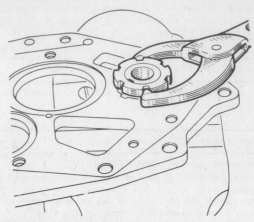

Fig. 6.26 Removing the pivoting interlock nut (Sec 9)

Fig. 6.24 Rear cover, type 395 gearbox (Sec 9)

1   5th speed locking bolt
2   Rear cover bolts

Secondary shaft

19 Hold the shaft by the 1st gear in a soft-jawed vice, and proceed as described in Section 3, paragraphs 21 to 31.

Differential unit

20 Proceed as described in Section 3, paragraphs 35 to 37.

Rear cover

21 Tap out the two roll pins retaining the selector finger to the shaft. Remove the clip and pull out the shaft. Retain the collars, springs and finger.
22 Prise out the cover seals.

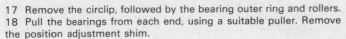

17 Remove the circlip, followed by the bearing outer ring and rollers.
18 Pull the bearings from each end, using a suitable puller. Remove the position adjustment shim.

General

23 The gearbox is now completely dismantled. Reference should be made to Section 4 for the principles of examination and renovation.

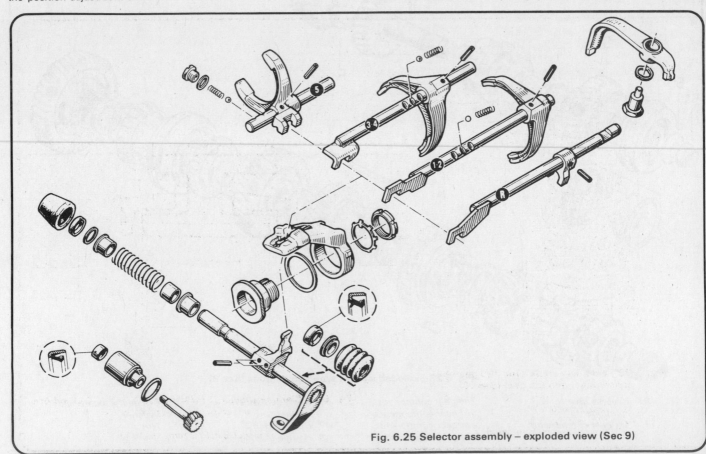

Fig. 6.25 Selector assembly – exploded view (Sec 9)

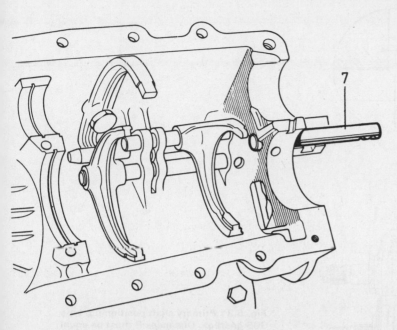

Fig. 6.27 Selector fork arrangement, type 395 gearbox (Sec 9)

7   5th gear selector fork

10 Gearbox, type 395 – reassembly

Secondary shaft
1   Proceed as described in Section 5, paragraphs 1 to 14.
2   Support the assembly vertically, by gripping the 1st gear in a soft-jawed vice.

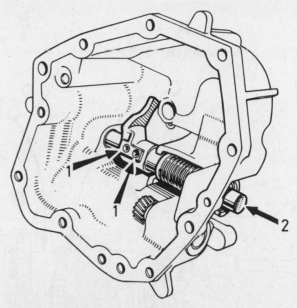

Fig. 6.28 Rear cover, type 395 gearbox – roll pin locations (Sec 9)

1   Roll pin   *2   Shaft*

3   Fit the spacer plate, the 5th gear, the wavy washer and speedometer drive worm. Tighten but do not lock the worm (see Section 3, paragraph 20).
4   Referring to Fig. 6.30, check the dimension indicated between both the 3rd and 4th speed synchro rings. When checking, the synchro ring must be stuck to the hub cone, and the gear pressed against the hub.

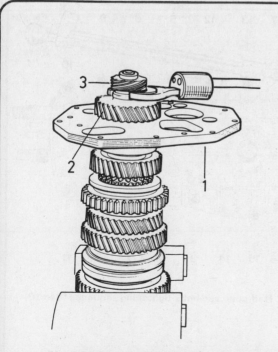

Fig. 6.29 Secondary shaft, type 395 gearbox – spacer plate and 5th gear (Sec 10)

1   Spacer plate
2   5th gear
3   Speedometer drive worm

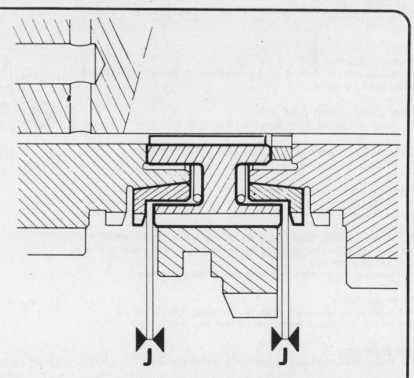

Fig. 6.30 Clearance check, 3rd/4th synchro hub-to-3rd and 4th synchro rings (Sec 10)

J = 0.008 in (0.2 mm) minimum

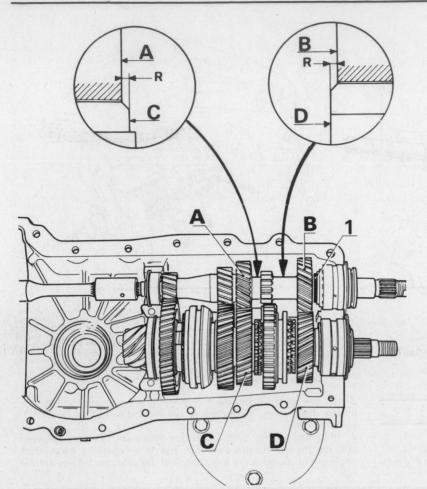

Fig. 6.31 Primary shaft positioning, type
395 gearbox. Distances R must be equal
(Sec 10)

1   Position adjustment washer

Primary shaft
5   Reassemble in reverse of the procedure described in Section 9,
paragraphs 16 to 18.

Final drive pinion protrusion
6   Refer to Section 5, paragraphs 18 and 19.

Differential bearings – fitting and adjustment
7   Proceed as described in Section 5, paragraphs 20 to 25.

Primary shaft positioning
8   Fit the secondary shaft into the RH half gearcase, with the
speedometer worm nut, wavy washer, 5th gear and spacer plate
removed.
9   Fit the primary shaft, and check that steps (R) are equal (Fig.
6.31). If not, adjust by replacing washer (item (1)) with a washer of the
appropriate thickness.
10 Remove the primary and secondary shaft assenblies.

Selector forks
11 Proceed as described in Section 5, paragraphs 29 to 33.
12 Fit the 5th speed selector shaft (see Fig. 6.27).

Reverse gear
13 Proceed as described in Section 5, paragraph 34.

Rear cover
14 Fit new oil seals into the case.
15 Position the spacers, springs, circlip and selector finger, and slide
the control shaft through them and into position (Figs. 6.25 and 6.27).
16 Fit the roll pins.

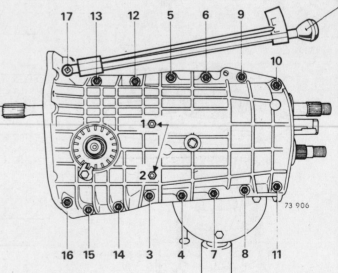

Fig. 6.32 Half gearcase bolts tightening sequence (Sec 10)

Half casings
17 Proceed as described in Section 5, paragraphs 37 and 38.
18 Smear the spacer plate gasket with non-setting jointing compound, fit the plate and gasket together with the locating dowels, and
fit and tighten the three bolts.
19 Tighten the half-casing bolts in the correct order, to the specified
torque figure (Fig. 6.32).

20 To the primary shaft, fit the spacer washer, needle bearing with sleeve, and the 5th speed gear.
21 Assemble together the synchro hub, sliding gear and fork, and fit them to the primary shaft, followed by the wavy washer and synchro nut.
22 To the secondary shaft, fit the 5th speed gear, wavy washer and speedometer worm nut.
23 Fit the roll pin to the 5th speed selector fork.
24 Engage both 5th and reverse gears, and tighten the speedo worm nut and primary shaft nuts to the specified torque figures. Lock them both.
25 Referring to Fig. 6.34, push the 5th speed synchro ring hard against the taper and the gear hard against the hub. Check the given dimension.
26 With all gears in neutral, smear the rear cover gasket with jointing compound and fit the cover and gasket with the rocking lever entered into the slot in the shafts. Fit and tighten the cover bolts.
27 Fit the 5th gear locking ball, spring washer and plug. Use jointing compound on the plug threads.
28 Fit the differential ring nuts as described in Section 5, paragraphs 43 and 44.

Crownwheel and pinion – backlash checking
29 Proceed as described in Section 5, paragraphs 45 to 47.

Clutch housing
30 Proceed as described in Section 5, paragraphs 48 to 51.

11 Rear cover and associated items, type 395 gearbox – removal and refitting with the gearbox in the vehicle

1 Drain the gearbox oil, and remove the gear control linkage and speedometer cable.
2 Proceed as described in paragraphs 3 and 4, Section 9.
3 Dismantle the cover as described in Section 9, paragraphs 21 and 22.
4 Reassemble as described in Section 10, paragraphs 14 to 16. Ensure that any burrs present on the selector shaft which might damage the new oil seal are removed, using fine emery cloth.
5 Refit all parts in the reverse order of dismantling.

12 Fifth speed synchro, type 395 gearbox – removal and refitting with the gearbox in the vehicle

1 Drain the gearbox oil, and remove the gear control linkage and speedometer cable.
2 Proceed as described in Section 9, paragraphs 3 to 7.
3 When renewing components, note that the hub and sliding gear are matched.
4 To refit, insert the selector fork in the sliding gear. Fit the gear assembled to the primary shaft, fit a new nut, and tighten to the correct torque with 5th and reverse gear engaged. Lock the nut.
5 Fit a new speedo worm nut, tighten to the correct torque, and lock.
6 Fit a new roll pin to the selector fork.
7 Proceed as described in Section 10, paragraphs 25 to 27.
8 Reconnect the control linkage and speedometer cable.
9 Refill the gearbox with oil.

13 Speedometer drive worm and wheel, all gearboxes – removal and refitting with the gearbox in the vehicle

1 Drain the gearbox oil, and remove the gear control linkage and speedometer cable.
2 Proceed as described in Section 9, paragraphs 3 and 4 (paragraph 3 on type 395 gearbox only).
3 Remove the speedometer wheel from the cover.
4 With the handbrake applied and 1st gear selected, unscrew and remove the speedometer worm nut.
5 To refit, fit and tighten a new worm nut, to the specified torque figure.
6 Place the new speedometer wheel in the cover.

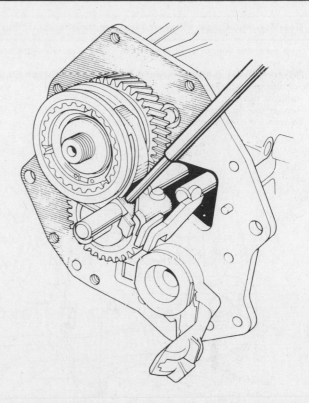

Fig. 6.33 Fitting the roll pin to the 5th speed selector fork (Sec 10)

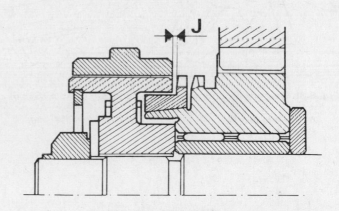

Fig. 6.34 Clearance check – 5th speed synchro ring-to-synchro hub (Sec 10)

$J = 0.008$ in (0.20 mm)

7 Refit the rear cover as described in Section 10, paragraphs 26 and 27 (locking ball and associated items on type 395 gearbox only).
8 Reconnect the control linkage and speedometer cable.
9 Refill the gearbox with oil.

14 Differential outlet oil seal, all gearboxes – removal and refitting with the gearbox in the vehicle

1 Disconnect the driveshaft at the gearbox as described in Chapter 8.
2 Mark the position of the ring nut on both nut and casing, take off the locking plate, and unscrew the ring nut. Count and record the number of turns necessary for removal.
3 Remove the old components and fit the new seal and O-ring as described in Section 5, paragraphs 43 and 44.
4 Refit in the reverse order of the removal procedure.

15 Gearbox mountings, all gearboxes – removal and refitting

1   Just take the weight of the gearbox, using a suitable jack and a piece of wood.
2   Referring to Fig. 6.35, remove the three bolts (A) and bolt (B), and remove the mounting.
3   Remove nut (C), and withdraw the rubber pad.
4   Repeat the operation on the other side of the vehicle.
5   To refit, reverse the removal procedure.

16 Gearchange control mechanism, all gearboxes – removal, refitting and adjusting

1   Working inside the car, remove the screws retaining the bellows, and the bolts securing the gearlever housing.
2   Remove the bolt (1) from the balljoint (see Fig. 6.36).
3   Remove the nut and bolt (2).

Refitting and adjusting (first type)
4   Refit in reverse of the removal procedure.
5   Referring to Fig. 6.36, loosen bolt (1).
6   Set lever (3) in the 3rd/4th speed plane.
7   Referring to Fig. 6.37, fit a shim 0.08 in (2 mm) thick on the 352 gearbox or 0.4 in (10 mm) thick on the 395 gearbox between lever (1) and housing (2).
8   Tighten the bolt (1 in Fig. 6.36), and recheck the clearance.

Refitting and adjusting (second type)
9   Refit in reverse of the removal procedure.
10 Select neutral.
11 Referring to Fig. 6.38, unlock nut (1).
12 Set lever (2) in the 3rd/4th speed plane.
13 Proceed as described in paragraph 7.
14 Referring to Fig. 6.38, tighten the nut and recheck the clearance.

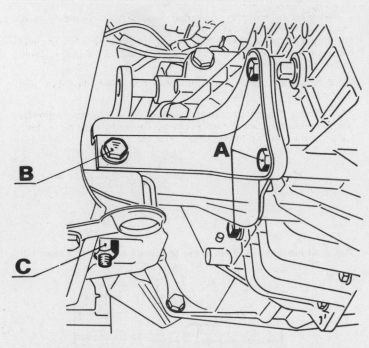

Fig. 6.35 Gearbox mountings, type 352 and 395 gearbox (Sec 5)

A   Bolt to gearbox   C   Nut to rubber pad
B   Bolt to sub-mountings

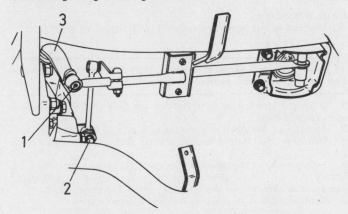

Fig. 6.36 Control linkage layout, first type (Sec 16)

1   Bolt   3   Lever
2   Nut and bolt

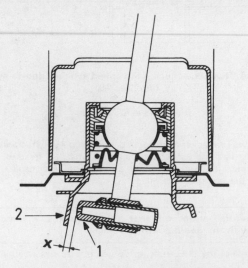

Fig. 6.37 Control lever adjustment detail (Sec 16)

1   Lever   X   Shim position
2   Housing

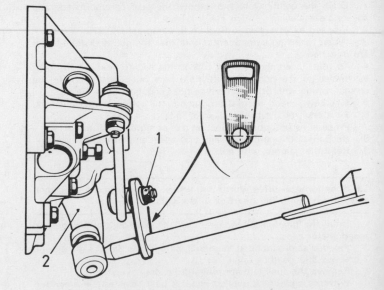

Fig. 6.38 Control linkage layout, second type (Sec 16)

1   Nut   2   Lever

17 Fault diagnosis – manual transmission

1   Faults can be sharply divided into two main groups: some definite failure with the transmission not working, and noises implying some component worn, damaged, or out of place.

2   The failures can sometimes be tracked down by commonsense and remembering the circumstances in which they appeared.

3   If there is a definite fault within the transmission then it has to be removed and dismantled to repair it, so further diagnosis can wait till the parts can be examined.

4   If the problem is a strange noise, the decision must be taken whether in the first place it is abnormal, and if so whether it warrants action.

5   Noises can be traced to a certain extent by doing the test sequence as follows.

6   Find the speed and type of driving that makes the noise. If the noise occurs with engine running, car stationary, clutch disengaged, gear engaged, the noise is not in the transmission. If it disappears after the clutch is engaged in neutral, halted, it is the clutch.

7   If the noise can be heard faintly in neutral, clutch engaged, it is in the gearbox. It will presumably get worse on the move, especially in some particular gear.

8   Final drive noises are only heard on the move. They will only vary with speed and load, whatever gear is engaged.

9   Noise when pulling is likely to be either the adjustment of preload of the differential bearings, or the crownwheel and pinion backlash.

10 Gear noise when free-wheeling is likely to be the relative positions of crownwheel and pinion.

11 Noise on corners implies excessive tightness or excessive play of the bevel side gears or idler pinions in the differential.

12 In general, whining is gear teeth at the incorrect distance apart. Roaring or rushing or moaning is bearings. Thumping or grating noise suggests a broken gear tooth.

13 If subdued whining somes on gradually, there is a good chance the transmission will last a long time to come, but check the transmission oil level!

14 Whining or moaning appearing suddenly, or becoming loud, should be examined quickly.

15 If thumping or grating noises appear, stop at once. If bits of metal are loose inside, the whole transmission, including the casing, could quickly be wrecked.

16 Synchromesh wear is obvious. You just 'beat' the gears and crashing occurs.

17 Difficulty in engaging gears can be caused by an incorrectly adjusted selector control mechanism, so check this before assuming the problem is within the gearbox.

Chapter 7 Automatic transmission

For modifications, and information applicable to later models, see Supplement at end of manual

Contents

Specifications

General

Transmission type 4139
Gears .... 3 forward, and 1 reverse
Gear ratios:
    First .... 2.39 : 1
    Second .... 1.48 : 1
    Third .... 1.02 : 1
    Reverse .... 2.05 : 1
    Final drive .... 9 x 32 (3.56 : 1)

Lubricant

Type .... Dexron type ATF (Duckhams Uni-Matic or D-Matic)
Capacity:
    Total .... 10.5 Imp pts (6.3 US qts, 6.0 litres)
    Drain and refill .... 5.5 Imp pts (3.2 US qts, 3.0 litres)
    Converter .... 3.0 Imp pts (1.9 US qts, 1.8 litres)

Torque wrench settings

	lbf ft	Nm
Selector arm nut	27	37
Bolts, converter-to-driveplate	21	28

1 General description

The automatic transmission, when fitted, provides fully automatic gearchanging without the use of a clutch. An override system of manual gear selection remains available to the driver.

The transmission consists of three main assemblies, namely the torque converter, the final drive and the gearbox.

The torque converter takes the place of a conventional clutch, transmitting the engine torque smoothly and automatically to the gearbox. Increased torque is provided for starting off.

The gearbox contains an epicyclic gear train, providing three forward gear ratios and one reverse, and the mechanical, hydraulic and electrical gear train control elements.

The epicyclic gear train consists of an assembly of helical gears which provides for different ratios to be obtained, depending upon the pressure of the hydraulic feed to the receivers. The gear assembly consists of 2 sunwheels, 3 pairs of planet wheels joined by a planet wheel carrier, and an involute gear ring.

A geared oil pump located at the rear of the unit supplies oil at the required pressures to the converter, the brakes and clutches, and for gear lubrication.

The hydraulic distributor ensures the regulation of oil pressure to suit the engine load, and the pressure feed or release to the brakes and clutches. Ratio changes are effected by the operation of two solenoid ball valves, instructed by the governor and computer. Circuit pressure is controlled by the capsule and pilot valve, thus determining pressure to the receiver and controlling gear changing quality.

A freewheel transmits torque in the same direction as the roadwheels, but does not permit engine braking in first gear when D or 2 are selected. The clutches and brakes lock or release the gear train components in various ways depending upon hydraulic feed pressure,

and thereby provide the different ratios.

Gearchange instruction is given by the governor and computer unit, the exact moment of change varying according to vehicle speed and engine torque.

The governor is in fact an alternator, and provides power to the computer, the amount being dependent upon vehicle speed and engine loading.

The computer supplies electrical pulses to the solenoid ball valves, depending upon the position of the selector lever and upon the current received from the governor.

The kickdown switch earths one of the computer circuits, causing instant selection of a lower gear in some circumstances. The switch is operated by the throttle pedal at the extreme of travel.

The solenoid ball valves open or close hydraulic passages, to permit gear changing.

The selector lever, centrally placed inside the car, has 6 positions as follows:

P (or park): Transmission in neutral, and the driving wheels mechanically locked.

R (or reverse): Reverse gear position. When the ignition switch is on, the reversing lights will automatically illuminate.

N (or neutral): Transmission in neutral.

D (or drive): Gears engage automatically.

2 : Second gear hold.

1 : First gear hold.

Note that whenever the selector lever is moved between D, P and R, the vehicle must be stationary, the footbrake applied, and the accelerator pedal raised. The mechanism must also be unlocked, by squeezing together the top of the selector lever.

To start the engine, the selector lever must be in either the P or N positions, for safety reasons. The starter will not function in other positions.

To move away, place the selector lever in the D position and drive away on the accelerator pedal.

In special circumstances, such as on very hilly and twisting roads, the selection of 2 will prevent frequent gear changing, and will provide engine braking when moving downhill, whilst retaining automatic changing between 1 and 2. Similarly, if 1 is selected, second and third gears are not obtainable.

When the vehicle is not required, engage P.

On normal roads, the most economical use is provided by driving with the selector lever at D, and with light accelerator pressure to give gearchanges at low engine speeds. Do not use positions 1 or 2.

When driving fast, gearchanging will take place at higher speeds. To obtain snap acceleration, such as when overtaking, smartly press the throttle pedal to the floor. This will cause the kickdown switch to

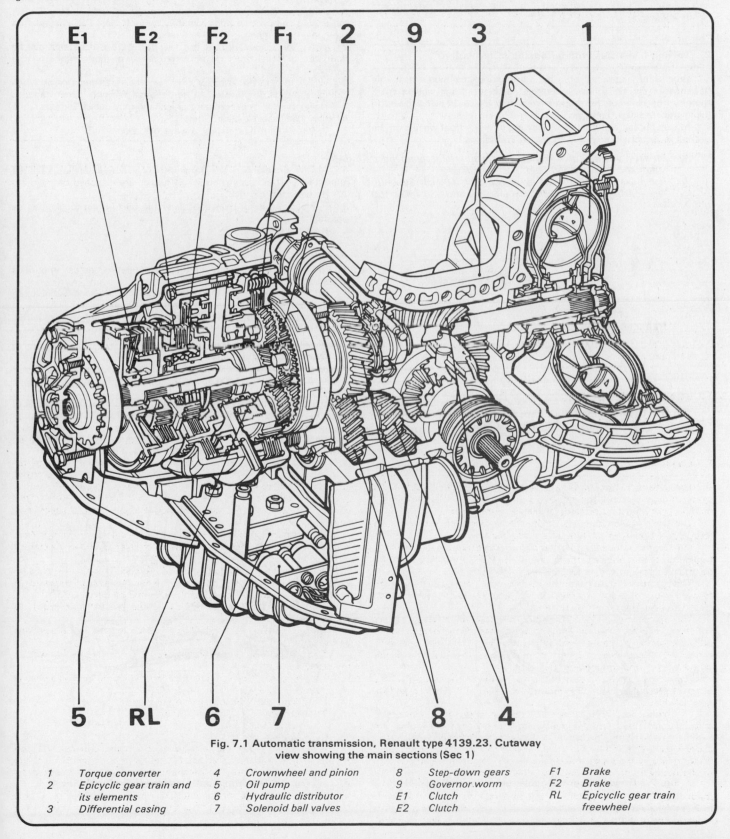

Fig. 7.1 Automatic transmission, Renault type 4139.23. Cutaway view showing the main sections (Sec 1)

1	Torque converter	4	Crownwheel and pinion	8	Step-down gears	F1	Brake
2	Epicyclic gear train and its elements	5	Oil pump	9	Governor worm	F2	Brake
		6	Hydraulic distributor	E1	Clutch	RL	Epicyclic gear train freewheel
3	Differential casing	7	Solenoid ball valves	E2	Clutch		

operate, and give an immediate change down to a lower gear.

In cold weather, wait for between ½ and 2 minutes, depending upon the temperature, before moving the selector lever. This will prevent stalling of the engine.

The complexity of the automatic transmission unit makes it largely unsuitable for working upon by the home mechanic, and any problems arising should be discussed with a Renault agent. Trouble-free running and long life will only be obtained if the unit is serviced correctly, and is not abused.

2 Towing – vehicles with automatic transmission

1   When the engine is off, the pump supplying lubricant to the transmission is no longer operative, and the front wheels must therefore be clear of the ground when the vehicle is being towed to avoid transmission damage.

2   In exceptional circumstances, towing with the front wheels on the ground is permissible on the following conditions:

(a) *An extra quantity of 4 Imp pts (2 litres) of the recommended automatic transmission fluid must be added*

(b) *The maximum permissible towing speed is 18 mph (30 km/h)*

(c) *The maximum permissible distance towed is 30 miles (50 km)*

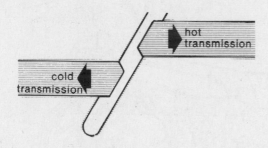

Fig. 7.2 Dipstick upper (hot transmission) and lower (cold transmission) fluid levels (Sec 3)

The surplus oil must be drained off, after towing.

3   When towing as in paragraph 2, the selector lever must be in N.

3 Automatic transmission fluid – checking, draining and refilling

Checking

1   With the vehicle on level ground, and the selector lever in P, start the engine and wait for approximately two minutes. This ensures that the converter is filled.

2   When the transmission is hot, ie after a drive of at least half an hour, the oil level must be between the two upper marks on the dipstick.

3   When the transmission is cold, ie when starting up, or when changing the oil, the level must be between the two lower marks.

4   Never overfill, or overheating or leakage may occur. Top up via the dipstick tube, using a clean funnel.

5   When wiping the dipstick, always use a non-fluffy rag.

Draining

6   Always drain when the transmission is hot, to ensure that the impurities held in suspension in the hot oil are disposed of.

7   Remove the dipstick.

8   Remove the drain plug and allow the oil to drain for as long as possible.

9   Refit the drain plug.

Refilling

10 Refill via a funnel, using one that has a filter to trap any impurities in the oil.

11 With the funnel in the dipstick tube, pour in the specified quantity of the recommended automatic transmission fluid.

12 Start up, check the level (see paragraph 3), and top up as required.

4 Governor control cable – adjustment

1   Free the locknut (1) and adjust the stop (2) to the mid position (see Fig. 7.3).

2   Depress the throttle pedal fully, and adjust the cable by means of stop (6), until play is just eliminated (see Fig. 7.4).

3   With the throttle pedal fully down, adjust the cable at the governor end to give a clearance (J) of 0.3 to 0.5 mm (0.012 to 0.020 in) between the stop peg and the quadrant (see Fig. 7.5).

4   Tighten the locknuts.

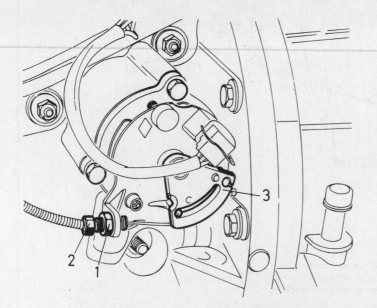

Fig. 7.3 Governor control cable (governor end) (Sec 4)

| 1 | Locknut | 3 | Adjuster |
| 2 | Stop | | |

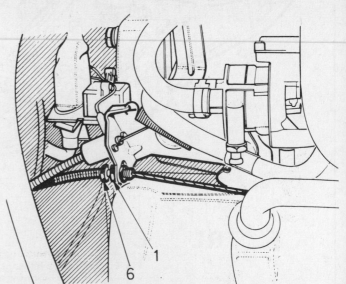

Fig. 7.4 Governor control cable (engine end) (Sec 4)

| 1 | Locknut | 6 | Stop |

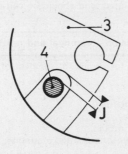

Fig. 7.5 Detail of governor control cable adjustment (Sec 4)

3	Quadrant	J	See text
4	Stop peg		

5 Gear selector lever – removal, refitting and adjustment

Removal

1   Place the selector lever to N.
2   Remove the circlip (1) at the computer control arm (Fig. 7.6).
3   Remove the cover (2) by taking out the securing bolts.
4   Remove the securing nut (3) on the fork end (4) and the nut (5) on the control rod bracket (6) (Fig. 7.7).
5   Inside the car, slide out the selector lever gate, and remove the lever casing.

Refitting and adjustment

6   Ensure that the computer is still correctly set in the N position, ie the fourth notch.
7   Ensure that the bellows (7) and cover (2) are placed over the rod (8).
8   Fit the lever (4).
9   Fit the rod (8) through the bearing at the control lever end.

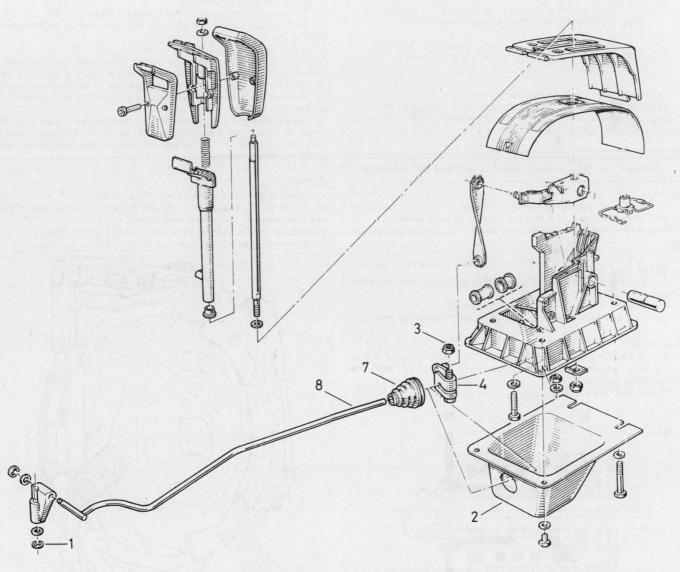

Fig. 7.6 Selector lever arrangement (Sec 5)

1	Circlip	3	Nut	7	Bellows	8	Selector rod
2	Cover	4	Fork end				

8   Adjustment of the throttle cable, kickdown switch and governor control cable are inter-related, and should be carried out together.

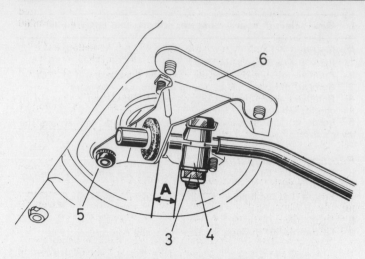

Fig. 7.7 Selector rod (transmission end) (Sec 5)

3	Nut	6	Bracket
4	Fork end		$A = \frac{19}{32}$ in (15 mm)
5	Nut		

10 Reconnect the rod (8) at the transmission end.
11 Reconnect the arm at the control lever end, and lightly secure.
12 Check dimension A (Fig. 7.7), reset if necessary to $\frac{19}{32}$ in (15 mm), and tighten fixings (3) and (5).

---

6   Kickdown switch – testing, removal, refitting and adjustment

Testing
1   To check the switch operation, connect a 12 volt test bulb between the switch terminal and the battery positive terminal.
2   Press the accelerator hard down. If the bulb lights, the switch is in order.

Removal and refitting
3   Remove the accelerator cable, and disconnect the wire from the switch.
4   Remove the two screws, and then the switch.
5   To refit, reverse the removal procedure, and adjust the switch.

Adjustment
6   Adjust the throttle cable with enough initial play to give $\frac{1}{8}$ to $\frac{5}{32}$ in (3 to 4 mm) movement in the stop sleeve (3 in Fig. 7.8) with the throttle pedal fully depressed.
7   Ensure that the kickdown switch cover is in position.

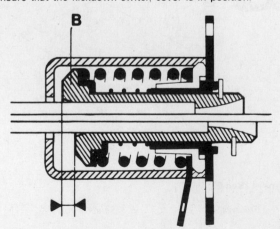

Fig. 7.8 Kickdown switch free play adjustment (Sec 6)

B   Stop sleeve

---

7   Automatic transmission – removal and refitting

1   Refer to Chapter 1, for details of removal of transmission and engine combined.

Removal (without engine)
2   Disconnect the battery.
3   Drain the transmission fluid.
4   Remove the starter motor bolts and mounting, and withdraw it to the rear.
5   Disconnect the driveshafts at the transmission (see Chapter 8).
6   Remove the cables at the governor and computer.
7   Disconnect the vacuum capsule pipe.
8   Disconnect the speedometer cable and the reversing light switch wires.
9   Place the selector lever to N.
10 Remove the circlip (1) at the computer (see Fig. 7.6).
11 Remove the three bolts securing the converter to the driveplate, rotating the crankshaft as necessary.
12 Disconnect the exhaust downpipe.
13 Support the transmission on a jack, but ensure that the load is properly spread by using a substantial piece of wood of adequate size.
14 Remove the transmission mountings, one at each side-member.
15 Remove the engine-to-transmission fixing bolts, and carefully lift out the unit with the help of an assistant.
16 Ensure that the converter does not become separated from the remainder of the unit. Keep it in position by bolting a suitable piece of steel through one of the engine-to-transmission bolt holes, restraining the converter.

Refitting
17 Remove the plate fitted to retain the converter.
18 Fit the unit to the engine, refit and tighten the bolts, and refit the mounting pads at the vehicle side-members.
19 Reconnect the exhaust pipe, and refit the starter motor.
20 When assembling the converter to the driveplate, ensure that the

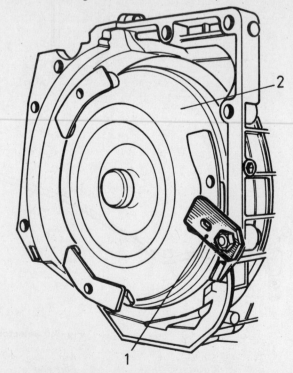

Fig. 7.9 Retaining the converter, using a piece of steel plate (Sec 7)

1   Retaining plate   *2   Converter*

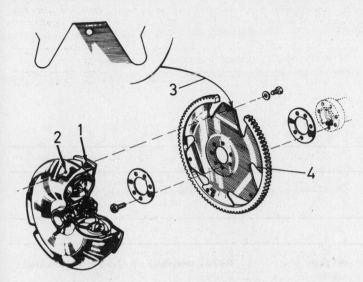

Fig. 7.10 Reassembly details for the converter and driveplate (Sec 7)

1 *Boss on converter*
2 *Timing hole opposite boss*
3 *Sharp angled machined arm*
4 *Driveplate*

boss on the converter which is in line with the timing hole is fitted opposite the sharp angled machined arm on the driveplate (identified by a paint splash).

21 Tighten the fixing bolts a little at a time and in rotation, ensuring proper location, until the correct torque figure is reached.

22 Refit the TDC pick-up support plate, and adjust the pick-up if necessary (see Chapter 6).

23 Reverse the procedure described in paragraphs 2 to 10.

24 Adjust the governor and accelerator control cables (see Sections 4 and 6).

25 After refilling the transmission unit, run the engine, and check and top up as necessary.

8 Fault diagnosis – automatic transmission

1 Most faults in the automatic transmission can be traced to incorrect adjustment of the cables or selector linkage, or to an incorrect fluid level.

2 Diagnosis of faults other than those described in paragraph 1 requires specialised knowledge and equipment, and should be referred to a Renault agent.

3 Non-functioning of the starter motor is frequently traceable to a defect in the wiring or earthing to the transmission, or to incorrect selector lever adjustment.

4 If a serious fault in the transmission is suspected, on no account remove the transmission before the fault has been professionally diagnosed. The diagnostic equipment used by Renault dealers is designed for use with the transmission in the vehicle.

Chapter 8 Driveshafts

Contents

Specifications

Driveshafts

	R1340	R1341 (manual)	R1341 (automatic)
Wheel end joint	GE86	GE86	GE86
Gearbox end joint	GI62	GI76	GI76
Joined by	Tube	Solid shaft	Tube
Damper fitted	No	Yes	Yes

Replacement driveshaft arrangements acceptable
Shaft arrangement:

	R1340	R1341 (manual)	R1341 (automatic)
GE86 + tube + GI62	Yes	No	No
GE86 + shaft and damper + GI76	No	Yes	No
GE86 + tube and damper + GI76	No	Yes	Yes

Torque wrench settings

	lbf ft	Nm
Top balljoint nut	38	51
Balljoint nut, on steering arm	26	35
Stub axle nut	120	163
Roadwheel nuts	53	71

1 Driveshafts – general description

The shafts transmit the drive from the transmission unit to the front roadwheels. Each shaft has joints at the ends to allow for the relative motion between the transmission unit and the suspension.

Shaft location is by splines at each end. Roll pins retain the shaft at the transmission end, and a nut and washer at the wheel end.

A rubber bellows protects the shaft driving joints from water and dirt.

More than one type of shaft is in use, depending upon the vehicle model. When ordering a new shaft, always take the old one and compare them to ensure correct replacement parts are supplied.

Very little work is possible upon the driveshaft assemblies by the home mechanic. Signs of wear or damage are vibration, clunking and ticking sounds, especially when cornering. A close eye should be kept on the rubber bellows, since a split here can swiftly lead to the ruin of the joint assembly.

2 Driveshafts – removal and refitting

Removal
1 Prise out the wheel centre, and loosen the hub nut.
2 Loosen the roadwheel nuts.
3 Compress the suspension, and insert the home-made tool illustrated in Fig. 11.7 (see Chapter 11).
4 Raise the vehicle, and remove the roadwheel.

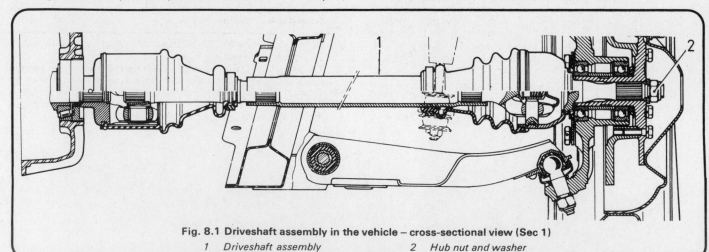

Fig. 8.1 Driveshaft assembly in the vehicle – cross-sectional view (Sec 1)

1 Driveshaft assembly   *2 Hub nut and washer*

2.6 Roll pin, tapped out to permit withdrawal of shaft

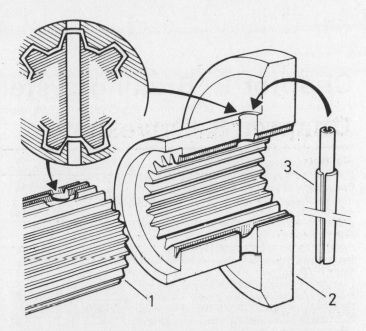

Fig. 8.2 Lining up the driveshaft and gearbox shaft holes (Sec 2)

1   Gearbox shaft   3   Roll pin
2   Driveshaft

2.11 Roll pin holes aligned, preparatory to fitting of the pin

5   Free off the nuts securing the balljoints on the steering arm and at the upper stub axle carrier (see Chapter 11). Using a suitable extractor, free the balljoint tapers.
6   Under the vehicle, tap out the roll pins securing the driveshaft at the gearbox end, using a suitable sized pin punch (photo).
7   Remove the brake caliper (see Chapter 9), leaving the flexible hose connected. Do not allow the caliper to hang on the hose.
8   Remove the nuts from the two balljoints (see paragraph 5), separate the joints and pull the driveshaft off the gearbox.
9   Loosely reconnect the steering arm balljoint, holding it in place with the nut, and using a suitable hub extractor, or alternatively a soft drift, push the shaft inwards and out of the hub.
10 Disconnect the steering arm balljoint, and remove the driveshaft.

Refitting

11 Grease the gearbox splines with Molykote BR2 or an equivalent grease, align the driveshaft roll pin hole with that in the gearbox shaft, and push the driveshaft on (photo).
12 Tap home the new roll pins. Seal the holes with a suitable non-setting sealer.
13 Place the driveshaft through the hub, and loosely fit the hub washer and nut.
14 Fit the balljoint taper pins to the upper stub axle carrier and to the steering arm, and fully tighten the nuts. Nip the hub nut up as tightly as is possible at this stage.
15 Refit the brake caliper (see Chapter 9).
16 Refit the roadwheel, lower the vehicle, and tighten the hub nut to the correct torque figure. Also tighten the roadwheel nuts.
17 Compress the suspension and remove the tool from the suspension.
18 Pump the brake pedal several times to settle the caliper piston to its normal position.

3 Driveshafts – overhauling

1   No repair is possible if the driveshaft assembly is worn, and a complete new unit must be obtained.
2   The bellows can be renewed, but the operation is felt to be beyond the scope of the home mechanic due to the special tools specified for the work. The advice of the local Renault dealer should be sought if bellows renewal becomes necessary. Prompt renewal of a damaged bellows may at least save the cost of a new driveshaft!

Chapter 9 Braking system

For modifications, and information applicable to later models, see Supplement at end of manual

Contents

Specifications

System type .... Hydraulic, dual circuit, servo-assisted

Front brakes
Type .... Bendix or Girling disc
Discs:
    Diameter .... 8.976 in (228 mm)
    Thickness:
        New .... 0.394 in (10 mm)
        Minimum .... 0.354 in (9 mm)
Brake pad thickness, including backing:
    New .... 0.590 in (15 mm)
    Minimum .... 0.276 in (7 mm)

Rear brakes
Drum internal diameter:
    New .... 7.096 in (180.25 mm)
    Maximum after regrinding .... 7.136 in (181.25 mm)
Lining width .... 1.575 in (40.0 mm)
Lining minimum thickness .... 0.020 in (0.5 mm) above rivet heads

Master cylinder
Make and type .... Teves or Bendix tandem
Bore .... 0.748 in (19 mm)
Stroke .... 1.181 in (30 mm)

Handbrake
Type .... Mechanical, operating on rear wheels only
Minimum lever travel .... 9 notches (essential, to permit self-adjusting mechanism to function)

Brake fluid type/specification .... Hydraulic fluid to SAE J1703 (Duckhams Universal Brake and Clutch Fluid)

Torque wrench settings

	lbf ft	Nm
Bleed screws	6	8
Metal pipe unions	10.5	14
Unions, metal pipe-to-flexible pipe	9	12
Hose union on calipers	15	20.5
Bolts, caliper bracket-to-stub axle carrier	49	66.5
Bolts, disc-to-hub	19	26
Stub axle nut (front wheels)	120	163
Roadwheel nuts	53	71
Guide bolts (Girling caliper)	26	35

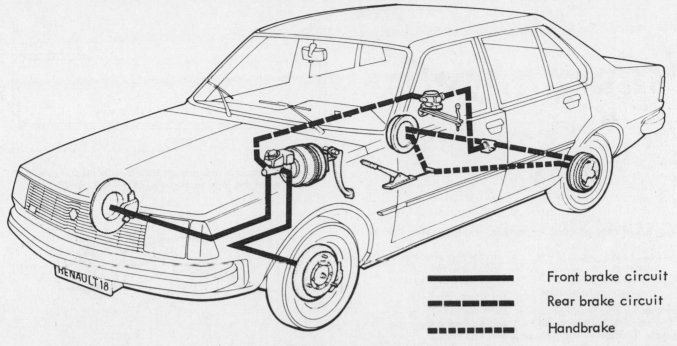

Fig. 9.1 Braking system layout (Sec 1)

Front brake circuit
Rear brake circuit
Handbrake

1 General description

A servo-assisted hydraulic braking system is employed. Disc brakes are fitted at the front of the car and self-adjusting drum brakes at the rear. The handbrake operates mechanically on the rear wheels only.

The hydraulic system is of the dual circuit type, so that even in the event of failure of a component in one circuit, braking effort will still be available from the other.

Hydraulic equipment may be of Bendix, Girling or Teves manufacture. Components and units are not necessarily interchangeable.

A brake pressure limiting valve is fitted, its purpose being to limit the hydraulic pressure to the rear brakes during heavy braking.

2 Master cylinder – description, removal and refitting

Description

1   The master cylinder has a primary piston applying pressure to the front brakes, and a secondary piston applying pressure to the rear, the two systems being independent of one another. The pistons resume their initial positions by virtue of the return springs, once the brake pedal is released.

2   If a failure occurs in one circuit, braking is still provided by the other. However, it should be noted that pedal travel will increase.

Removal

3   Drain the master cylinder, preferably using a syringe. Take care not to get any fluid on the car paintwork, as it acts as an effective paint stripper.

4   Remove the reservoir and the rubber rings.

5   Disconnect the three metal pipes.

6   Remove the two nuts securing the cylinder to the servo, and remove the cylinder.

Refitting

7   Check the length of the pushrod which is proud of the servo face, and adjust if necessary to give a dimension X as shown in Fig. 9.14.

8   Refit the cylinder to the servo, and tighten the securing nuts.

9   Reconnect the brake pipes (see Fig. 9.2).

10   Refit the rubber rings and the fluid reservoir.

11   Refill the system, and bleed (Section 21).

3 Master cylinder, Teves type – dismantling, overhauling and reassembly

Dismantling

1   Remove the cylinder from the vehicle (see Section 2).

2   Mount the cylinder body in a soft-jawed vice.

3   Using a suitable implement, compress the piston and spring assembly. Remove the stop screw, snap-ring and stop washer.

4   Release the piston/spring assembly carefully, and take the primary piston out.

5   Gently blow out the secondary piston with air.

Overhauling and reassembly

6   Examine the cylinder bore. If scoring is present, renew the complete cylinder assembly.

7   Preferably fit all new rubber parts. Ease all cups gently on to the

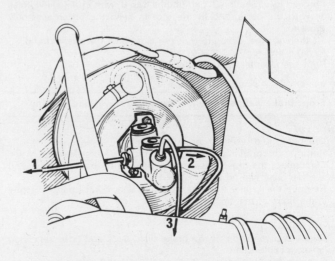

Fig. 9.2 Connections to the master cylinder (Sec 2)

1   RH front wheel   3   Brake pressure limiter
2   LH front wheel

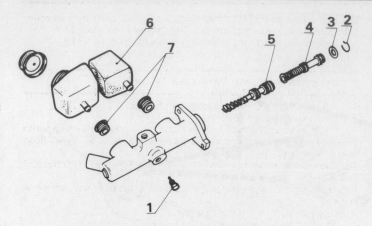

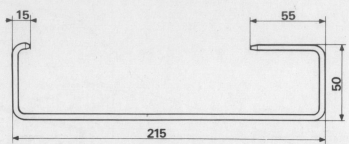

Fig. 9.4 Piston compressing tool – Bendix master cylinder.
Dimensions in mm; diameter of rod is 6 mm (Sec 4)

Fig. 9.3 Master cylinder, Teves type – exploded view (Sec 3)

1 Stop screw	*5 Secondary piston assembly*
2 Snap-ring	*6 Fluid reservoir*
3 Stop washer	*7 Reservoir sealing rings*
4 Primary piston assembly	

pistons, using only the fingers.
8   Dip all parts in clean brake fluid, and insert the secondary piston
assembly, followed by the primary piston assembly.
9   Reverse the procedure given in paragraphs 2 and 3.
10 Refit the cylinder (see Section 2).

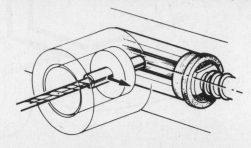

Fig. 9.5 Removing the secondary piston roll pin – Bendix master
cylinder (Sec 4)

**4   Master cylinder, Bendix type – dismantling, overhauling and
reassembly**

Dismantling
1   Manufacture the tool shown in Fig. 9.4.
2   Remove the master cylinder (see Section 2). Fit the special tool
into the ends of the cylinder to retain the pistons.
3   Grip a 3.5 mm drill in a vice, bring the master cylinder up to it, and
enter the drill in the secondary piston roll pin, working via the
secondary port. Rotate the cylinder to cause the drill to enter and grip
the pin, and pull the cylinder away, leaving the pin on the drill (Fig.
9.5).
4   Repeat the procedure given in paragraph 3, for the primary piston.
5   Remove the tool, allowing the pistons to be withdrawn.

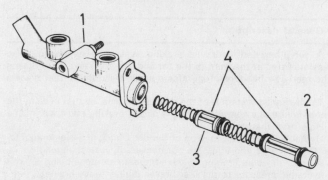

Fig. 9.6 Bendix master cylinder – exploded view (Sec 4)

1 Body	*3 Secondary piston assembly*
2 Primary piston assembly	*4 Roll pin slots*

Overhauling and reassembly
6   Proceed as in Section 3, paragraphs 6 to 8, with t1he roll pin holes
in line with the ports, and fit the tool to keep the piston assembly
compressed.
7   Fit the roll pins, ensuring that the slots face the pushrod end.
8   Remove the tool.
9   Refit the cylinder, as described in Section 2.

5   Front brake calipers, Bendix type – removal and refitting

Removal
1   Drain the fluid reservoir, preferably using a syringe.
2   Remove the retaining clips and tap out the keys sideways.
3   Disconnect the brake hose at the union with the metal pipe, and
remove the hose clip on the bracket.
4   Remove the caliper from its mounting and, if necessary, unscrew
the brake hose at the caliper.

Refitting
5   Check the condition of the brake disc, pads and hose, renewing
any unserviceable items.
6   Remove the bleed screw, fill the caliper with fluid via the hole for
the hose, and refit the bleed screw without tightening.
7   Fit the hose to the caliper (if removed) using a new washer.
8   Press the caliper piston into the bore, and place the lower end of
the caliper into the caliper bracket, followed by the top end (Fig. 9.7).

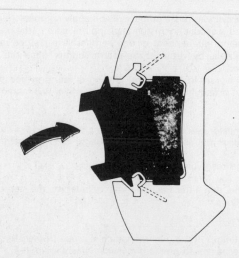

Fig. 9.7 Refitting a Bendix caliper in the caliper bracket (Sec 5)

9   Fit the caliper lower retaining key.
10 Place a screwdriver in the upper key slot, press down on the caliper, and slip the upper key into place. Take the screwdriver out, and tap the key home.
11 Fit new spring clips to retain the keys.
12 Refit the brake hose to the bracket, ensure that it is not twisted, and fit the clip.
13 Whilst holding the hose, fit and tighten the metal pipe union.
14 Bleed the brakes (Section 21).
15 Press the brake pedal several times, to correctly position the pads.

6   Front brake calipers, Girling type – removal and refitting

Removal
1   Proceed as in Section 5, paragraphs 1 to 4, except that the caliper is held by two bolts instead of keys. Remove the bolts (photos).

Refitting
2   Proceed as described in Section 5, paragraphs 5 to 7.
3   Push the piston into the bore.
4   Place the caliper in position over the pads, and fit the lower bolt. Push the caliper into place and insert the top bolt.

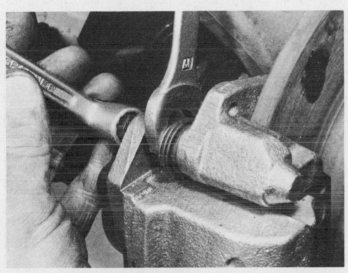
6.1a Removing a caliper retaining bolt

6.1b Removing the caliper

5   Tighten the lower bolt fully, whilst holding the guide with an open-ended spanner to prevent turning. Then tighten the upper bolt.
6   Proceed as described in Section 5, paragraphs 12 to 15.

7   Front brake calipers, Bendix type – overhauling

1   Remove the caliper as described in Section 5.
2   Remove the rubber dust cover.
3   Place a piece of wood between the piston and caliper opposite flanges to prevent damage, and apply air pressure to the brake hose hole to push out the piston.
4   Use a smooth, flexible steel strip, eg a feeler gauge, to lift the seal in the piston bore up and out.
5   Examine the piston and piston bore, and discard if any scoring is evident.
6   Clean all parts in methylated spirit.
7   Reassemble in reverse order, using new rubber parts.

8   Front brake calipers, Girling type – overhauling

Piston and seals
1   Proceed as described in Section 7. If a caliper bore proves to be scored or rusty, renew the caliper assembly. If only the piston is defective, a new piston may be fitted.

Sliding guides
2   Correct operation of the caliper will be impaired if the guides are corroded.
3   To remove the guides, first remove the caliper (see Section 6).
4   Remove the dust covers and guides (see Fig. 9.8).
5   Fit new dust covers to the caliper bracket.
6   Lubricate the new guides with the special grease provided with them. Fit them to the caliper bracket. Note that the hexagon guide with a rubber sleeve is fitted to the top fo the bracket, opposite the bleed screw, and the steel guide with two flats is fitted at the bottom.
7   Refit the caliper as described in Section 6.
8   Take note that under no circumstances may the clamping bolts (Fig. 9.8, item 10) be removed or loosened.

9   Front brake pads, Bendix type – changing

1   Brake pads must be renewed as complete axle sets, when any one lining has worn down to the minimum specified thickness. The pads

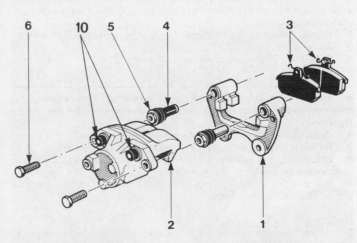
Fig. 9.8 Front brake caliper, Girling type – exploded view (Sec 8)

1   Caliper bracket
2   Caliper
3   Brake pads
4   Guide pins
5   Dust covers
6   Caliper retaining bolts
10 Caliper assembly bolts (not to be loosened)

can be inspected without removing the caliper.
2   Raise the front of the vehicle, and remove the roadwheels.
3   Disconnect the wire from the pad wear transmitter.
4   Remove the retaining clips and tap out the keys sideways.
5   Tip the caliper and remove it, but do not allow the weight to hang on the flexible hose.
6   Remove the brake pads and springs.
7   Take the dust covers from the pistons, and clean the covers and piston ends with methylated spirit. Lubricate the piston with Spagraph grease or equivalent.
8   Check the condition of the brake disc, pads and hose, and renew any unserviceable items.
9   Refit the dust covers.
10 Push the piston fully back into the bore.
11 Refit the springs, and then the pads which must be completely free to slide. Note that on service pads, the retaining peg must be fitted at the top.
12 Refit the caliper as described in Section 5. Press the brake pedal several times, to correctly position the pads.

10 Front brake pads, Girling type – changing

1   Proceed as in Section 9, paragraphs 1 to 3 (photo).
2   Do not remove the cap screws. Remove the guide bolts whilst holding the guide to prevent turning. *Note that the guide bolts must be changed whenever pads are renewed.*
3   Lift out the caliper, followed by the pads. Do not allow the caliper to hang on the brake hose.
4   Check the condition of the brake disc and hose, and renew any unserviceable items.
5   Fit the new pads to the caliper bracket (photos).
6   Refit the caliper as described in Section 6, paragraphs 4 and 5.
7   Press the brake pedal several times, to correctly position the brake pads.

11 Front brake caliper bracket – removal and refitting

Removal
1   Raise the vehicle, and remove the roadwheel.
2   Remove the caliper (see Section 5 or 6), leaving the hose connected. Do not let the caliper hang on the hose.
3   Take out the brake pads.
4   Remove the two caliper bracket bolts, and then the bracket (photo).

Refitting
5   Place the caliper bracket in position, and after smearing the

10.1 Brake pad wear transmitter wire

10.5a Fitting a brake pad

10.5b The pads in position

11.4 The caliper assembly (caliper bracket bolts arrowed)

threads of the retaining bolts with thread locking compound, fit the bolts using new locking washers.

6   Refit the caliper as described in Section 5 or 6.

7   Refit the roadwheels. Press the brake pedal several times, to correctly position the pads.

12   Brake discs – examination, removal and refitting

Examination

1   Examine the disc for excessive or uneven wear. If this is present, the disc must be renewed or replaced.

2   If a dial gauge is available, check for run-out of the disc, ie lateral movement. Measuring on an 8,583 in diameter (218.0 mm), the maximum acceptable run-out is 0.04 in (0.1 mm). Do not confuse disc run-out with hub bearing wear.

Removal and refitting

3   Remove the brake caliper bracket, as described in Section 11.

4   Remove the hub and disc assembly (see Chapter 11).

5   Separate the disc from the hub by removing the six bolts and washers.

6   To refit, reverse the removal procedure.

13   Rear brake drums – removal, refitting and bearing adjustment

Removal

1   Release the handbrake, and free off the handbrake cable adjustment.

2   Remove the sealing plug from the access holes, and use a screwdriver through the holes to push the handbrake lever, until the stop peg is freed from the brake shoe. Push the handbrake internal lever further to the rear.

3   Tap off the grease cap.

4   Take out the split pin, remove the lockplate, and undo and remove the stub axle nut and washer.

5   Take off the drum, using a puller if it is tight. Recover the outer bearing, which will fall out.

Refitting and bearing adjustment

6   Grease the bearings and the inside of the hub with a recommended lubricant, using about ¾ oz (20 gm).

7   Fit the drum, position the outer bearing, and fit the washer and nut.

8   Tighten the nut to 22 lbf ft (30 Nm), and then turn it back a little. Rotate the drum, and tighten or loosen the nut until the endfloat is zero to 0.0012 in (0.03 mm), or if no dial gauge is available, until endfloat is just perceptible.

9   Fit the locking plate and split pin.

10   Put ¼ oz (10 gm) of recommended grease in the end cap, and tap it home.

11   Press the brake pedal several times, to settle the brake shoes.

12   Adjust the handbrake (Section 22).

13   Refit the sealing plug to the backplate.

14   Rear brake linings, Bendix type brakes – removal and refitting

Removal

1   Raise the rear of the vehicle and remove the wheels.

2   Remove the brake drum (Section 13) (photos).

3   Place a strong rubber band or flexible wire round the operating cylinder, to prevent the pistons coming out.

4   Remove the upper return spring.

5   Disconnect the handbrake cable at the internal operating lever.

6   Press the bottom of the shoe steady springs, to release them.

7   Push the toothed lever fully inwards, and lift the brake shoes out from the backplate.

8   Free the link (1), Fig. 9.9, from the front shoe.

9   Move the toothed sector (4) to its original position.

10   Tilt the leading (front) shoe to 90° from the backplate, free the bottom return spring, and remove the shoe.

11   Remove the trailing (rear) shoe.

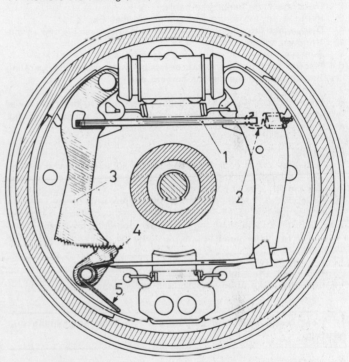

Fig. 9.9 Rear brake, Bendix type – drum removed (Sec 14)

1 Link	4 Toothed sector
2 Spring	5 Spring
3 Lever	

14.2a Rear brake mechanism (drum removed)

14.2b Brake cylinder and upper mechanism (drum removed)

14.2c Handbrake cable attachment and self-adjusting mechanism

Refitting

12 With the lower spring attached to both shoes, offer up the shoes with the front one at 90° to the backplate.
13 Push the toothed lever fully inwards, and offer up the link, straightening up the shoe as the link is fitted.
14 Reverse the procedure in paragraphs 3 to 6.
15 Check the automatic adjustment by measuring dimension 'A' (see Fig. 9.10), which should be approximately 0.040 in (1 mm) when the handbrake operating lever is resting against the shoe. If the dimension is incorrect, the link spring (2) (Fig. 9.9) and both shoe return springs must be renewed.
16 Refit the brake drum as described in Section 13.

15 Rear brake linings, Girling type brakes – removal and refitting

1 Raise the rear of the vehicle, remove the wheels, and then the brake drum (see Section 13).
2 Disconnect the handbrake cable.
3 Remove the top return spring and clip (see Fig. 9.11).
4 Disconnect the spring (3), and remove the lever (4), the spring, and the lever thrust washer.
5 Remove the two steady spring assemblies (6).
6 Place a strong rubber band, or flexible wire, round the operating cylinder, to prevent the pistons coming out.
7 Remove the link assembly (7).
8 Cross the brake shoes one behind the other, to bring the lower return spring from behind the fulcrum. Remove the shoes and spring, and separate them.

Refitting

9 Note the colour coding applied to the adjustable link assembly. That for the LH side of the vehicle has a grey threaded plunger, whilst that for the RH side is coded yellow.
10 Reverse the procedure given in paragraphs 4 to 8.
11 Adjust the link assembly length until the lining diameter is about 7 in (178 mm), by pushing the handbrake operating lever back to the shoe, and turning the ratchet wheel until the required dimension is obtained.
12 Fit the top return spring and clip.
13 Reconnect the handbrake cable.
14 Refit the brake drum as described in Section 13.

16 Rear brake operating cylinders – removal, overhauling and refitting

Removal

1 Remove the brake drum (see Section 13).
2 Remove the upper return spring, and pull the brake shoes apart.
3 Disconnect and plug the metal brake pipe at the wheel cylinder.
4 Remove the two cylinder-to-backplate fixing bolts, and remove the cylinder.

Overhauling

5 Clean the cylinder externally in methylated spirit, and dismantle it.
6 Examine the pistons and bore, and discard the complete unit if rust or scoring exist.
7 Always use new rubber parts when reassembling. Fit them using the fingers only.
8 Dip all items in clean brake fluid, fit them as illustrated in Fig. 9.12 or 9.13, and check that the parts slide correctly in the cylinder.

Refitting

9 Refit in the reverse order (see paragraphs 1 to 4), and bleed the hydraulic system.

17 Brake servo unit – checking, removal and refitting

Checking

1 Whilst the braking system will continue to function with an inoperative servo, considerably more pedal pressure will be required. If this is the case, a loss of vacuum may be the cause.

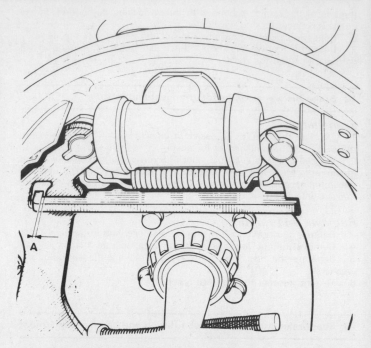

Fig. 9.10 Setting the automatic rear brake adjustment (Bendix brakes) (Sec 14)

A = 0.040 in (1 mm)

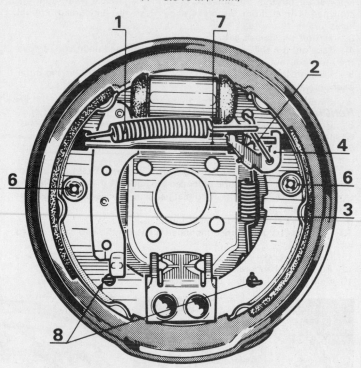

Fig. 9.11 Rear brake, Girling type – drum removed (Sec 15)

1	Upper return spring	6	Shoe steady spring assembly
2	Clip	7	Thrust link
3	Adjusting lever spring	8	Lower return spring
4	Adjusting lever		

2 To confirm a loss of vacuum, a proper gauge is required. However, the likely causes of loss will be either a faulty check valve, which can be renewed, or if this is ineffective the pushrod diaphragm will be faulty in which case a complete servo must be obtained. Also check the vacuum hoses.

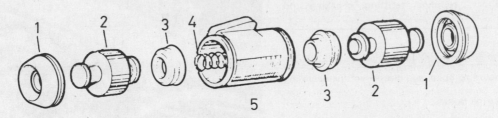

Fig. 9.12 Girling rear wheel cylinder – exploded view (Sec 16)

1 Covers   3 Seals   4 Spring   5 Body
2 Pistons

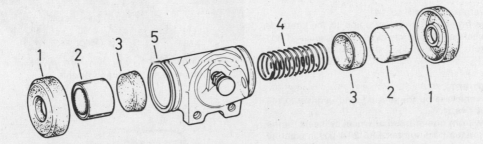

Fig. 9.13 Bendix rear wheel cylinder – exploded view (Sec 16)

1 Covers   3 Seals   4 Spring   5 Body
2 Pistons

Removal
3 Remove the master cylinder as described in Section 2.
4 Disconnect the vacuum hose.
5 Remove the brake pedal clevis pin.
6 Remove the nuts securing the servo to the bulkhead, and then the servo complete with the spacer plate.

Refitting
7 Check the pushrod dimensions on both sides of the servo, as shown in Fig. 9.14, and adjust as necessary by adjusting the pushrod nut or clevis.
8 Fit the master cylinder and spacer plate to the servo, offer up to the bulkhead, and secure.
9 Fit the pedal clevis pin.
10 Reconnect the servo vacuum hose, and the metal pipes to the master cylinder.
11 Top up the hydraulic system, and bleed.

18 Brake servo unit – air filter and check valve renewal

Air filter
1 Take out the brake pedal clevis pin, and remove the pedal.
2 Referring to Fig. 9.14, release the locknut on the clevis. Remove the clevis.
3 Prise out the filter, using a scriber, and clean or renew as necessary.
4 Press the filter into position, and refit the clevis.
5 Reset the clevis position, and tighten the locking nut.
6 Refit the clevis pin and brake pedal.

Check valve
7 Remove the vacuum hose from the servo.
8 Twist the check valve, and pull it out.
9 To refit, reverse the removal procedure.

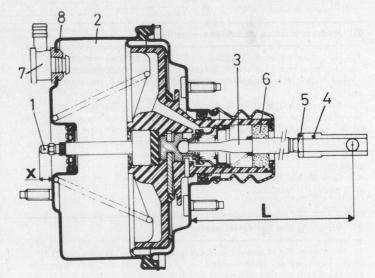

Fig. 9.14 Servo unit assembly – sectional view (Sec 17)

1 Pushrod   6 Air filter
2 Servo body   7 Check valve
3 Pushrod   8 Check valve sealing washer
4 Clevis   X = 9 mm (0.354 in)
5 Locknut   L = 126 mm (4.960 in)

19 Brake pressure limiter – removal, refitting, checking and adjustment

Removal
1   Slacken the flexible hose at the limiter (photo).
2   Disconnect the metal pipe at the limiter.
3   Remove the limiter mounting bolts, and disconnect the operating rod at the chassis.
4   Unscrew the limiter at the hose.

Refitting
5   Refit the hose to the limiter, and place it in position to check that the hose is not twisted. If it is twisted, unscrew the hose at the three-way union and offset it one notch, then re-tighten.
6   Refit in reverse of the removal procedure, and bleed the hydraulic system (Section 21).
7   Adjust the limiter (see paragraphs 9 to 14).
8   Note that, if the hose from the three-way union to the limiter is renewed, the copper washers should be checked for thickness. If any considerable variation from 0.059 in (1.5 mm) exists, renew the washers.

Checking and adjustment
9   First ensure that the vehicle is on the ground, with a driver in the driving seat, and the boot empty.
10 Remove a bleed screw from one of the rear wheel cylinders, fit the special pressure gauge (Renault part number FRE 214-02) in place of the screw, and bleed the circuit and gauge.
11 Press the pedal several times, and check the cut-off reading given by the gauge. Correct readings are as follows:

Fuel tank	Cut-off pressure
Full	30 to 34 bars (435 to 490 lbf/in^2)
Half full	27 to 31 bars (390 to 450 lbf/in^2)
Empty	24 to 28 bars (345 to 405 lbf/in^2)

12 Adjust the cut-off pressure if necessary, by screwing the adjusting nut up to increase, or down to decrease.
13 Check the pressure gauge reading several times, and if satisfactory remove the gauge.
14 Bleed the brakes.

20 Hydraulic pipes and hoses – general

1   Hydraulic pipes and hoses should be examined periodically, the metal pipes being checked for signs of severe corrosion, and the rubber hoses for cracks. Both should be checked for any signs of chafing.
2   Renew any defective rubber hoses with new parts.
3   Metal pipes can sometimes be purchased complete and ready to fit. Alternatively, it will be necessary to have replacements made by an engineering concern who possess the necessary tools. When ordering, it is advisable to provide the manufacturer with the old pipe as a pattern.
4   Care should be taken to ensure that the correct metric pipe fittings and ends are supplied.

21 Bleeding the hydraulic system

1   If any of the hydraulic components in the braking system have been removed or disconnected, or if the fluid level in the master cylinder has been allowed to fall appreciably, it is inevitable that air will have been introduced into the system. The removal of all this air from the hydraulic system is essential if the brakes are to function correctly, and the process of removing it is known as bleeding.
2   There are a number of one-man, do-it-yourself, brake bleeding kits currently available from motor accessory shops. It is recommended that one of these kits should be used wherever possible as they greatly simplify the bleeding operation and also reduce the risk of expelled air and fluid being drawn back into the system.
3   If one of these kits is not available then it will be necessary to gather together a clean jar and a suitable length of clear plastic tubing which is a tight fit over the bleed screw, and also to engage the help

19.1 Brake pressure limiter and flexible hose

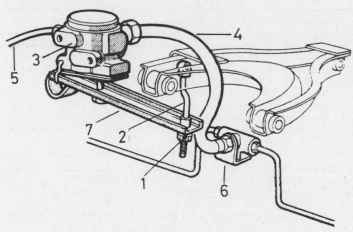

Fig. 9.15 Brake pressure limiter (Sec 19)

1 Nut for adjuster link	5 Metal pipe
2 Adjuster link	6 Three-way connector
3 Limiter body	7 Limiter lever
4 Flexible hose	

of an assistant.
4   Before commencing the bleeding operation, check that all rigid pipes and flexible hoses are in good condition and that all hydraulic unions are tight. Take great care not to allow hydraulic fluid to come into contact with the vehicle paintwork, otherwise the finish will be seriously damaged. Wash off any spilled fluid immediately with cold water.
5   If hydraulic fluid has been lost from the master cylinder, due to a leak in the system, ensure that the cause is traced and rectified before proceeding further or a serious malfunction of the braking system may occur.
6   To bleed the system, clean the area around the bleed screw at the wheel cylinder to be bled. If the hydraulic system has only been partially disconnected and suitable precautions were taken to prevent further loss of fluid, it should only be necessary to bleed that part of the system. However, if the entire system is to be bled, start at the wheel furthest away from the master cylinder.
7   Remove the master cylinder filler cap and top up the reservoir. Periodically check the fluid level during the bleeding operation and top up as necessary.
8   If a one-man brake bleeding kit is being used, connect the outlet tube to the bleed screw and then open the screw half a turn. If possible position the unit so that it can be viewed from the car, then depress

the brake pedal to the floor and slowly release it. The one-way valve in the kit will prevent dispelled air from returning to the system at the end of each stroke. Repeat this operation until clean hydraulic fluid, free from air bubbles, can be seen coming through the tube. Now tighten the bleed screw and remove the outlet tube.

9   If a one-man brake bleeding kit is not available, connect one end of the plastic tubing to the bleed screw and immerse the other end in the jam jar containing sufficient clean hydraulic fluid to keep the end of the tube submerged. Open the bleed screw half a turn and have your assistant depress the brake pedal to the floor and then slowly release it. Tighten the bleed screw at the end of each downstroke to prevent expelled air and fluid from being drawn back into the system. Repeat this operation until clean hydraulic fluid, free from air bubbles, can be seen coming through the tube. Now tighten the bleed screw and remove the plastic tube.

10   If the entire system is being bled the procedures described above should now be repeated at each wheel, finishing at the wheel nearest to the master cylinder. Do not forget to recheck the fluid level in the master cylinder at regular intervals and top up as necessary.

11   When completed, recheck the fluid level in the master cylinder, top up if necessary and refit the cap. Check the 'feel' of the brake pedal which should be firm and free from any 'sponginess' which would indicate air still present in the system.

12   Discard any expelled hydraulic fluid as it is likely to be contaminated with moisture, air and dirt which makes it unsuitable for further use.

22.4 The handbrake cable adjuster

22 Handbrake – cable adjustment

1   Correct adjustment is essential, so that the self-adjusting braking system functions normally. Adjustment should be carried out only when work on the linings or on the cables themselves has been carried out.

2   All four wheels must be on the ground (or on a lift).

3   Release the handbrake at the lever.

4   Release the adjuster locknut, and turn the adjusting nut until the cable deflection is approximately $\frac{3}{4}$ in (20 mm) in the centre of the cable run (photo, and Fig. 9.17).

5   Check that the handbrake lever travel is at least 9 notches.

23 Handbrake lever, rod and cable – removal and refitting

Lever

1   Remove both centre safety belt anchor bolts.

2   Disconnect the handbrake under the car.

3   Disconnect the handbrake warning switch.

4   Remove the lever fixing bolts from the floor, and pull the lever out.

5   To further dismantle, take out the ratchet pin, and the pin in the toothed quadrant.

6   Unscrew the lever plunger knob, and remove the detail parts beneath it.

7   To refit, assemble the detail items under the plunger knob in the

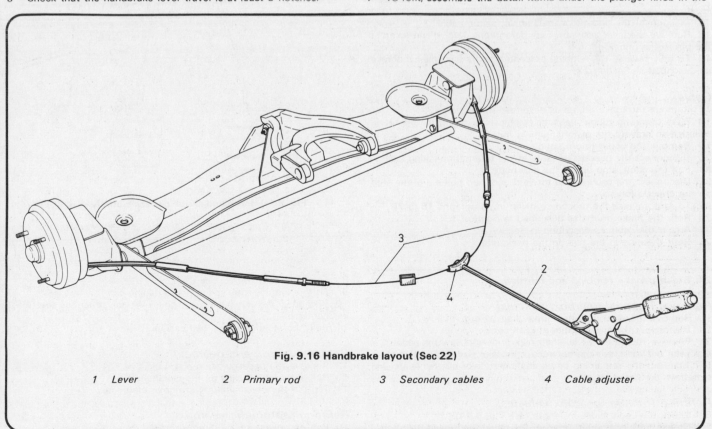

Fig. 9.16 Handbrake layout (Sec 22)

| 1 Lever | 2 Primary rod | 3 Secondary cables | 4 Cable adjuster |

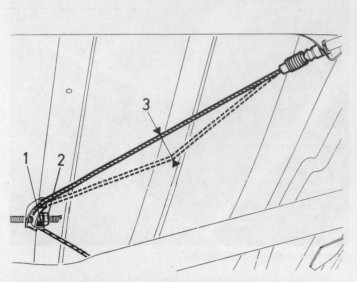

Fig. 9.17 Adjustment of the handbrake cable (Sec 22)

1   Adjuster   *3   Cable deflection*
2   Adjuster locking nut

23.22 A cable stop on the chassis

following order: flat washer, rubber washer, flat washer, spring and plunger.
8   Refit the ratchet and toothed quadrant pins.
9   Refit the bellows.
10 Offer the assembly to the floor, bolt in place, and reconnect the handbrake warning light switch.
11 Refit the seat belt anchor bolts.
12 Reposition the bellows, beginning at the front.
13 Reconnect the rod and cable, and adjust the mechanism (see Section 22).

Rod
14 Disconnect the underfloor mechanism at the cable.
15 Remove the lever assembly (see paragraphs 1 to 4) and take out the pin to the underfloor rod.
16 To refit, reverse the removal procedure and adjust the handbrake as described in Section 22.

Cable
17 Remove the brake drums as described in Section 13.
18 Place a strong rubber band or flexible wire round the operating cylinder, to prevent the pistons coming out.
19 Remove the upper return spring.
20 Disconnect the handbrake cable at the internal operating lever.
21 Pull the cable stop out of the backplate.
22 Disconnect the cable at the forward end, and pull out of the stop on the chassis (photo).
23 To refit, reverse the procedure given in paragraphs 18 to 20.
24 Refit the brake drums as described in Section 13.
25 Adjust the cable as described in Section 22.
26 Refit the plug in the hole in the backplate.

24 Brake pedal – removal and refitting

Removal (manual gearbox vehicles)
1   Remove the steering wheel lower half casing.
2   Disconnect the clutch cable at both ends.
3   Remove the clutch pedal shaft clip, and withdraw the pedal.
4   Take out the clevis pin between the brake pedal and servo.
5   Take out the remaining pedal shaft clip, take out the shaft, and withdraw the brake pedal.

Refitting (manual gearbox vehicles)
6   Check the servo pushrod length (see Fig. 9.14).
7   Grease the pedal shaft, offer up the pedal, and insert the shaft.

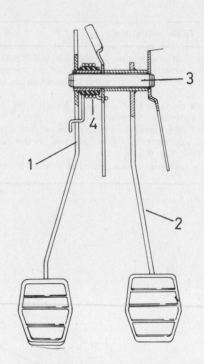

Fig. 9.18 Brake and clutch pedal assembly (Sec 24)

1   Clutch pedal   *3   Pedal shaft*
2   Brake pedal   *4   Spring*

8   Fit the pedal-to-servo clevis pin.
9   Fit the clutch pedal, spring and retaining clip.
10 Reconnect the clutch cable and check the adjustment as described in Chapter 5.
11 Refit the steering wheel half casing.

Removal (automatic gearbox)
12 Remove the pedal-to-servo clevis pin.
13 Remove the two clips on the pedal shaft.
14 Tap out the pedal shaft and remove the pedal.

Refitting (automatic gearbox)
15 Refit in reverse of the removal procedure.

25 Fault diagnosis – braking system

Symptom	Reason(s)
Excessive pedal travel	Self-adjusting mechanism to rear brakes not functioning Disc pads or linings badly worn
Poor braking, although pedal is firm	Disc pads or linings worn Discs or drums badly scored Seized piston in caliper or wheel cylinder Brake pads or linings of incorrect grade fitted Brake pads or linings contaminated Servo inoperative
Uneven braking, with vehicle swerving to one side	Seized piston in caliper or wheel cylinder Pads or linings contaminated Mixture of friction materials fitted Tyre pressures incorrect
Spongy brake pedal	Air in the hydraulic system Mounting bolts loose on one of the brake system components New disc pads or shoes not bedded-in
Pedal travels to floor, with minimal resistance and braking	Hydraulic system leak causing pressure loss Master cylinder not sustaining pressure

Chapter 10 Electrical system

For modifications, and information applicable to later models, see Supplement at end of manual

Contents

Specifications

General
System type .... 12 volt, negative earth
Battery type .... Lead acid

Alternator
Make .... SEV-Marchal or Paris-Rhone
Drive .... Belt from crankshaft pulley
Maximum output .... 52A at 14 V, 8000 alternator rpm

Starter motor
Make .... Ducellier or Paris-Rhone
Type .... Pre-engaged

Windscreen wipers
Wiper arms (except rear, Estate) .... Champion CCA4
Wiper blades (except rear, Estate) .... Champion X-4103

Fuse identification

Fuse no – complete plate	Fuse no – single plate	Rating (amps)	Circuits protected
1	1	8	Flasher unit, stop-lights
2	–	5	Not used
3	–	5	Not used
4	–	5	Not used
5	–	5	Not used
6	2	8	Cigar lighter, interior lights
7	–	5	Not used
8	3	16	Windscreen wiper/washer
9	–	5	Not used
10	4	5	Console illumination
11	–	16	LH front window motor
12	5	5	Instrument panel
13	–	16	RH front window motor
14	6	5	Automatic transmission
15	8	16	Reversing lights, heated rear window
16	–	5	Not used
17	7	16	Heater blower motor

Torque wrench settings

	lbf ft	Nm
Oil pressure switch	19	25
Fan thermal switch	30	40

1 General description

The electrical system is a 12 volt negative earth type. The major components consist of a battery, an alternator for charging purposes, and a starter motor.

Where components are connected into the electrical system, the greatest care should be taken to see that correct polarities are observed. Failure to observe this precaution may result in irreparable damage to the items concerned.

2 Battery – general description and maintenance

1   Two battery types are in use. That with a black container should be held only by a clamp and tie-bolts, and **not** by the base. The type with a blue container should be held by the lug on the base, although a clamp and tie-bolts are permissible (photo).
2   Top up the battery as necessary, so that the plates are just covered. Do not overfill, and use only distilled water.
3   Keep the battery top, terminals and lead connectors clean. Periodically grease the terminals with petroleum jelly.
4   If battery charging is to be carried out, the battery should first be disconnected. The same applies if repairs involving electric arc welding are to be carried out.
5   When disconnecting the battery leads, always detach the negative lead first. When reconnecting the leads, always attach the positive lead first. Further cautionary notes concerning the battery are given in Safety first! at the front of this manual.

2.1 The battery (base-clamped type)

3 Starter motor – removal and refitting

Removal – vehicle type R1340
1   Disconnect the battery.
2   Disconnect the wires to the starter motor and solenoid.
3   Remove the three bolts from the starter motor to clutch housing.
4   Remove the rear mounting bolt.
5   Remove the engine mounting bolt securing the rear bracket, and tilt the bracket a minimum of 90°.
6   Pull back the motor, and lower it to touch the exhaust pipe.
7   Turn the starter through 90° and remove it.

Removal – vehicle type R1341
8   Proceed as in paragraphs 1 and 2.
9   Remove the air filter and elbow.
10   Remove the three bolts securing the starter motor to the engine block.
11   Remove the rear mounting bolt.
12   Remove the mounting LH bolt.
13   Twist the rear bracket to the horizontal position.
14   Pull the starter motor back a little, tilt through 90°, and remove it.

Refitting – vehicle type R1340
15   With the starter at the side of the engine mounting, nose uppermost, turn it through 90° to make contact with the exhaust pipe. Push through the engine mounting, and forwards into place.
16   Reverse the procedure in paragraphs 1 to 5.

Refitting – vehicle type R1341
17   Position the starter motor between the alternator and engine mounting, tilt it through 90°, and place in position.
18   Reverse the procedure given in paragraphs 8 to 13.

4 Starter motor, Ducellier – dismantling, overhauling and reassembly

Dismantling
1   Remove the starter motor (see Section 3).
2   Remove the motor rear shield.
3   Remove the armature end bolt.

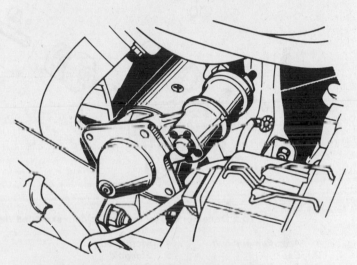

Fig. 10.1 The starter motor turned through 90° to remove it (Sec 3)

4   Remove the terminal.
5   Remove the rear bearing.
6   Take off the motor body.
7   Remove the solenoid securing body.
8   Remove the bearing pin from the solenoid-to-pinion fork.
9   Remove the armature and solenoid.

Overhauling
10   Examine the commutator condition, and if necessary either have it skimmed by a specialist and the segments undercut, or fit a replacement.
11   Examine the brushes and the pinion assembly, and change them if necessary.
Pinion assembly
12   With the armature removed, tap the stop collar with a suitable-sized piece of tube, thereby freeing the clip (Fig. 10.3).
13   Remove the pinion assembly.
14   Fit the replacement pinion.
15   Fit the circlip, and push the stop collar over it.
16   Reassemble the motor, and adjust the fork position as described in paragraphs 21 to 23.
Brushes
17   With the starter body removed, unsolder the brushes and solder new brushes in their place (Fig. 10.4).
Solenoid
18   With the body and armature removed, take out the solenoid.
19   Remove the bolt securing the fork to the solenoid core, and

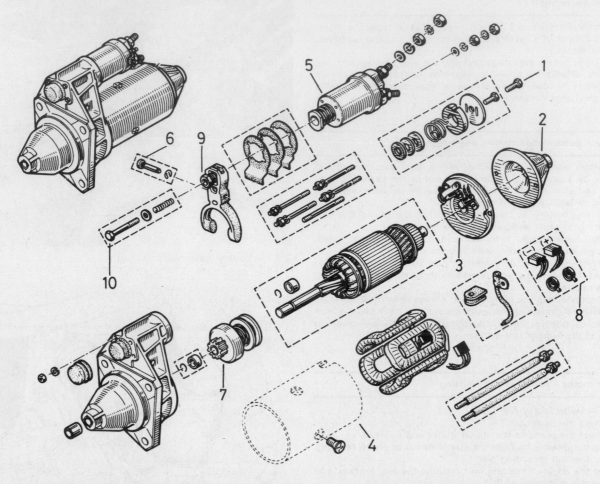

Fig. 10.2 Ducellier starter motor, typical – exploded view (detail differences will exist between types) (Sec 4)

1	Armature end bolt	4	Motor body	7	Pinion assembly
2	Cap	5	Solenoid	8	Brushes
3	Rear bearing	6	Solenoid fork bearing pin	9	Adjusting nut, solenoid
				10	Bolt, solenoid

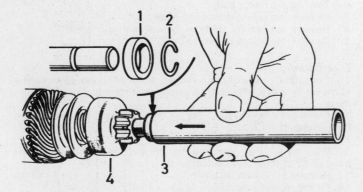

Fig. 10.3 Removing the pinion assembly (Sec 4)

1	Stop collar	3	Tube
2	Clip	4	Pinion assembly

remove the fork.

20 To reassemble, secure the fork to the solenoid by refitting and fully tightening the bolt.

21 Refit the solenoid (see paragraph 25) and adjust the assembly fork by first removing the plug at the solenoid front end.

22 Ensure that minimal clearance exists between the bolt and adjusting nut, and that the pinion assembly is resting against the armature.

23 Push the solenoid bolt fully inwards, and check the clearance between the pinion and stop collar which should be as shown in Fig. 10.6. Turn the adjusting nut as necessary to obtain the correct clearances (Fig. 10.2, items 9 and 10).

Reassembly

24 Lubricate the starter front bush.

25 Fit the armature and solenoid into the starter nose, replace and tighten the solenoid securing nuts, and refit the fork pin.

26 Fit the commutator and washers, first the steel washer and then the fibre one.

27 Oil the rear bush.

28 Refit the body.

29 Refit the rear bearing, the spring, and the washer, noting the slots. Fit and tighten the fixing bolt.

30 Refit the shield.

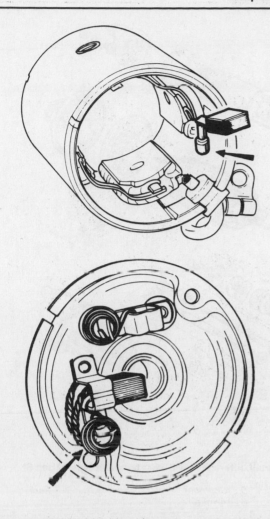

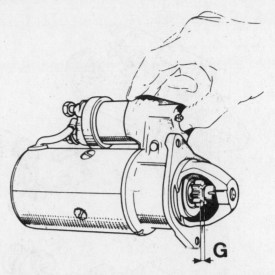

Fig. 10.6 Checking the pinion-to-stop collar clearance (Sec 4)

$G = 0.020$ to 0.060 in $(0.5$ to 1.5 mm$)$

3   Remove the clip (where fitted), the rear bearing, and the body, noting washer arrangements.
4   Remove the pinion operating fork pin.
5   Remove the solenoid securing nuts.
6   Remove the armature and solenoid.

Overhauling
Pinion assembly
7   Proceed as in Section 4, paragraphs 12 to 15.
8   Reassemble the motor, and adjust the fork position by first removing the solenoid cap.
9   Connect the solenoid to a battery to energise it, and check that the clearance between the pinion and the stop is as shown in Fig. 10.7.
10 To correct the clearance, turn the adjusting screw (1) as necessary (Fig. 10.9).

Fig. 10.4 Starter motor brushes, with the soldering points arrowed (Sec 4)

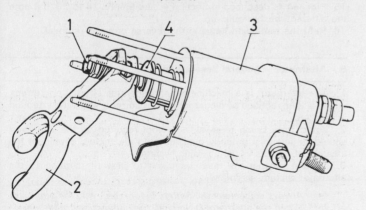

Fig. 10.5 Solenoid and fork assembly, removed from the starter motor (Sec 4)

| 1 | Fork securing bolt | 3 | Solenoid body |
| 2 | Fork | 4 | Core |

5   Starter motor, Paris-Rhone – dismantling, overhauling and reassembly

Dismantling
1   Remove the starter motor (see Section 3).
2   Disconnect the lead.

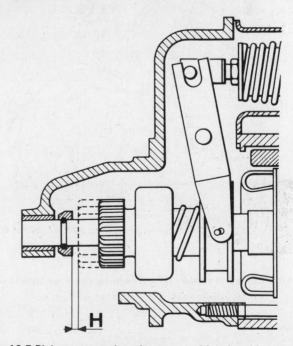

Fig. 10.7 Pinion-to-stop ring clearance, with solenoid energised (Sec 5)

$H = 0.059$ in $(1.5$ mm$)$

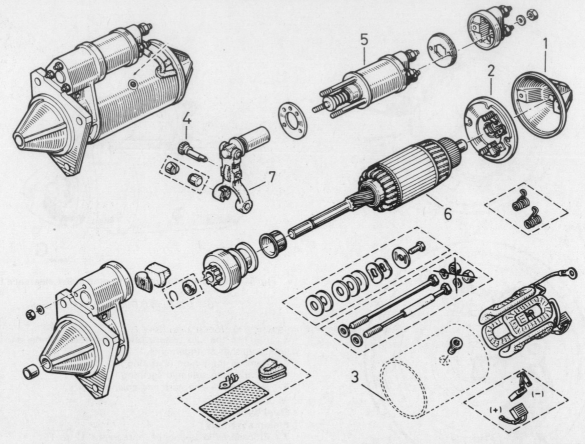

Fig. 10.8 Paris-Rhone starter motor, typical – exploded view (detail differences will exist between types) (Sec 5)

1	Cap	3	Body	5	Solenoid	7	Fork
2	Bearing plate	4	Fork bearing pin	6	Armature		

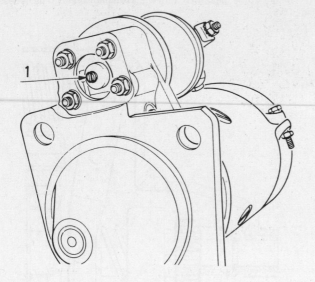

Fig. 10.9 Adjusting screw (1) for pinion-to-stop clearance (Sec 5)

Brushes
11 Proceed as in Section 4, paragraph 17.
Solenoid
12 Disconnect the lead, and remove the solenoid securing nuts.
13 Withdraw the solenoid, noting the spring and seal.
14 To refit, reverse the removal procedure.

Reassembly
15 Proceed as described in Section 4, paragraphs 24 to 28, but note the different washer build-up.
16 Refit the remaining items in the reverse order to removal.

---

6   Alternator – general description

1   The alternator is a machine which generates alternating current, which is then rectified by diodes into direct current. Very little attention is required.
2   Apart from brush renewal, any work required on the alternator should be left to a properly qualified auto electrician, as should the location of defects in the charging system.
3   Certain precautions must be taken to prevent damage to the alternator system, as follows:

    (a) Always disconnect the battery before removing the alternator
    (b) When the alternator is running, the connections must always be properly made, ie the positive terminal connected to the battery, and the alternator and battery negative terminals earthed
    (c) Never disconnect the battery or the regulator when the alternator is running
    (d) Ensure that the regulator earth connection is always properly made, when running

---

7   Alternator – removal, refitting and brush renewal

Removal
1   Disconnect the battery, and then the wiring to the alternator.

Fig. 10.10 Alternator, Paris-Rhone type – exploded view (Sec 7)

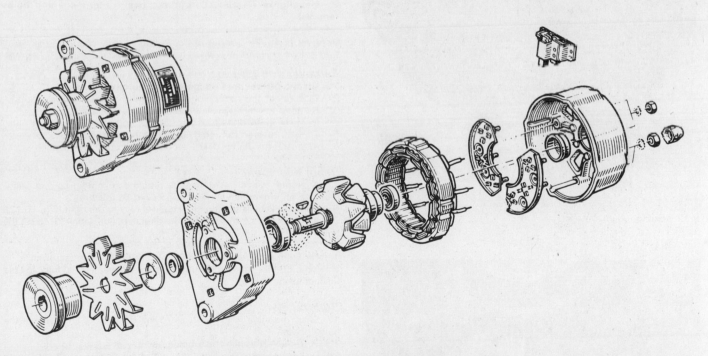

Fig. 10.11 Alternator, SEV-Marchal – exploded view (Sec 7)

2   On the R1341 vehicle only, remove the lower main alternator mounting from below.

3   On all models, remove the tensioning bolt, and the drivebelt.

4   Remove the mounting bolts (or the mounting, as relevant, see paragraph 2) and remove the alternator.

Refitting

5   Reverse the removal procedure. Adjust the drivebelt as described in Chapter 2.

Brush renewal

6   Change the brushes complete with their carrier, by removing the

two retaining screws and withdrawing the assembly. Fit a new assembly.

8   Voltage regulator – removal and refitting

1   The voltage regulator is located on the engine side of the main bulkhead, above the ignition coil.

2   To remove, disconnect all wires, noting their positions, after first disconnecting the battery.

3   Remove the fixing nuts, followed by the regulator.

4   To refit, reverse the removal procedure.

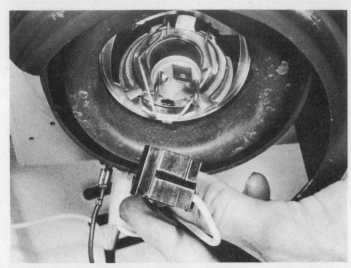

9.2 Remove the bulb connector block

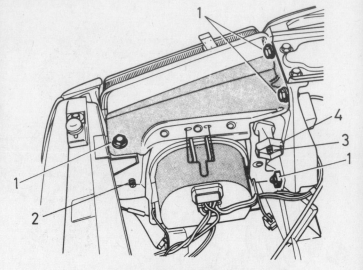

Fig. 10.12 Headlamp assembly (Sec 9)

1   *Light unit securing bolts (in 4 places)*
2   *Beam direction setting screw*
3   *Height setting screw*
4   *Loading adjustment screw*

9   Headlights – removal, refitting, beam alignment and bulbs renewal

Note: *Holts Amber Lamp is useful for temporarily changing the head-light colour to conform with the normal usage on Continental Europe*

Removal and refitting
1   Lift the bonnet, and disconnect the battery.
2   Disconnect the connector block (photo).
3   Remove the direction indicator unit (see Section 11).
4   Remove the 4 bolts (see Fig. 10.12), and remove the light unit.
5   To refit, reverse the removal procedure.
6   Align the headlamp beam, as necessary.

Beam alignment
7   A proper optical instrument is required to obtain correct align-ment. In an emergency, however, proceed as follows.
8   With the vehicle empty, turn knob (4) fully to the right (Fig. 10.12).
9   Adjust screw (2) to set the beam direction, and screw (3) to set the height.
10 Set knob (4) as necessary to suit the vehicle loading.
11 On Estate models, finally adjust using the load corrector knob on the side of the steering column, when the vehicle is fully loaded (including the driver).

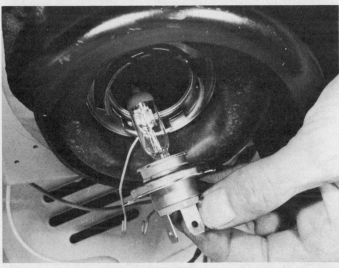

9.12 Headlight bulb removal

Bulb renewal
12 Snap the clips from the back of the main bulb, and remove the bulb (photo).
13 To remove the pilot bulb holder, simply pull it out (photo).
14 Do not touch a headlight bulb with bare fingers, but hold it with a clean cloth. If a bulb is accidentally touched, clean it carefully with methylated spirit. Greasy finger marks will blacken the bulb and may shorten its life.

10 Rear lamps – bulb changing and dismantling

Saloon, bulb changing
1   Referring to Fig. 10.13, press the two lugs together to release the plastic cover.
2   Lift off the plastic cover, thus gaining access to the bulbs.

Estate, bulb changing and lamp dismantling
3   Remove the screws securing the lamp to the body (photo).
4   Pull out the lamp body, thus releasing the two snap-in lugs from their slots.

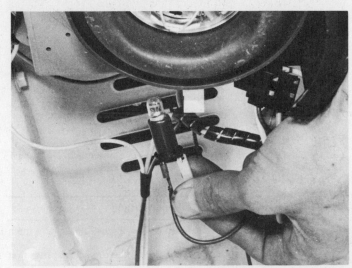

9.13 Pilot bulb removal

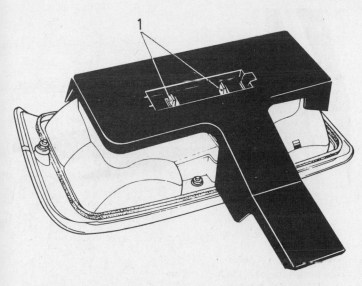

Fig. 10.13 Rear lamp plastic cover (Saloon) (Sec 10)

1   Cover release lugs

5   Release the bulb holder assembly by freeing the lugs at each end (photos).

Saloon, rear lamp dismantling

6   To remove the lens, unscrew and remove the four nuts.
7   Release the bulb holder from the plastic cover by pressing the lugs apart where these protrude through the back of the bulb holder. Disconnect the junction blocks.

11 Direction indicators and bulbs, front – removal and refitting

1   Pull out the bulb holder (photo).
2   Depress the clip, and pull the unit away (photo).
3   To refit, align the unit and press it into position until it clips home.
4   Refit the bulb holder.
5   To remove a bulb from the holder, simply turn it and pull it out.

12 Windscreen wiper arms, mechanism, motor and spindles – removal and refitting

Wiper arms

1   Lift the plastic cover and remove the locknut (photo).
2   Pull the arm from the tapered spindle, press the small lever on the blade, and remove the blade (photos).
3   To refit, make sure that the motor is parked, align the arm, and refit the locknut. Snap the blade into place.
4   Moisten the screen. Switch on to check for correct sweeping action.

Wiper mechanism

5   Remove the wiper arms.
6   Remove the two wiper spindle nuts.
7   Disconnect the earth wire and junction block from the motor supply lead.
8   Remove the mechanism plate fixing bolt.
9   Push the mechanism inwards to free the spindles, and move it sideways to withdraw.
10   To refit, first ensure that the short and long cranks are in line, and then refit in reverse of the removal procedure.

Motor

11   Remove the mechanism, and then remove the crank nut.

10.3 Rear lamp, Estate – screw removal

10.5a Releasing the bulb holder lugs

10.5b The bulb holder

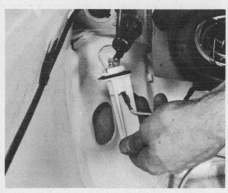

11.1 Removing a bulb holder from the front direction indicator lamps

11.2 Removing a direction indicator unit

12.1 A wiper spindle and nut assembly

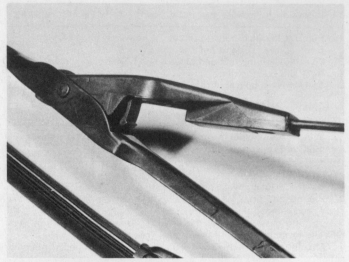

12.2a A wiper arm and blade, showing the release lever

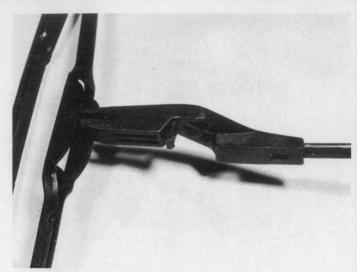

12.2b Removing the blade from the arm

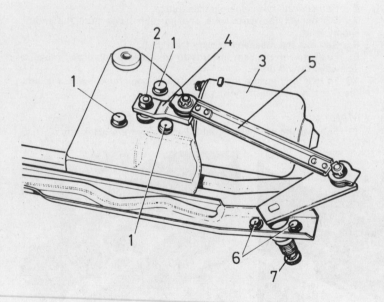

Fig. 10.14 Wiper mechanism (Sec 12)

1	Motor fixing bolts	5	Long crank
2	Crank nut	6	Spindle bolts
3	Motor	7	Spindle
4	Short crank		

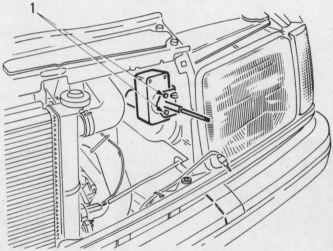

Fig. 10.15 Headlamp wiper assembly (where fitted) (Sec 13)

1   Assembly securing nuts

12 Remove the three fixing bolts, and remove the motor (see Fig. 10.14).
13 To refit, reverse the removal procedure, ensuring that the long and short cranks are in line.

Wiper spindle (SEV type)

14 Remove the mechanism.
15 Remove the two bolts, items (6) in Fig. 10.14, and remove the spindle.
16 To refit the spindle, reverse the removal procedure.

Wiper spindle (Bosch type)

17 Proceed as in paragraphs 15 to 17, but note that the spindle is riveted instead of being bolted. Drill out the rivets. When refitting, use new rivets or suitable nuts and bolts.

13 Headlamp wash/wipe system (where fitted) – removal and refitting

1   Disconnect the battery.
2   Disconnect the junction block, the feed wires to the motor, and the washer fluid pipe.
3   Remove the screw securing the wiper blade, and remove the blade.
4   Remove the front grille (see Chapter 12).
5   Remove the nuts securing the wiper assembly, and withdraw the assembly (Fig. 10.15).
6   To remove the motor, take out the three motor cover screws, and remove the cover followed by the motor (Fig. 10.16).
7   To refit, reverse the removal procedure.

14 Rear wiper arm and drive mechanism (Estate) – removal and refitting

Arm

1   The arm is retained on the tapered and splined spindle with a nut.

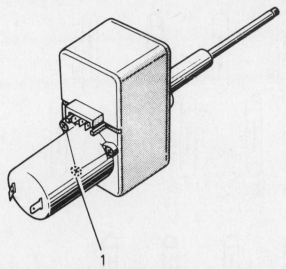

Fig. 10.16 Wiper assembly (headlamps) showing motor fixing screws (1) (Sec 13)

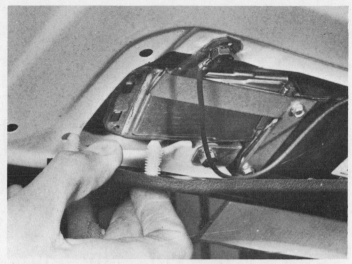

14.4 The trim panel on the tailgate of the Estate being removed

2　To remove the arm, remove the nut, lift the blade up and away from the screen, and then pull it off the pivot. Do not attempt to remove the arm without first lifting it from the screen. Remove the blade as described in Section 12, paragraph 2.

3　To refit, reverse the removal procedure, first ensuring that the blade lies horizontal.

Drive mechanism

4　To remove, carefully prise out the snap fasteners which retain the trim panel covering the mechanism on the inside of the tailgate (photo).

5　Remove the wiper arm as described in paragraphs 1 to 3. Remove the spindle nut.

6　Remove the wiring and washer pipe connections.

7　Remove the screws securing the mechanism to the tailgate lid, and lift out the mechanism (photo).

8　To refit, reverse the removal procedure.

15 Accessories plate – description, removal and refitting

Description

1　Two types of plate are used, depending upon the vehicle version, the more complex plate being designed to cover all the optional fittings. The fuse arrangements are thus different, depending upon the plate employed (see Fig. 10.17).

2　The plate is located under the dashboard on the driver's side.

Removal and refitting

3　Disconnect the battery.

4　Remove screws (1), Fig. 10.18.

5　Disconnect the wiring (photo).

6　Remove the plate and bracket, and separate them if necessary.

7　To refit, reverse the removal procedure.

8　To renew a blown fuse, simply open the cover, pull out the old fuse, and press in a new fuse of the same rating. If a fuse blows repeatedly, investigate the wiring to the component(s) concerned and the component themselves. Do not bypass fuses with foil or wire, nor fit a fuse of a higher rating, or serious damage may result.

16 Instrument panel and bulbs – removal and refitting

Removal

1　Disconnect the battery.

2　Remove the steering wheel half casings.

3　Remove the two panel screws.

4　Disconnect the speedometer cable and the three junction blocks.

5　Undo the two retaining screws at the base of the panel, then tilt the panel forward at the top by careful use of a screwdriver. Release the top and push the bottom inwards.

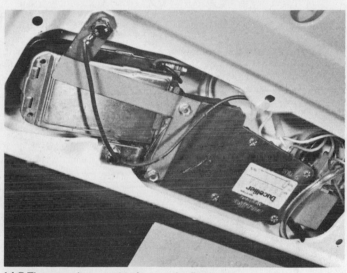

14.7 The rear wiper mechanism on the Estate

15.5 Removing the wiring connectors from the accessory panel

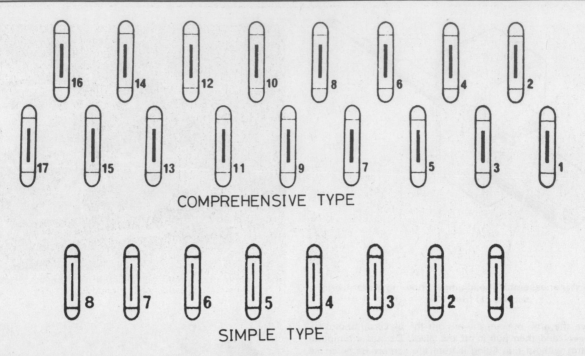

Fig. 10.17 Fuse arrangement on accessories plate, showing comprehensive and simple types. For fuse identification see Specifications (Sec 15)

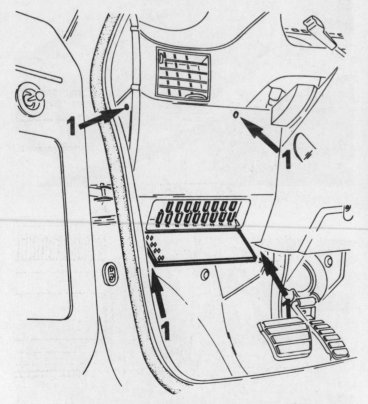

Fig. 10.18 Accessories plate (LH drive vehicle shown) (Sec 15)

1   Fixing screws

6   Remove the panel horizontally (photo).
7   To remove an instrument bulb, twist the holder and pull it out. Take out the bulb (photo).

Refitting
8   Refit any bulbs and holders removed (photo).

16.6 Remove the panel

9   Offer up the panel, tilting it rearwards as necessary to engage the two pegs.
10   Reconnect the speedometer cable and junction blocks.
11   Refit the remaining items in reverse of the removal procedure.

___

17 Speedometer cable – disconnecting at the instrument panel

1   Remove the accessories plate (Section 15).
2   Remove the steering wheel lower half casing (see Chapter 11).
3   Locate the cable behind the instrument panel, press the clips inwards where the cable joins the speedometer head, and at the same time pull the cable away, thus removing it.
4   To refit, just push the cable onto the back of the speedometer head.
5   Refit the remaining items in reverse of the removal procedure.

16.7 Removing an instrument panel bulb

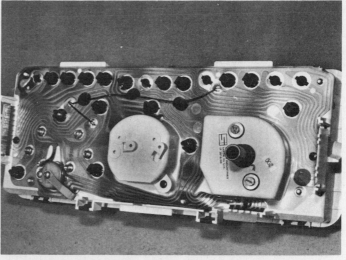

16.8 The instrument panel, ready for refitting

18 Electro-mechanical door locks – general description

1   This section should be read in conjunction with Sections 18 and 20, in Chapter 12.
2   The system provides for simultaneous locking of the four doors, either from outside by turning the key in either front door, or from inside by pressing the door-locking switch on the central console.
3   A red tell-tale on each door, when visible, indicates that the doors are locked.
4   An electrical circuit, consisting of an inertia switch and a thermal cut-out, causes all the doors to become unlocked in the event of an impact of more than a certain magnitude occurring. Very light impacts will not cause this to happen.
5   Note that for certain territories, automatic rear door locking can be cancelled by the rocker switch above the door handle.
6   To the mechanical lock (described in Chapter 12, Section 18) is added a changeover switch, connected to the key in the slot, and an electro-magnetic actioner, the operation of which is governed by the two-way switch, ie by the key position, or alternatively by the position of the console switch.
7   The inertia switch, described in paragraph 4, can be reset by depressing the resetting button, in the event of an accidental unlocking situation occurring. The cut-out prevents the inertia switch circuit remaining alive too long, and is reset by pressing the cut-out button below, and to the offside of, the steering wheel.

19 Electric window winder motor (where fitted) – removal and refitting

1   Remove the winder mechanism as described in Chapter 12.
2   Remove the screw and nut securing the motor and pull it away from the reduction gear.
3   Test the motor by connecting it to a 12 volt battery. No repairs are possible to a faulty motor.
4   To remove the reduction gear from the mechanism, unscrew the securing nuts and separate the items.
5   To refit, reverse the removal procedure.

20 Steering column controls assembly – removal and refitting

1   Disconnect the battery.
2   Remove the steering wheel and the half casings.
3   Remove the fixing bolt and screw.
4   Disconnect the junction block, and remove the controls assembly (photo).
5   To refit, reverse the removal procedure.

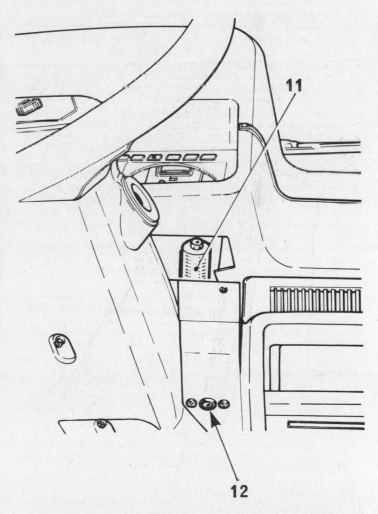

Fig. 10.19 Safety devices for electro-magnetic locks (Sec 18)

11 Resetting button for inertia switch   12 Thermal cut-out button

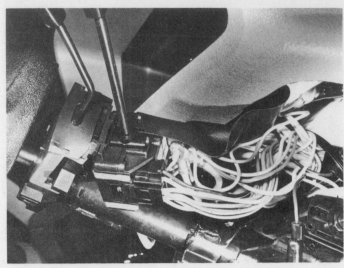

20.4 The controls assembly and junction block

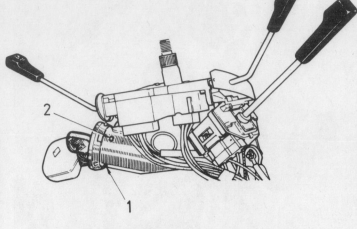

Fig. 10.20 The ignition/starter switch (Sec 21)

1   Screw   2   Retainer

21 Steering column switches – removal and refitting

Ignition/starter switch

1   Proceed as in Section 20, paragraphs 1 and 2.
2   Disconnect the junction block.
3   Turn the key to 'G' and withdraw it.
4   Remove the screw (1), press in the retainer (2), Fig. 10.20, and push the switch out from the rear.
5   Fit the key, turn it to the 'Stop' position, and remove the key again.
6   Remove the two screws at the rear, and slide the switch apart.
7   To refit, reverse the removal procedure, noting that retainer (2) should be pressed with a scriber as the switch is pressed home.

Direction indicator, lighting and wiper-washer switches

8   Separation of the direction indicator and wiper-washer switch is not practicable, and if a defect occurs, the complete assembly must be renewed.
9   To remove the lighting switch, remove the steering column controls assembly as described in Section 20.
10   Tap out the pin (1), Fig. 10.21, using a $\frac{9}{64}$ in (3.8 mm) diameter pin punch, and with a short length of suitable sized tube as a support.
11   Refitting is the reverse of the removal procedure.

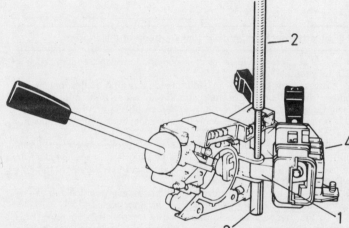

Fig. 10.21 Removing the lighting switch from the switch assembly (Sec 21)

1   Pin   3   Tube
2   Pin punch   4   Lighting switch

22 Oil pressure, cooling fan and coolant temperature switches – general

Oil pressure switch

1   This is fitted on the LH side of the engine and oil filter on the R1340, and on the RH side of the engine on the R1341.
2   The switch should operate at pressures above 5 lbf/in² (0.35 kgf/cm²), causing the oil pressure warning light to extinguish.
3   If a faulty switch is suspected, check by substituting a new one.

Cooling fan thermal switch

4   The switch, when fitted, will be found at the base of the radiator.
5   Check the switch by connecting it in circuit with a 12 volt battery and a sidelight bulb. No illumination should occur.
6   With the switch base immersed in water at 88°C (190°F), the bulb should light after a 'warming-up' period.
7   On pouring cold water in, the bulb should go out again at 79°C (174°F).
8   A defective switch must be renewed.

Coolant temperature sender switch

9   This switch will be found screwed into the body of the water pump.

10   If the temperature gauge is suspected of being inaccurate, first check the wiring from the sender and the instrument panel printed circuit. Testing of the sender unit is by substitution of a known good unit.

23 Horns – general

1   The horns are secured to the body of the vehicle, thereby providing an earth return for the current. They cannot be serviced and must be renewed if faulty.
2   If a horn fails to work, check that the earthing is secure, and that the feed wire is live when the horn push is pressed. (Check using a 12 volt bulb). If the feed is live, but the horn will not sound, it must be renewed.

24 Cigar lighter – general

1   When the knob of the cigar lighter is pressed inwards, the resistance element in the lighter heats up.
2   If the lighter does not function, first check the fuse and the feed wire. Check the earth wire. If all these are correct, the lighter will have to be renewed.

25 Cooling fan motor assembly (where fitted) – removal and refitting

1   Disconnect the battery and the fan motor junction block.
2   Remove the three securing nuts and withdraw the fan assembly.
3   To remove the motor, remove the centre securing nut, followed by the fan blades.
4   Drill out the rivets securing the motor, and remove the motor.
5   To refit, reverse the removal procedure, employing new blind rivets to secure the motor.

26 Fuel tank gauge unit – removal and refitting

1   Disconnect the battery.
2   Disconnect the wires from the gauge and from the luggage compartment light.
3   Remove the cardboard covering the tank.
4   Turn the gauge unit, using a screwdriver, and remove it. Cover the aperture in the tank with a sheet of polythene or similar if a new unit is not to be fitted immediately.
5   Refitting is the reverse of the removal procedure. A defective unit must be renewed, no repair being possible.

27 Interior lamp – general

1   This lamp has a three-position switch, with the centre being the 'Off' position. The left and right positions provide one permanently-on position, and one where the lamp is on only when a front door is open.
2   The cover simply snaps open, with a hinge at the switch end. The bulb then snaps in and out for replacement purposes (photo).

28 Radio and tape player installation

1   The console fitted to Renault 18 models contains locations for a radio of standard size and for an oval loudspeaker 9 cm by 15 cm (3.5 in by 5.9 in approx). Vehicles fitted with the 'radio pre-equipment' option have in addition a roof aerial fitted, along with wiring for a loudspeaker in each door and basic interference suppression measures. All cars have a plug in the headlining covering the hole for the aerial.
2   When purchasing a radio, make sure that it is the correct polarity for your car – negative earth. Units with dual polarity must be set correctly. Follow the radio manufacturer's instructions.
3   Earth and feed wires will be found in the console, and (on models so equipped) speaker leads and the aerial lead. It may be necessary to fit suitable connectors to the earth and feed wires – these should be supplied with the radio. Disconnect the battery before starting work.
4   To fit a roof aerial, remove the blanking plug from between the sun visors and drill a hole of the appropriate size (usually 10 mm) for the base of the aerial. Scrape the paint away from the underside of the hole to give a good earth contact. Remove the passenger side sun visor and glove compartment, also the left-hand glove compartment bracket, and feed the aerial wire along the top crossmember, down the passenger side windscreen pillar, out of the hole under the dash and into the console.
5   If the above sounds like too much work, you can fit a wing-mounted aerial in the position shown in Fig. 10.24, and drill a couple of bolts in the scuttle and the windscreen bottom crossmember to put the wire through. Always fit grommets when a wire has to be fed through a hole, and don't forget to paint bare metal surfaces to avoid future rusting.
6   The need for interference suppression will vary according to the type and quality of radio, the signal strength in the locality, and not least on the driver's sensitivity to unwanted noise! As mentioned above, basic suppression measures may already have been taken.
7   Suppression of components such as the coil, alternator and voltage regulator is by fitting condensers as shown. Additionally, if the radio is not fitted with an in-line choke, one should be fitted in the supply lead as close to the radio as possible. Suitable chokes and condensers are available from your local car accessory shop, or (maybe cheaper!) from a radio component shop.

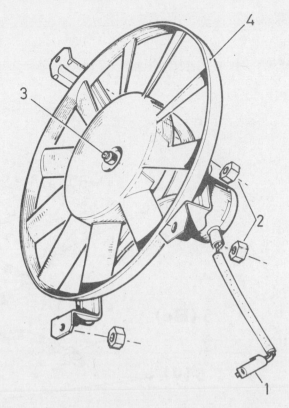

Fig. 10.22 Cooling fan motor assembly (Sec 25)

1   Junction block
2   Assembly securing nuts
3   Fan blades securing nut
4   Fan blades

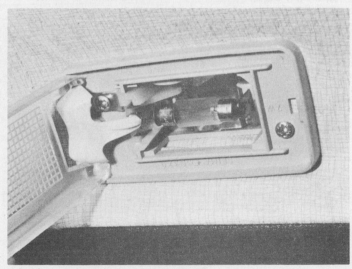

27.2 The interior lamp (cover snapped open)

8   Further suppression may be necessary to the ignition HT leads – fit a single suppressor at the centre (king) lead where it enters the distributor cap. Earth straps (braids) may also be effective between the bonnet and the body, between the exhaust pipe and floor, and between the front anti-roll bar and the side-member. These are not routine measures, however, and should only be undertaken if necessary.
9   Tyre static can be cured by having anti-static powder put into the tyres. Brake static – more noticeable on hot dry days – is quite expensive to cure, though special kits are available, and you may decide to live with it. Turn signals are not normally suppressed. Electric

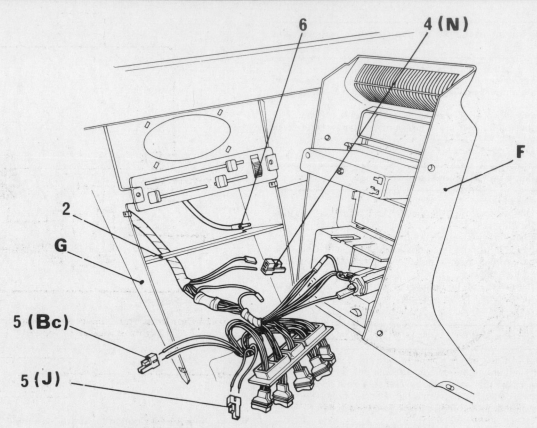

Fig. 10.23 Console removed, showing radio wiring harness (Sec 28)

2	Bulkhead	5(Bc) White terminal, for	6 Aerial lead (when	F Console front
4(N)	Black terminal, for	LH speaker (when fitted)	fitted)	G Console body
	feed and earth	5(J) Yellow terminal, for		
		RH speaker (when fitted)		

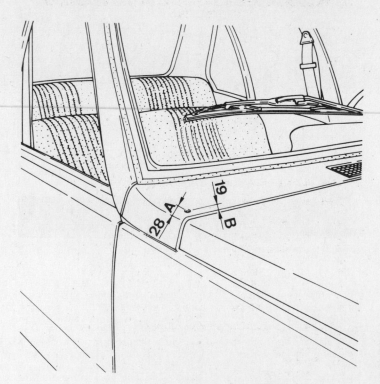

Fig. 10.24 Correct position for wing-mounted aerial – use LH wing on RHD cars. Dimensions in mm (Sec 28)

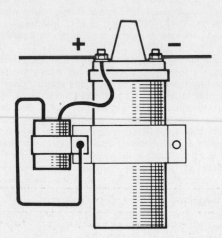

Fig. 10.25 Correct way to connect a suppressor capacitor to the coil (Sec 28)

motors (windscreen wipers, washers, etc) can be suppressed using condensers as for the coil.

10 In conclusion, it is pointed out that it is relatively easy and cheap to eliminate 95 per cent of all noise, but to eliminate the final 5 per cent is time and money consuming. It is up to the individual to decide if it is worth it. Remember also that one cannot have concert hall performance from a cheap radio.

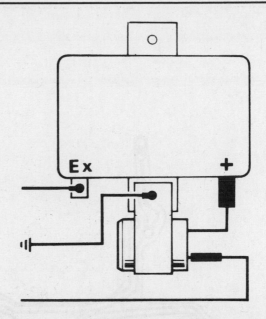

Fig. 10.27 Correct way to connect a suppressor capacitor to the
voltage regulator (Sec 28)

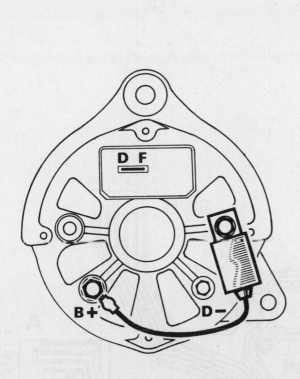

Fig. 10.26 Correct way to connect a suppressor capacitor to the
alternator (Sec 28)

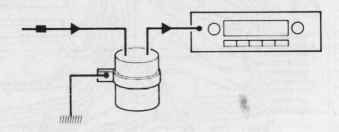

Fig. 10.28 Connect an in-line choke to the radio power supply
(Sec 28)

29 Wiring diagrams and wire identification

1   Each wire in the diagrams has a four-section code. An example of
such a code is 133 N2-41. In this case:

133 is the wire number
N is the wire colour (see below)
2 is the wire diameter (see below)
41 is the unit to which the wire goes

2   Wire colour code letters are as follows:

Be	Beige	N	Black
Bc	White	Or	Orange
B	Blue	R	Red
C	Clear	S	Pink
G	Grey	V	Green
J	Yellow	Vi	Violet
M	Maroon		

3   Wire diameter code numbers are as follows:

Code no	mm
1	7/10
2	9/10
3	10/10
4	12/10
5	16/10
6	20/10
7	25/10
8	30/10
9	45/10

4   Wire harness identification is as follows:

A	Front
B	Rear
K	Starter
L	Interior light
P	Door locking
R	Engine
U	Headlight

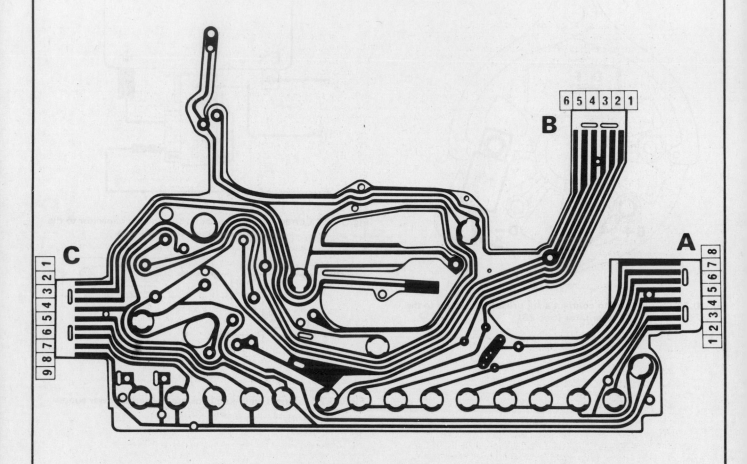

Fig. 10.29 Instrument panel printed circuit diagram

Connector A:
1   Not in use
2   Not in use
3   Not in use
4   Main beam warning light
5   Not in use
6   Rear screen demister warning light
7   Handbrake On warning light
8   Oil pressure warning light

Connector B:
1   Not in use
2   Brake pad wear warning light
3   + after ignition switch
4   Coolant temperature gauge
5   Fuel gauge
6   Lighting

Connector C:
1   Tachometer make and break
2   Permanent + (clock)
3   Direction indicators
4   Earth
5   Rear foglights
6   Hazard warning lights tell-tale
7   Choke On warning light
8   Not in use
9   + after ignition switch

Key to main wiring diagrams. Refer also to Section 29

1 LH front sidelight and direction indicator	66 Rear screen demister
2 RH front sidelight and direction indicator	67 Luggage compartment light
7 LH headlight main and dipped beams	68 LH rear light assembly
8 RH headlight main and dipped beams	69 RH rear light assembly
9 LH horn	70 Number plate light
10 RH horn	72 Reversing lights switch
11 Regulator	80 Junction block - front harness to engine harness
12 Alternator	81 Junction block - front harness to rear harness
14 RH earth	84 Junction block - front harness to auto-transmission
15 Starter	harness
16 Battery	85 Junction block - window winder harness
17 Engine cooling fan motor	90 Wire junction - air conditioning electro-magnetic
18 Ignition coil	clutch
20 Windscreen washer pump	91 Wire junction - brake pad wear warning light
21 Oil pressure switch	92 Wire junction - optional air conditioning
22 Thermal switch on radiator	98 Glove compartment illumination earth
24 LH front brake	99 Dashboard earth
25 RH front brake	101 Fuel tank mounting earth
26 Windscreen wiper plate	102 Regulator mounting earth
27 Brake master cylinder	106 Rear foglights switch
28 Heating-ventilating fan motor	110 Engine cooling fan motor relay
30 Connector No 1 - instrument panel	117 Heating-ventilating fan rheostat illumination
31 Connector No 2 - instrument panel	121 Glove compartment illumination
32 Connector No 3 - instrument panel	131 Electro-magnetic locks cut-out
34 Hazard warning lights switch	132 Electro-magnetic locks inertia switch
35 Rear screen demister switch	133 LH front door lock changeover switch
36 Heating-ventilating fan motor rheostat	134 RH front door lock changeover switch
37 LH window winder changeover switch	135 LH front door electro-magnetic lock solenoid
38 RH window winder changeover switch	136 RH front door electro-magnetic lock solenoid
40 LH door pillar switch	137 LH rear door electro-magnetic lock solenoid
41 RH door pillar switch	138 RH rear door electro-magnetic lock solenoid
42 LH window winder	140 Junction block - electro-magnetic lock harness
43 RH window winder	143 Wire junction - main lighting switch harness
44 Accessories plate (fuse box)	144 Wire junction - interior light harness
45 Junction block - front harness to accessories plate	146 Thermal switch
46 Junction block - front harness to accessories plate	148 Tailgate fixed contact
47 Junction block - front harness to accessories plate	150 LH front door loudspeaker
48 Junction block - front harness to accessories plate	151 RH front door loudspeaker
49 Junction block - front harness to accessories plate	152 Electro-magnetic door locks switch
52 Stop-lights switch	153 Car radio loudspeaker wires
53 Ignition-starter switch	158 Auto-transmission selector lever illumination
56 Cigar lighter	171 Rear screen washer-wiper switch
57 Car radio feed	172 Diagnostic socket
58 Windscreen washer-wiper switch	173 Wire junction - fuel tank unit
59 Combination lighting switch	174 RH headlight wiper-washer motor
60 Direction indicator switch	175 LH headlight wiper-washer motor
62 LH interior light	176 Headlight washer-wipers time switch relay
63 RH interior light	177 Headlight washer pump
64 Handbrake	178 Headlight washer pump (RHD)
65 Fuel gauge tank unit	179 Wire junction - windscreen washer - headlights washer

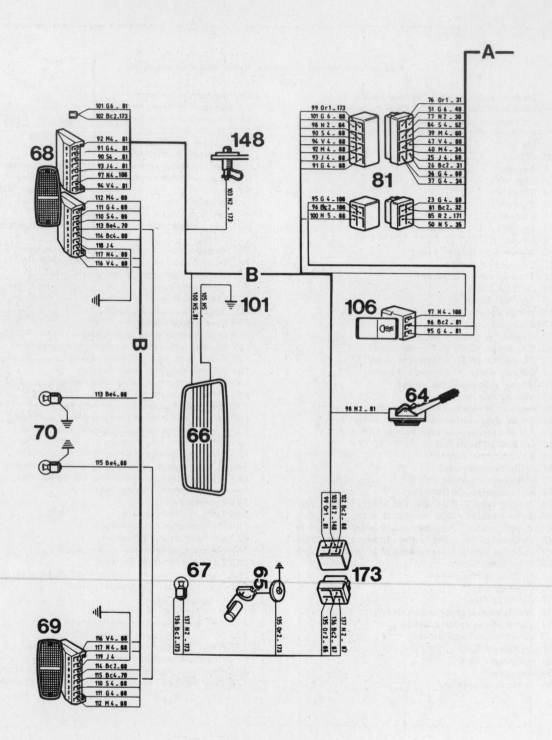

Fig. 10.30 Wiring diagram – R1340 (typical). Not all items are fitted to all models.

139

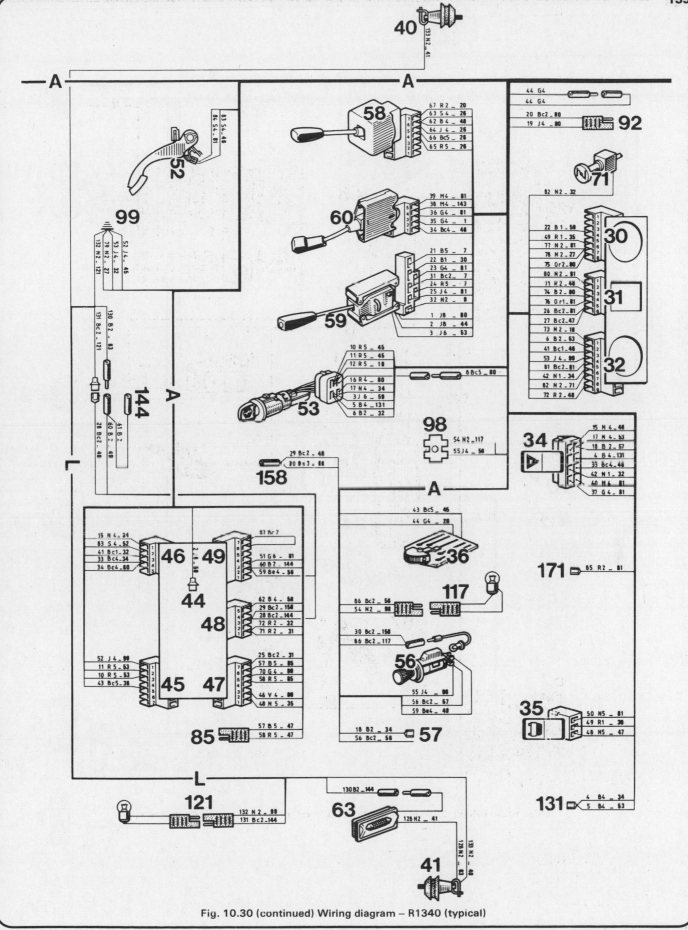

Fig. 10.30 (continued) Wiring diagram – R1340 (typical)

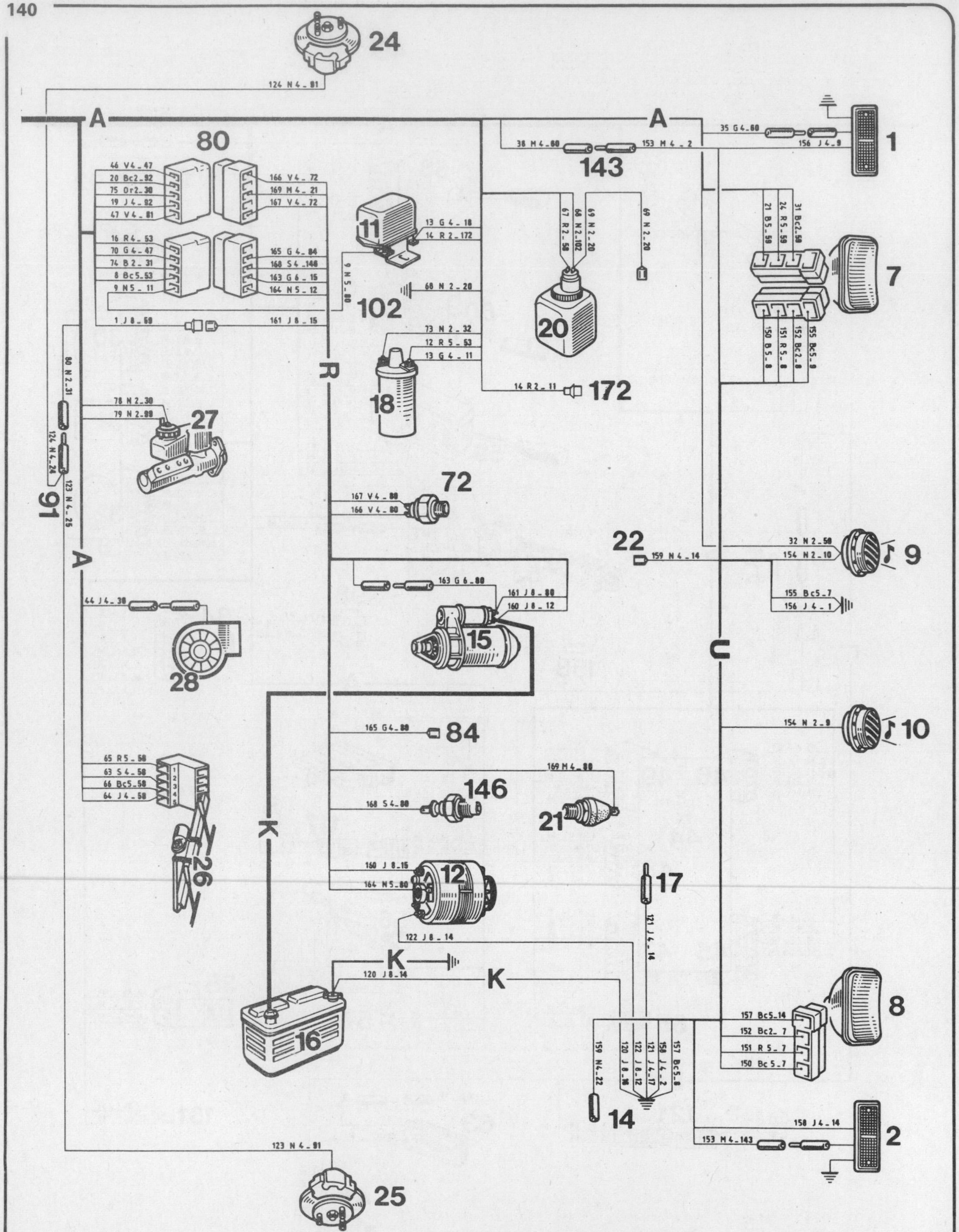

Fig. 10.30 (continued) Wiring diagram – R1340 (typical)

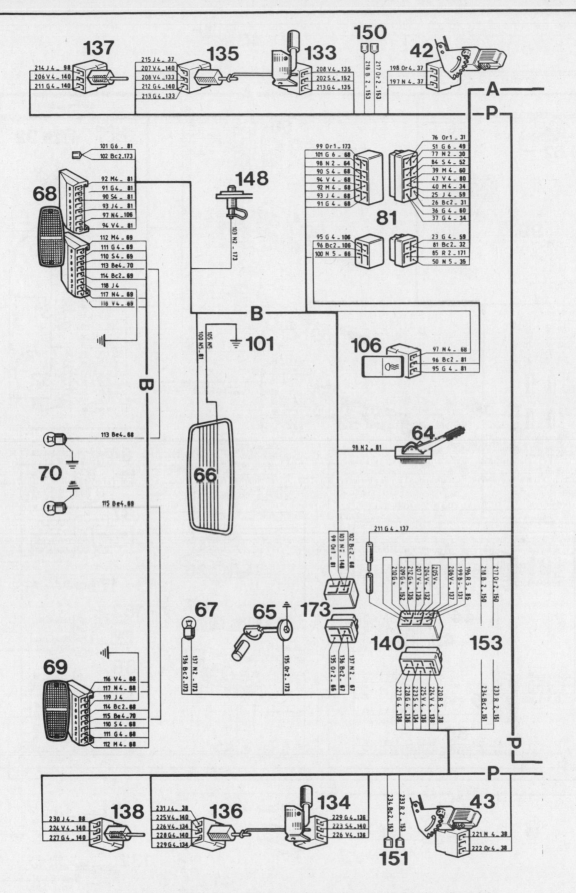

Fig. 10.31 Wiring diagram – R1341 (typical). Not all items are fitted to all models.

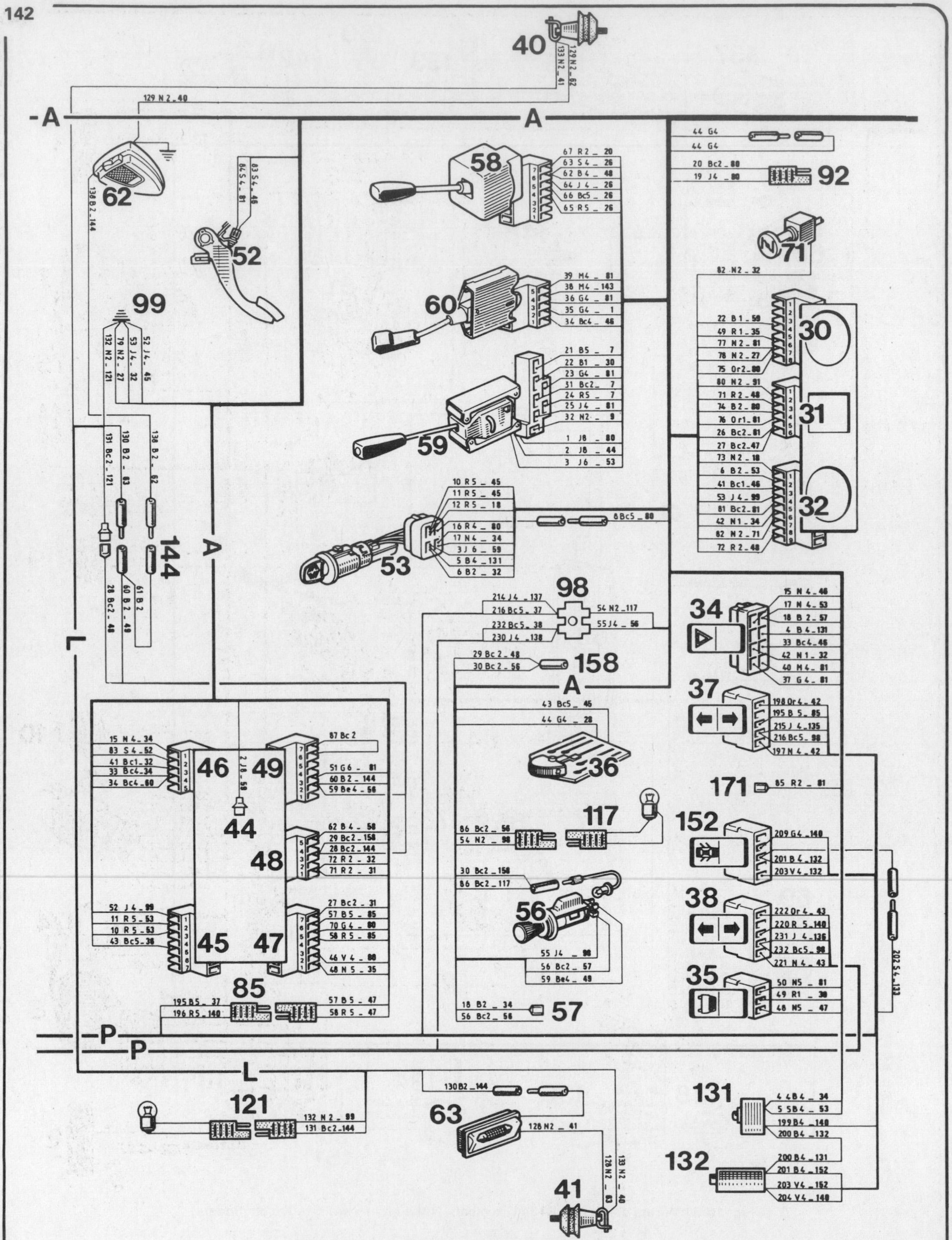

Fig. 10.31 (continued) Wiring diagram – R1341 (typical)

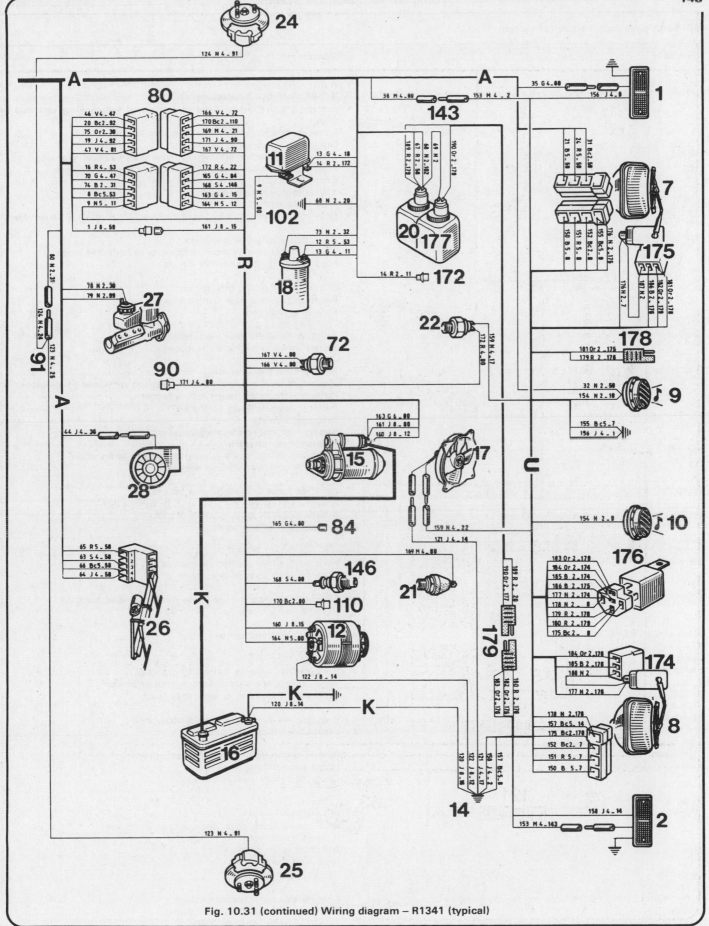

Fig. 10.31 (continued) Wiring diagram – R1341 (typical)

30 Fault diagnosis – electrical system

Symptom	Reason(s)
Starter fails to turn engine	Battery discharged
	Battery defective internally
	Battery terminal leads loose or earth lead not securely attached to body
	Loose or broken connections in starter motor circuit
	Starter motor solenoid faulty
	Starter motor pinion jammed in mesh with flywheel gear ring
	Starter brushes badly worn, sticking, or brush wires loose
	Commutator dirty, worn or burnt
	Starter motor armature faulty
	Field coils earthed
	Gear selector lever not engaged in 'P' or 'N' (automatic transmission)
Starter turns engine very slowly	Battery in discharged condition
	Starter brushes badly worn, sticking or brush wires loose
	Loose wires in starter motor circuit
Starter spins but does not turn engine	Starter motor pinion fork sticking
	Pinion or flywheel gear teeth broken or worn
	Battery discharged
Starter motor noisy or excessively rough engagement	Pinion or flywheel gear teeth broken or worn
	Starter motor retaining bolts loose
Battery will not hold charge for more than a few days	Battery defective internally
	Electrolyte level too low or electrolyte too weak due to leakage
	Plate separators no longer fully effective
	Battery plates severely sulphated
	Drivebelt slipping
	Battery terminal connections loose or corroded
	Alternator not charging
	Short in lighting circuit causing continual battery drain
	Regulator unit not working correctly
Ignition light fails to go out, battery runs flat in a few days	Drivebelt loose and slipping or broken
	Alternator brushes worn, sticking, broken or dirty
	Alternator brush springs weak or broken
	Internal fault in alternator

Horn/s

Symptom	Reason(s)
Horn operates all the time	Horn push either earthed ot stuck down
	Horn cable to horn push earthed
Horn fails to operate	Cable or cable connection loose, broken or disconnected
	Horn has an internal fault
Horn emits intermittent or unsatisfactory noise	Cable connections loose

Lights

Symptom	Reason(s)
Lights do not come on	If engine not running, battery discharged
	Wire connections loose, disconnected or broken
	Light switch shorting or otherwise faulty
Lights come on but fade out	If engine not running, battery discharged
	Connections loose or corroded
Lights work erractically – flashing on and off, especially over bumps	Battery terminals or earth connection loose
	Lights not earthing properly
	Contacts in light switch faulty

Wipers

Symptom	Reason(s)
Wiper motor fails to work	Blown fuse
	Wire connections loose, disconnected or broken
	Brushes badly worn
	Armature worn or faulty
	Field coils faulty
Wiper motor works very slowly and takes excessive current	Commutator dirty, greasy or burnt
	Armature bearings dirty or unaligned
	Armature badly worn or faulty

Symptom	Reason(s)
Wiper motor works slowly and takes little current	Brushes badly worn Commutator dirty, greasy or burnt Armature badly worn or faulty
Wiper motor works but wiper blades remain static	Wiper motor gearbox parts badly worn Faulty linkage

Chapter 11 Suspension and steering

For modifications, and information applicable to later models, see Supplement at end of manual

Contents

Specifications

General

Suspension type – front .... Independent, wishbones and coil springs, with anti-roll bar
Suspension type – rear .... Trailing rigid axle, A-frame and coil springs, with anti-roll bar
Shock absorbers .... Telescopic all round
Steering type .... Rack and pinion, power-assisted optional extra on some models

Front wheel geometry

Camber angle, unladen .... $0° \pm 30'$
Castor angle, unladen .... $30' \pm 30'$
Toe-out, unladen .... 0 to $\frac{1}{8}$ in (0 to 3 mm)
King pin inclination .... $9° 30' \pm 30'$. Equal on both sides to within $1°$

Rear wheel geometry

Camber angle, unladen .... $0°$ to $0° 30'$
Toe-out, unladen .... 0 to $\frac{1}{16}$ in (0 to 1.5 mm)

Rear wheel bearings

Type .... Taper roller
Endfloat .... 0 to 0.0012 in (0 to 0.03 mm)

Springs

	Front (auto)	Front (manual)	Rear
Colour	Yellow	Blue	Green
Type number	326	434	965
Wire diameter	0.547 in (13.9 mm)	0.547 in (13.9 mm)	0.502 in (12.75 mm)
Number of coils	6.9	6.9	5.75
Free length	23.82 in (605 mm)	23.23 in (590 mm)	16.14 in (410 mm)

Manual steering

Retractable steering shaft length .... 12.902 ± 0.04 in (327.7 ± 1 mm)

Power-assisted steering

Pump drivebelt tension .... $\frac{5}{32}$ to $\frac{3}{16}$ in (4 to 5 mm) at midpoint of longest run
Fluid type .... Dexron type ATF (Duckhams Uni-Matic or D-Matic)
Fluid capacity (from dry) .... 2.2 Imp pints (1.3 US qts, 1.25 litres) approx

Wheels and tyres

Wheel size and type:
R1340 TL 4.50 B13 3FH36
R1340 GTL and R1341 .... 5.00 B13 3FH36
Tyre size:
R1340 TL .... 145 SR 13
R1340 GTL and R1341 .... 155 SR 13

	Normal use		Heavy load or high speed	
Tyre pressures in lbf/in² (bar):	Front	Rear	Front	Rear
Saloon with 145 SR 13	25 (1.7)	28 (1.9)	28 (1.9)	31 (2.1)
Saloon with 155 SR 13	23 (1.6)	26 (1.8)	26 (1.8)	29 (2.0)
Estate	25 (1.7)	35 (2.4)	26 (1.8)	38 (2.6)

Torque wrench settings

	lbf ft	Nm
Steering		
Steering box fixing bolts	19	26
Tie-rod (track rod) locking nut	26	35
Tie-rod (track rod) ball-pin nut	26	35
Steering wheel nut	33	45
Column universal joint locknuts	26	35
Column flexible coupling bolts	11	15
Front suspension		
Shock absorber upper fixing nut	11	15
Shock absorber lower fixing nut	44	60
Shock absorber lower pivot nut	60	80
Upper suspension arm pivot nut	74	100
Upper suspension arm balljoint nuts	30	40
Upper suspension arm ball-pin nut	37	50
Lower suspension arm pivot nut	80	110
Lower suspension arm ball-pin nut	37	50
Anti-roll bar clamps-to-chassis nuts	11	15
Anti-roll bar end-to-link nuts	11	15
Stub axle nut	120	160
Bearing closure plate bolts	48	65
Roadwheel nuts	52	70
Castor tie-rod-to-suspension arm nut	30	40
Rear suspension		
Side arm front pivot fixing nut	26	35
Side arm rear pivot fixing nut	40	55
Contro A arm front pivot fixing nuto	80	110
Centre A-arm rear pivot fixing nut	30	40
Centre A-arm rear pivot clamp nuts	11	15
Roadwheel nuts	52	70
Shock absorber upper securing nut	11	15
Shock absorber lower securing nut	22	30

1 General description

Rack and pinion steering is fitted. Internal balljoints in the rack are covered by rubber gaiters, and connect by short track rods to the balljoints on the stub axle carrier steering arm. Power-assisted steering is available as an optional extra on certain models.

The independent front suspension basically consists of an upper and lower suspension arm on each side, pivoted on the chassis at the inner end by rubber bushes. At the outer end of the arms, balljoints are provided and connect with the stub axle carrier. The stub axle carriers carry the bearings for the front wheel hubs. Suspension control is by coil springs, an anti-roll bar, and double-acting shock absorbers with built-in pump and rebound stops.

Rear suspension is a trailing beam arrangement supported by an arm on each side, and one in the centre, all pivot points being rubber bushed. An anti-roll bar permanently joins the side arms. At each end of the axle beam, a stub axle carries the hub/drum assemblies on taper roller bearings. Suspension control is as for the front layout.

When ordering spares, care should be taken to ensure that the correct item is obtained. Many spares in the Renault range look alike but are nevertheless different dimensionally, a fact which may not be immediately obvious.

2 Steering wheel – removal and refitting

1   Remove the half casings from the steering column by taking out the retaining screws. Disconnect the battery.
2   Snap out the steering wheel embellisher.
3   Remove the centre nut retaining the steering wheel.
4   Pull off the wheel, using a suitable extractor tool.
5   To refit, reverse the procedure. Ensure that the steering rack is properly centred, and fit the wheel with the spokes evenly placed.

6   Centre pop the nut and shaft after fitting, for locking purposes.

3 Steering shaft flexible coupling – removal and refitting

1   Remove the steering rack assembly as described in Section 6.
2   Drill out the rivets which retain the coupling to the rack flange (photo).

3.2 Flexible steering coupling in position in the vehicle

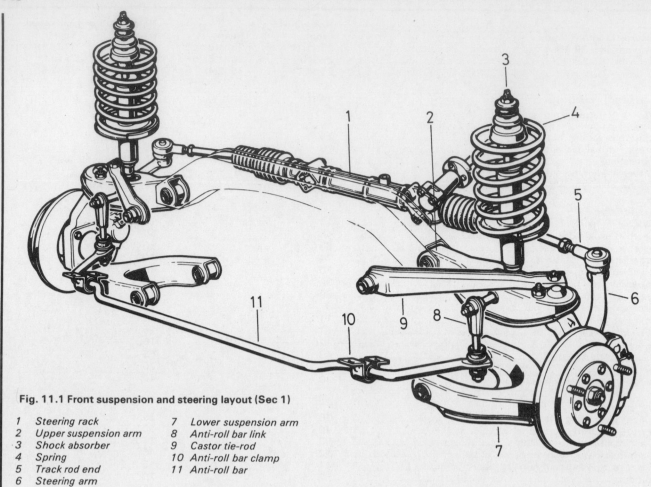

Fig. 11.1 Front suspension and steering layout (Sec 1)

1 Steering rack
2 Upper suspension arm
3 Shock absorber
4 Spring
5 Track rod end
6 Steering arm

7 Lower suspension arm
8 Anti-roll bar link
9 Castor tie-rod
10 Anti-roll bar clamp
11 Anti-roll bar

Fig. 11.2 Rear suspension layout (Sec 1)

1 Centre arm (A-frame)
2 Axle beam
3 Coil spring
4 Shock absorber

5 Brake drum
6 Side arm
7 Handbrake cable
8 Anti-roll bar

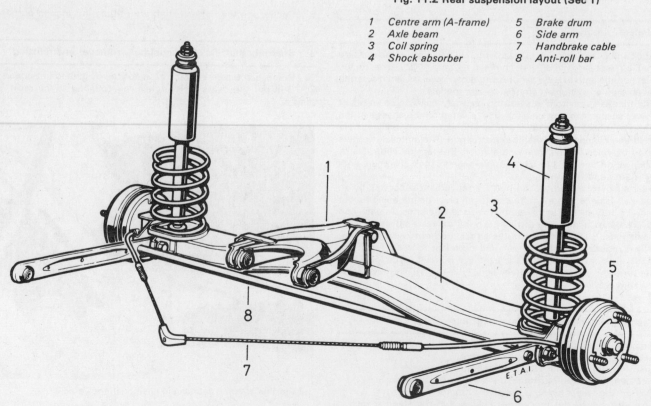

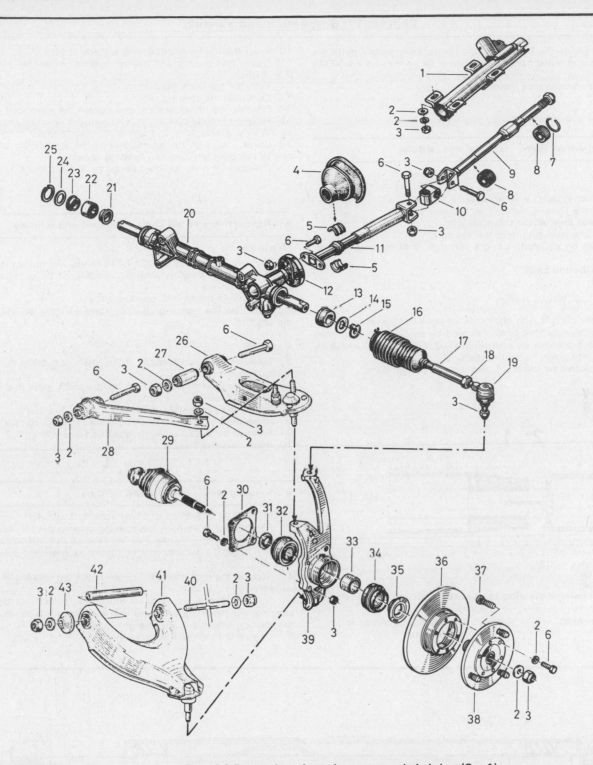

Fig. 11.3 Front axle and steering gear – exploded view (Sec 1)

1	Steering wheel shaft housing	*12*	Flexible coupling	*22*	Anti-knock bearing	*33* Spacer
2	Washers	*13*	Lock stop	*23*	Rubber bush	*34* Outer bearing
3	Nuts	*14*	Thrust washer	*24*	Outer thrust washer	*35* Ring
4	Bulkhead grommet	*15*	Lockplate	*25*	Circlip	*36* Disc
5	Bulkhead half sleeves	*16*	Rack bellows	*26*	Upper suspension arm	*37* Wheel stud
6	Bolts	*17*	Axial balljoint assembly	*27*	Sleeve	*38* Hub
7	Top bush snap-ring	*18*	Track rod (tie-rod) locking nut	*28*	Castor tie-rod	*39* Stub axle carrier
8	Bushes			*29*	Driveshaft outer joint	*40* Pivot stud
9	Steering wheel shaft	*19*	Track rod end	*30*	Bearing closure plate	*41* Lower suspension arm
10	Universal joint block	*20*	Steering rack housing	*31*	Ring	*42* Sleeve
11	Retractable steering shaft	*21*	Rubber bush	*32*	Inner bearing	*43* Shim

3   To refit, secure the coupling to the flange, using bolts 7 mm dia x 30 mm long, and Nyloc nuts. The nuts must be on the rack side of the flange.
4   Refit the steering rack assembly.

4   Steering wheel shaft – removal and refitting

Removal

1   Proceed as in Section 2, paragraphs 1 to 4.
2   Remove the lighting switch (see Chapter 10).
3   Unbolt and remove the universal joint.
4   Remove the snap-ring from the top bush.
5   Tap down on the shaft, using a soft drift, until the bottom bush comes out.
6   Remove the top bush.

Refitting

7   Grease the bushes with Hatmo or an equivalent grease.
8   Offer up the new split bush, followed by an old bush which has been reduced in diameter by approximately 0.080 in (2 mm), and draw the bush up into place, using the shaft.
9   Partially withdraw the shaft, and remove the old bush.

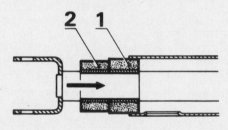

Fig. 11.4 Drawing a steering shaft lower bush into place (Sec 4)

1   New bush   2   Old bush, reduced in
                                     diameter

10 Insert the top bush, employing a piece of tube.
11 Ensure that the top and bottom bushes lie between the indents (Fig. 11.5).
12 Refit the top bush snap-ring.
13 Centralise the steering rack.
14 Offer up the universal joint, and tighten the lower bolt and nut.
15 Fit the universal joint upper bolt, turn the steering wheel to right or left by a quarter turn, and tighten the bolt and nut.
16 Check that, with the roadwheels in the straight-ahead position, one of the universal joint ball heads is upright.
17 Continue by reversing the operations in paragraphs 1 and 2.

5   Retractable steering shaft – removal and refitting

Removal

1   Remove the steering column half casings. Disconnect the battery.
2   Remove the steering shaft universal joint by withdrawing the bolts.
3   Disconnect the flexible couplings.
4   Withdraw the bulkhead grommet, and then the retractable steering shaft.

Refitting

5   Measure the overall length of the shaft. If outside the specified limits, renew the shaft.
6   With the grommet positioned on the shaft, pass it through the bulkhead.
7   Reconnect the flexible coupling.
8   Raise the front of the vehicle, and set the steering rack to centre.
9   Proceed as in Section 4, paragraphs 14 to 17.
10 Reconnect the battery.

6   Steering rack – removal and refitting

1   Proceed as in Section 8, paragraphs 1, 3 and 4.
2   Remove the bolts from the steering column flexible coupling.
3   Mark the position of the steering rack on the body crossmember so that it can be refitted at the correct height, then unscrew the mounting bolts.
4   Remove the rack assembly complete via the opening in the inner wing panel.
5   To refit, reverse the removal procedure, but do not tighten the steering rack mounting bolts until the steering rack is centralised, and the steering wheel is in the straight-ahead position.
6   Tighten the four rack securing bolts.

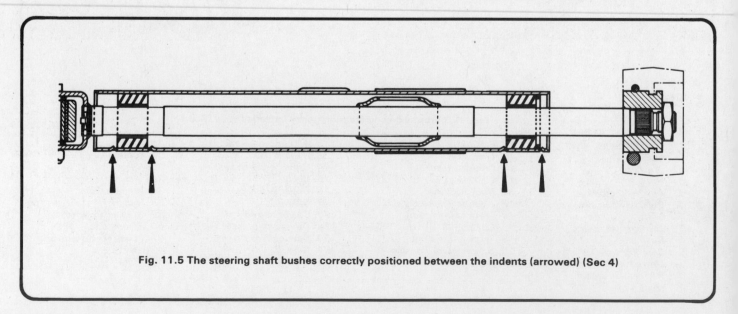

Fig. 11.5 The steering shaft bushes correctly positioned between the indents (arrowed) (Sec 4)

7   Refit the roadwheels, lower the vehicle, and tighten the wheel nuts.

8   Have the following checked by a Renault agent, who will possess the necessary specialised equipment:

(a)   *The steering box height setting*
(b)   *The front wheel toe-in*

7   Steering rack – dismantling

1   Dismantling of the steering rack by the average owner is not advised, and in the event of wear a replacement rack should be fitted or the advice of a Renault agent sought.

2   To renew a rack bellows, remove the relevant tie-load outer balljoint (Section 8).

3   Remove the bellows clips, where fitted, and ease off the bellows.

4   To refit, reverse the removal procedure.

8   Tie-rod (track rod) outer balljoints – removal and refitting

1   Raise the front of the vehicle and remove the roadwheels.

2   Hold the tie-rod with an open-ended spanner, and release the tie-rod locking nut.

3   Remove the ball-pin locking nut (photo).

4   Use a suitable balljoint extractor to free the joint taper from the stub axle carrier. Alternatively, use a pair of hammers, one each side of the tapered hole in the stub axle carrier, and use carefully-directed blows to shock the taper free.

5   Hold the tie-rod, and unscrew the balljoint, noting the number of turns required to remove it for refitting purposes.

6   To refit, reverse the removal procedure, refitting the balljoint the correct number of turns. Tighten the locknut.

7   On completion, have the toe-in checked, preferably by a Renault agent with the correct equipment.

9   Front anti-roll bar – removal and refitting

1   Remove the nuts securing the clamps to the side-members (photo).

2   Remove the nuts securing the links to the upper arm (photo).

3   Withdraw the bar and links.

4   Examine all parts, and renew as necessary.

5   Offer up and refit the anti-roll bar and associated parts, but do not tighten the fixings.

6   With the wheels on the ground, load the vehicle to give the setting in Fig. 11.19 (see Section 20).

7   Tighten the anti-roll bar fixings to the specified torque figures.

10   Hub and disc assembly, stub axle carrier and front wheel bearings – removal and refitting

Hub/disc assembly and stub axle carrier – removal

1   Fit the special tool as described in Section 11, paragraph 22.

2   Raise the relevant side of the vehicle, place it on a stand, and remove the roadwheel (photo).

3   Remove the brake caliper and caliper bracket (see Chapter 9). Tie the caliper up out of the way. There is no need to disconnect the hydraulic hose.

4   Using a suitable bar across the wheel studs to prevent the hub from turning, preferably one with holes to accept two of the wheel studs, unscrew the stub axle nut. This nut is very tight. Retain the nut and thrust washer.

5   Pull off the hub and disc assembly. If it should prove tight, a slide hammer with a suitable adaptor, or other tool, may have to be obtained.

6   Disconnect all three balljoints, as described in Section 8, paragraph 4 (photos).

7   Remove the stub axle carrier, taking care not to pull the driveshaft out of the gearbox coupling. Note the bearing spacer.

8.3 Removing the ball-pin nut on the track rod end

9.1 A front anti-roll bar clamp

9.2 Links – anti-roll bar-to-upper arm

10.2 General view of hub, disc and stub axle carrier

10.6a Removing the upper arm balljoint nut

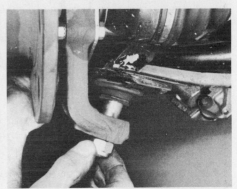

10.6b Removing the lower arm balljoint nut

Hub/disc assembly and stub axle carrier – refitting

8   Assemble the hub/disc and stub axle carrier together, preferably under a suitable press, by supporting the hub/disc assembly and then using a piece of tube of approximately $1\frac{11}{16}$ in (43 mm) outside diameter by $1\frac{13}{32}$ in (36 mm) inside diameter. Ensure that the bearing spacer is not forgotten.
9   Refit the assembly in reverse of the removal procedure, tightening all fixings to the correct torque figures.
10 Compress the suspension, and remove the special tool.
11 Operate the brake pedal to reposition the piston in the caliper.

Bearings – removal and refitting (stub axle carrier removed)

12 Remove the bolts from the bearing closure plate, and remove the plate.
13 Support the back of the hub carrier, and press out the inner bearing using a piece of tube or a drift.
14 Carefully fit the new bearing using a piece of tube or drift, ensuring that the lips of the grease seal are not damaged, and that no load is placed on the inner bearing ring.
15 Using a suitable bearing puller, withdraw the outer bearing from the hub/disc assembly. Note that it may be possible for the bearing to be levered off using two levers, each pivoting on a packing piece.
16 Fit a new bearing, ensuring that the sealed end is facing outwards, using a piece of tube slightly larger in inside diameter than the hub centre. Grease the seal lip and hub before fitting, and ensure that no press load is applied to the bearing outer track.
17 Apply some multi-purpose wheel bearing grease to the centre of the stub axle carrier, and assemble it to the hub/disc as described in paragraph 8.
18 Refit the bearing closure plate, washers and bolts.
19 Refit the assembly to the vehicle as described in paragraphs 9 to 11.

---

11 Front suspension upper arm – servicing and refitting

Removal

1   Raise the front of the vehicle.
2   Remove the relevant roadwheel.
3   Loosen the shock absorber bottom locknut.
4   Remove the fixing securing the castor tie-rod to the arm.
5   Remove the fixing securing the anti-roll bar link to the arm.
6   Remove the balljoint nut at the stub axle carrier, and free the taper using a suitable extractor. Alternatively, use 2 hammers as described

in Section 8, paragraph 4.
7   Remove the arm pivot pin.
8   Raise the arm, and unscrew the lower shock absorber mounting.
9   Remove the arm.

Refitting

10 Position the arm, and screw on the shock absorber lower mounting.
11 Position the balljoint pin in the stub axle carrier, and loosely refit the nut.
12 Refit, but do not tighten, the upper arm pivot pin.
13 Similarly, refit the shock absorber lower mounting pin, after smearing with Hatmo or an equivalent grease.
14 Reconnect, but do not tighten, the tie-rod.
15 Fit the special tool as described in paragraph 22.
16 Tighten the upper arm pivot pin, the shock absorber mounting pin, the tie-rod nut and the upper balljoint nut.
17 Remove the special tool, by compressing the suspension and taking it out.

Pivot bushes

18 To remove the bushes, remove the arm as described above.
19 Obtain a piece of tube 1.0 in (26 mm) dia and press out the old bush.
20 Press in the new bush using the same tube, to the dimension given in Fig. 11.6.

Upper arm balljoint

21 The joint may be changed with the arm either on or off the vehicle.
22 With the arm still on the vehicle, either Renault tool T.Av.509-01 or the home-made equivalent shown in Fig. 11.7 must be employed. Have two people put their weight on the relevant side of the vehicle, and insert the tool between the lower arm pivot pin and the shock absorber mounting pin (or the anti-roll bar link nut, if a damper is fitted to the driveshaft) (photos).
23 Raise the relevant side of the vehicle on a stand.
24 Remove the roadwheel.
25 Disconnect the balljoint as described in paragraph 6.
26 Remove the fixing bolts, drill out the rivets, and remove the balljoint.
27 To refit a balljoint, position the joint on the arm with the castor tie-rod in place. Secure the joint with the bolts supplied in place of the rivets, ensuring that the nuts and washers are at the top. Tighten to the specified torque.
28 Reconnect the balljoint, and tighten the securing nut to the specified torque.
29 Compress the suspension and remove the special tool.
30 Have the toe-in checked.

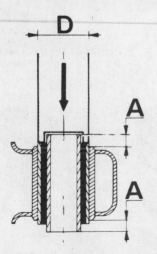

Fig. 11.6 Pressing out an upper arm bush using a piece of tube (Sec 11)

$A = \frac{15}{64}$ in (6 mm)
$D = 1.0$ in (26 mm) diameter

11.22a Special tool in place (driveshaft not fitted)

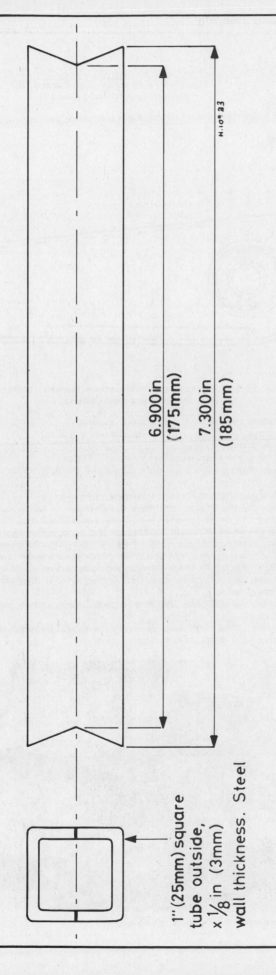

1'' (25mm) square
tube outside,
x ⅛ in (3mm)
wall thickness. Steel

6.900in
(175mm)

7.300in
(185mm)

H.1023

Fig. 11.7 Spacer too (Sec 11)

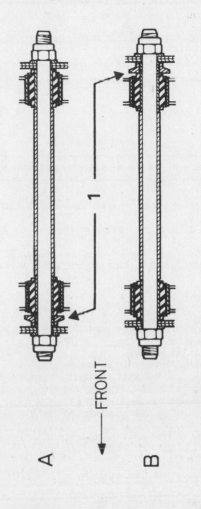

A ←── FRONT

A

B

1

Fig. 11.8 Front suspension lower arm pivot – sectional view (Sec 12)

1 Shim   A Manual steering   B Power steering

11.22b Special tool in place (driveshaft fitted)

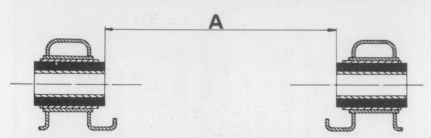

Fig. 11.9 Front suspension lower arm
bushes – positioning (Sec 12)

$A = 5\frac{15}{16}$ in (151 mm)

12 Front suspension lower arm – removal, steering and refitting

Removing

1   Raise the relevant side of the vehicle on a stand, and remove the roadwheel.
2   Disconnect the lower balljoint, as described in Section 11, paragraph 6.
3   Lower the anti-roll bar from the chassis fixings (see Section 9).
4   Remove the nut from the lower arm pivot pin, and withdraw the pin from the front.
5   Remove the arm.

Refitting

6   Insert the balljoint pin into the stub axle carrier, and retain it by screwing the nut lightly into place.
7   Raise the relevant side of the vehicle.
8   Grease the pivot pin, using Hatmo or an equivalent grease.
9   Position the arm, ensuring that the shim is correctly in place (Fig. 11.8), and insert the pin from the front. Do not tighten yet.
10 Fit the spacer tool as described in Section 11, paragraph 22.
11 Tighten the bottom balljoint nut, pivot pin nuts and anti-roll bar fixings.
12 Compress the suspension and take out the special tool.

Pivot bushes

13 Remove only one bush at a time, thus ensuring that the correct dimension is maintained between them.
14 Employ a piece of tube $1\frac{7}{32}$ in (31 mm) outside diameter to push out one of the bushes.
15 Press in a new bush until dimension A is achieved (Fig. 11.9), and ensure that the slots are as indicated in Fig. 11.10.
16 Proceed similarly with the second bush, ensuring that dimension A is retained.

Balljoint

17 The balljoint cannot be dismantled, and must be renewed if the bellows becomes damaged.
18 Remove the lower arm.
19 Drill out the rivets which retain the joint.
20 Fit the new balljoint, using the bolts supplied with it. Ensure that the bolt leads are on the bellows side of the arm.
21 Refit the arm, as described in paragraphs 6 to 12.

13 Front springs and shock absorbers – removal and refitting

1   The shock absorbers may be removed alone, or together with the springs. The springs may not be removed alone (photo).

Shock absorber removal, without spring

2   Preferably employ Renault tool Sus 808, with the base bracket sitting astride the top arm pin and the remainder of the tool disposed as indicated in Fig. 11.11. Alternatively, a suitable proprietary spring clamp system may be employed. *Great care should be taken during this operation, as serious injury may be sustained to the person if a spring should slip from the clamps.*
3   Raise the relevant side of the vehicle, place it on a stand, and remove the roadwheel.
4   With a jack under the lower balljoint, raise the body until it lifts off the stand.
5   Fit and make secure the spring clamping tool (see paragraph 2).
6   Remove the shock absorber lower mounting pin.

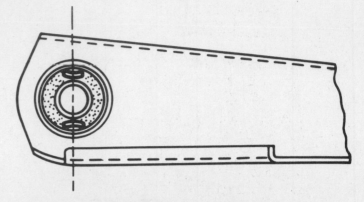

Fig. 11.10 Front suspension lower arm bushes – correct
orientation of the slots (Sec 12)

7   Remove the shock absorber lower locknut.
8   Remove the shock absorber top mounting (photo).
9   Lower the jack.
10 Remove the shock absorber by unscrewing it.
11 Push down on the top arm, and take the shock absorber out.

Shock absorber refitting, without spring

12 Place the shock absorber inside the spring.
13 Screw the shock absorber fully home in the lower mounting.
14 Pull the shock absorber top through the wing, ensuring that the mounting components are correctly refitted. Do not tighten yet.
15 Raise the suspension by means of a jack placed under the lower balljoint, and carefully remove the spring compression clamps.
16 Fit the spacer tool as described in Section 11, paragraph 22. Lower the jack.
17 Tighten the shock absorber bottom pivot pin and bottom mounting locknut.
18 Raise the jack sufficiently to permit the tool to be removed.
19 Tighten the shock absorber top mounting.

13.1 General view of front shock absorber

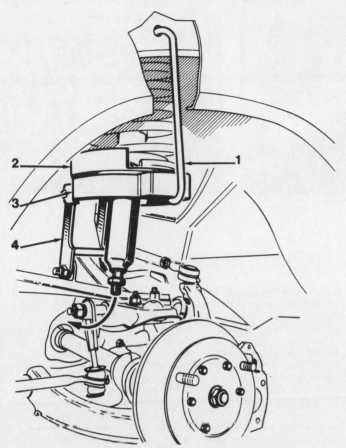

Fig. 11.11 Renault tool Sus 808 in use (Sec 13)

1   Hook from wing to base of   3   Pivoting base
    tool   4   Base bracket
2   Dummy cup

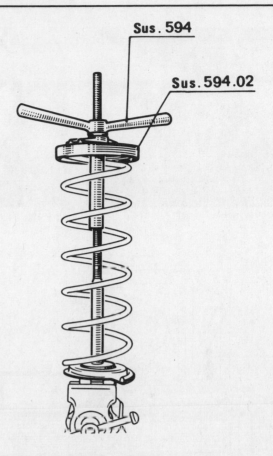

Fig. 11.12 Renault tool Sus 594 and Sus 594.02 in use (Sec 13)

23 Free the locknut at the bottom of the shock absorber. Lower the jack.
24 Remove the nut, and withdraw the shock absorber lower mounting pin.
25 Unscrew the shock absorber, and withdraw it with the spring and tool combined.

Shock absorber and coil spring – refitting together
26 Offer up the combined shock absorber, coil spring and tool, as removed (paragraph 25).
27 Reverse the sequence given in paragraphs 21 to 24 leaving the nut on the shock absorber lower mounting pin loose.
28 Proceed as in paragraphs 16 to 18.
29 Refit the roadwheel, and remove the stand and jack.

Coil spring – removal and refitting
30 Proceed as described in paragraphs 20 to 25.
31 Special tools, Sus 594 and Sus 594.02, shown in Fig. 11.12, or similar, are necessary to allow the spring to be safely decompressed. **Do not** use makeshift methods.
32 Fit the tool, tighten to permit the spring clamps to be released, and then unscrew the tool to decompress the spring.
33 To refit, reverse the directions given in paragraphs 30 to 32. Ensure that the bottom of the spring rests against the stop in the cap, and where tool Sus 809 is used, in line with the opening in the base of the tool.

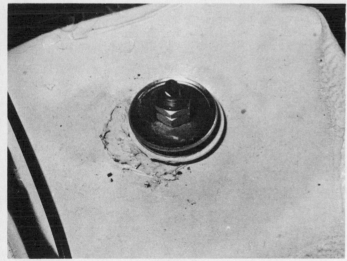

13.8 Front shock absorber top mounting

Shock absorber and coil spring – removal together
20 Proceed as in paragraphs 3 and 4.
21 Fit either Renault tool Sus 809 or a similar three-jaw spring compressor tool, with the 3 jaws hooked on the last but one coil from the top. Tighten each nut gradually until the spring pressure is contained by the clamps. *Make sure the compressor is secure.*
22 Disconnect the shock absorber at the top by removing the nuts.

14 Rear hub bearings – removal, refitting and adjusting

Removal
1   Raise and suitably support the rear of the vehicle.
2   Release the handbrake.
3   Remove the brake drum as described in Chapter 9, recovering the outer bearing (photo).
4   Remove the oil seal from the drum, by careful use of a screwdriver.

14.3 The rear hub outer bearing

14.5 The rear hub inner bearing

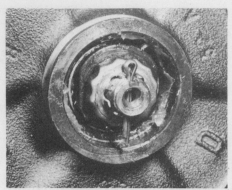

14.18 The nut lock and split pin on the rear hub

14 Fit the outer bearing, the washer, and the nut.
15 Adjust the bearing as described in paragraphs 16 onwards.

Bearing adjustment
16 Tighten the stub axle nut to a torque wrench setting of 22.5 lbf (30.5 Nm), and then turn it back by one-eighth of a turn.
17 Check the endplay if a dial gauge is available. Alternatively, ensure that endplay is *only just* detectable.
18 Fit the nut locking device, and then the split pin (photo).
19 Fill the hub cap with approximately $\frac{3}{8}$ ounce (10 gm) of wheel bearing grease, and fit it.
20 Adjust the rear brakes and handbrake (Chapter 9).
21 Refit the backplate plastic plug, preferably using a new part.

15 Rear suspension centre arm – removal, refitting and bush renewal

Removal
1 Raise the vehicle on a lift or stands.
2 Disconnect the link on the brake limiter at the arm (photo).
3 Remove the nuts securing the clamp at the forward end of the arm.
4 Remove the nuts securing the pivot pins at the rearward end of the arm.
5 Remove the clamp and pivot pins, followed by the arm.

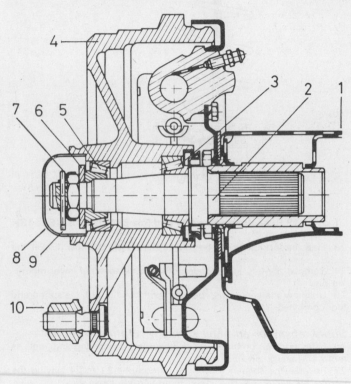

Fig. 11.13 Rear hub – sectional view (Sec 14)

1	Axle beam	6	Axle washer
2	Stub axle	7	Axle nut
3	Inner bearing	8	Hub cap
4	Brake drum	9	Split pin
5	Outer bearing	10	Roadwheel nut

5 Extract the inner bearing from the stub axle, using a suitable bearing puller (photo).
6 Tap the outer tracks of the bearing from the drum working through the hub centre, and using a suitable drift and hammer.

Refitting
7 Ensure that the seal deflector is in place.
8 Fit the inner bearing (excluding the outer track) to the stub axle. Tap carefully home using a suitable piece of tube.
9 Carefully tap the bearing outer track into place in the drum, using a suitable piece of tube or a socket.
10 Carefully tap a new oil seal into place.
11 Place about $\frac{3}{4}$ ounce (20 gr) of wheel bearing grease in the brake drum centre.
12 Smear a little grease on the bearings and oil seal. Take care not to contaminate the brake shoes or the friction surface of the drum.
13 Fit the brake drum over the stub axle.

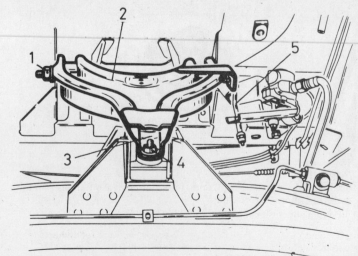

Fig. 11.14 Rear suspension centre arm (A-frame) (Sec 15)

1	Rear pivot nut	4	Clamp nut
2	Arm	5	Brake limiter link
3	Clamp		

15.2 General view of the rear suspension centre arm (A-frame), showing brake limiter link (arrowed)

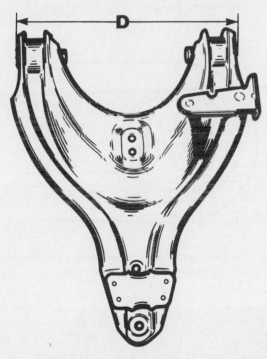

Fig. 11.15 Rear suspension centre arm (A-frame) showing dimension to be maintained when renewing bushes (Sec 15)

$$D = 9\tfrac{9}{16} \text{ to } 9\tfrac{37}{64} \text{ in } (243.0 \text{ to } 243.3 \text{ mm})$$

Refitting

6   Grease the pivot pins, using Hatmo or equivalent grease, and fit the arm by inserting the pins.
7   Refit but do not tighten the pivot pin nuts.
8   Refit but do not tighten the forward clamp nuts.
9   Ensure that the vehicle is fully lowered onto the roadwheels, and loaded to give the correct dimension as described in Section 20.
10   Tighten the nuts on the pivot pins and the centre bearing clamp.
11   Check the calibration of the brake limiter (see Chapter 9).

Bush renewal

12   Remove the arm as described above.
13   Press out one of the bushes employing a piece of tube $1\tfrac{1}{4}$ in outside diameter by $1\tfrac{1}{32}$ in inside diameter (31.5 mm by 26 mm).

14   Lightly grease the new bush, and press it into place until the correct dimension is achieved (see Fig. 11.15).
15   Press out the remaining old bush and fit a new one in the same way, maintaining the given dimension again.
16   Refit the arm.

16 Rear suspension side arms – removal refitting, and bush renewal

Removal

1   Raise the vehicle on a lift, or place it on stands.
2   Disconnect the secondary handbrake cables (see Chapter 9).
3   Remove the nuts securing the four pivot pins, knock the pins out, and remove the arm assembly (photos).

Refitting

4   Lubricate the pivot pins with Hatmo grease, or equivalent.
5   Offer up the arm assembly, fit the pins, and refit (but do not tighten) the securing nuts.
6   Reconnect the handbrake cables.
7   With the vehicle on the roadwheels, load it to give the dimension described in Section 20, and tighten the securing nuts on the side arms.

16.3a A front pivot on a rear suspension side arm

16.3b A rear pivot on a rear suspension side arm

Bush removal

8   The only bushes in the side arms which may be changed are those at the chassis end. Renew them one at a time, in order to assist in maintaining their relative positions.
9   Remove the arm assembly.
10 Press one bush out with the aid of a piece of tube $1\frac{5}{8}$ in outside diameter by $1\frac{5}{16}$ in inside diameter (41.5 mm by 33.5 mm).
11 Press in a new bush until the dimension given in Fig. 11.16 is obtained.
12 Press out the remaining old bush.
13 Fit a second new bush, again ensuring that the correct dimension is preserved.
14 Note the correct orientation of the slots in the bushes (Fig. 11.17).

---

17 Rear axle beam assembly – removal and refitting

Removal

1   Raise the rear of the vehicle, and lower it on to stands.
2   Remove the roadwheels.
3   Release the shock absorbers at the lower mountings, and push them upwards as far as possible.
4   Disconnect the flexible brake hose, running between the 3-way union and brake limiting valve, at one end only. Plug the open ends.
5   Pull the axle down, and withdraw the springs.
6   Disconnect the handbrake cables at the adjuster, and remove them from the retaining plates.
7   Support the axle on a jack.
8   Disconnect the brake limiting valve at the centre arm.
9   Remove the nuts, and then the pivot pins, from both the centre arm-to-chassis points, and from the side arms at the axle beam.
10 Remove the axle beam.

Refitting

11 Offer up the axle beam on a jack.
12 Lubricate the centre and side arm pivot pins with Hatmo grease or equivalent, and refit them together with their nuts. Do not tighten.
13 Reconnect the limiter to the centre arm.
14 Reconnect the brake hose.
15 Reconnect the handbrake cable.
16 Refit the roadwheels and lower the vehicle.
17 Load the vehicle as described in Section 20, and tighten the pivot pin nuts.
18 Refer to Chapter 9, and bleed the brakes, adjust the brake limiter, and adjust the handbrake.

---

18 Rear springs – removal and refitting

---

1   Remove the bolt securing the three-way union, and unclip the brake pipes adjacent to it.
2   Disconnect the shock absorber bottom mountings, collecting up the washers and rubber blocks.
3   Lift and support the vehicle at the rear.
4   Withdraw the springs.
5   To refit, move the axle beam sufficiently to allow the springs to be positioned with the lower end contacting the stop in the bottom cup.
6   Lower the vehicle.
7   Reconnect the shock absorbers at the lower end, and tighten to the specified torque.

---

19 Rear shock absorbers – removal and refitting

Removal

1   Open the boot lid, and remove the fuel tank cover plate. On Estate versions, remove the cap (photo).
2   Disconnect the shock absorber top mounting, using very thin spanners.
3   Raise the rear of the car and support it on stands.
4   Remove the relevant roadwheel.
5   Disconnect the shock absorber lower mounting.
6   Remove the bolt which retains the 3-way brake pipe union.
7   Compress the shock absorber.
8   Hold the axle beam out of the way, and withdraw the spring and

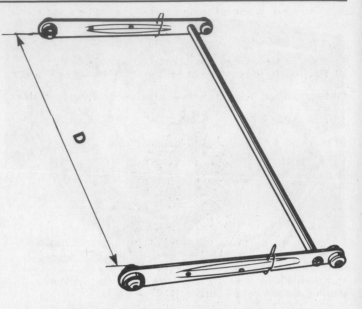

Fig. 11.16 Rear suspension side arms, showing dimension to be maintained when renewing bushes (Sec 16)

$$D = 37\frac{3}{4} \text{ in } (959 \text{ mm})$$

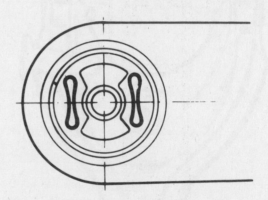

Fig. 11.17 Bushes in rear suspension side arms – correct orientation of slots (Sec 16)

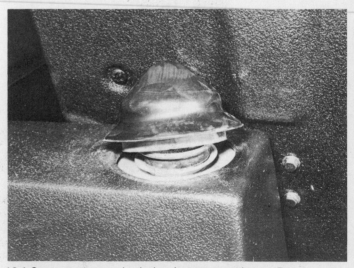

19.1 Cover cap on rear shock absorber top mounting, on Estate version

shock absorber, whilst collecting together the washers and rubber blocks.

Refitting

9   Place on the top stem one rubber block and spacer tube.
10 Place on the lower stem one flat washer followed by one rubber block.
11 With the axle beam held clear, offer up the spring and shock

absorber, placing the shock absorber top stem through the hole.
12 Place the spring so that the bottom end rests against the stop in the lower cup.
13 Secure the shock absorber at the top, with the washers properly fitted (see Fig. 11.18).
14 Extend the shock absorber downwards, and secure the lower end with the washers properly fitted (see Fig. 11.18).
15 Secure the 3-way union and refit the roadwheel.
16 Lower the vehicle.

20 Correct roadwheel-to-chassis relationship necessary, when tightening rubber bushed suspension pivots

Front of vehicle

1   Referring to Fig. 11.19.

   $H1$ = wheel centre height
   $H2$ = frame height, on line of wheel centres

2   The vehicle should be laden until the following formula is satisfied:

   $H1 - H2 = 3\frac{21}{32}$ in (93 mm)

3   The following items may then be tightened:

   (a) Upper and lower arm pivot pins
   (b) Shock absorber lower pivot pin
   (c) Castor tie-rod bracket
   (d) Anti-roll bar bearing
   (e) Castor tie-rod on upper arm

Rear of vehicle

4   Referring to Fig. 11.19:

   $H4$ = wheel centre height
   $H5$ = side arm front pivot height

5   The vehicle should be laden until the following formula is satisfied:

   $H4 - H5 = 1\frac{7}{8}$ in (48 mm)

6   The following items may then be tightened:

   (a) Side arm pivot pins
   (b) Centre arm pivot pins

21 Setting the steering to centre point

Refer to Fig. 11.20, and set dimension C to $2\frac{9}{16}$ in (65 mm) for manual gearboxes, or to $2\frac{13}{32}$ in (61 mm) for automatic transmission types.

22 Wheel balancing

1   This must be carried out by using special equipment. However, the points mentioned in the following paragraphs should be noted.
2   Renault balance weights are advised. They should be secured with yellow clips, except where thin rims are fitted. In this case, use white clips.
3   Never use an electric roadwheel rotator on the front wheels. It may, however, be used at the rear.

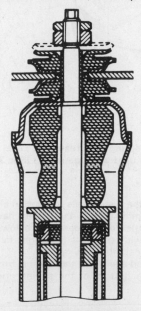
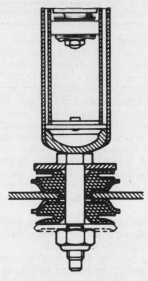

Fig. 11.18 Rear shock absorber – correct arrangement of washers at the top and bottom (Sec 19)

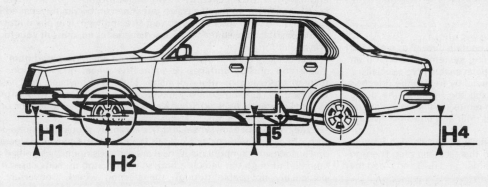

Fig. 11.19 Dimensions employed when tightening rubber-bushed suspension items (see Section 20)

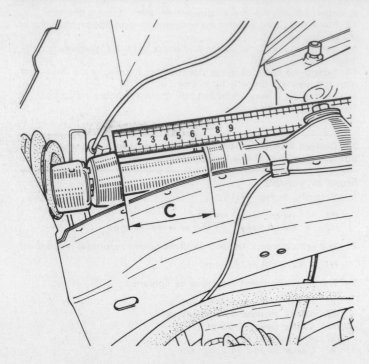

Fig. 11.20 Setting the steering rack to centre point (see Section 21)

4   Balance the front wheels by raising both wheels and securely placing the vehicle on stands. Do not raise one side only at the front for this check, or excessive strain will be placed upon the differential assembly when running the engine.

5   Rotate the front wheels using the engine, to the point of maximum out-of-balance, and employ a strobe light to find the position for the weight.

6   The rear wheels may be raised and balanced one at a time.

23 Power assisted steering – maintenance and overhaul

1   Power assisted steering is available as an optional extra on certain models. The pump providing power assistance is belt-driven from the crankshaft pulley. Procedures are as given for manual steering with the additions and differences noted below.

Maintenance

2   At the specified intervals, check the fluid level in the pump reservoir. The fluid should be up to the bottom of the strainer mesh. Top up with the specified fluid if necessary; look for leakage if frequent or heavy topping up is necessary. Maintain scrupulous cleanliness throughout.

3   Also at the specified intervals, check the tension and condition of the pump drivebelt. Adjustment is achieved in the same way as for the alternator drivebelt, by slackening the pump pivot and tensioner bolts and moving the pump as necessary.

Overhaul

4   The DIY mechanic is recommended to confine himself to renewing defective components in the power steering system, possibly on an exchange basis. Reconditioned units may be available from specialists.

5   Removal of the steering pump is achieved by freeing the drivebelt from the pulley, disconnecting and plugging the hoses at the pump (note the hose positions for refitting), then removing the tensioner and pivot bolts. Refit in the reverse order, and refill and bleed the system as described below.

6   Removal of the steering rack is essentially as described in Section 6. Additionally, disconnect the pipes from the rack and plug them to prevent fluid leakage. Refit in the reverse order, and refill and bleed the system as described below.

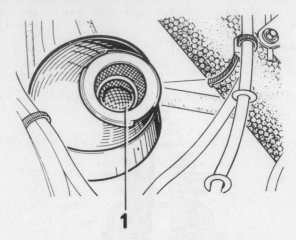

Fig. 11.21 Power steering fluid reservoir. Level should be up to bottom of strainer gauze (1) (Sec 23)

Bleeding

7   If the fluid level in the system falls so low that air is introduced into the pump, or after components of the system have been disturbed, bleeding should be carried out as described below.

8   Fill the reservoir with specified fluid right to the top. Turn the steering from lock to lock, then top up.

9   Start the engine and again turn the steering from lock to lock. Top up if necessary to the bottom of the strainer mesh, then refit the reservoir cap.

24 Wheels and tyres – general care and maintenance

Wheels and tyres should give no real problems in use provided that a close eye is kept on them with regard to excessive wear or damage. To this end, the following points should be noted.

Ensure that tyre pressures are checked regularly and maintained correctly. Checking should be carried out with the tyres cold and not immediately after the vehicle has been in use. If the pressures are checked with the tyres hot, an apparently high reading will be obtained owing to heat expansion. Under no circumstances should an attempt be made to reduce the pressures to the quoted cold reading in this instance, or effective underinflation will result.

Underinflation will cause overheating of the tyre owing to excessive flexing of the casing, and the tread will not sit correctly on the road surface. This will cause a consequent loss of adhesion and excessive wear, not to mention the danger of sudden tyre failure due to heat build-up.

Overinflation will cause rapid wear of the centre part of the tyre tread coupled with reduced adhesion, harsher ride, and the danger of shock damage occurring in the tyre casing.

Regularly check the tyres for damage in the form of cuts or bulges, especially in the sidewalls. Remove any nails or stones embedded in the tread before they penetrate the tyre to cause deflation. If removal of a nail *does* reveal that the tyre has been punctured, refit the nail so that its point of penetration is marked. Then immediately change the wheel and have the tyre repaired by a tyre dealer. Do *not* drive on a tyre in such a condition. In many cases a puncture can be simply repaired by the use of an inner tube of the correct size and type. If in any doubt as to the possible consequences of any damage found, consult your local tyre dealer for advice.

Periodically remove the wheels and clean any dirt or mud from the inside and outside surfaces. Examine the wheel rims for signs of rusting, corrosion or other damage. Light alloy wheels are easily damaged by 'kerbing' whilst parking, and similarly steel wheels may become dented or buckled. Renewal of the wheel is very often the only course of remedial action possible.

The balance of each wheel and tyre assembly should be maintained to avoid excessive wear, not only to the tyres but also to the steering and suspension components. Wheel imbalance is normally signified by vibration through the vehicle's bodyshell, although in many cases it is particularly noticeable through the steering wheel. Conversely, it should be noted that wear or damage in suspension or steering components may cause excessive tyre wear. Out-of-round or out-of-true tyres, damaged wheels and wheel bearing wear/maladjustment also fall into this category. Balancing will not usually cure vibration caused by such wear.

Wheel balancing may be carried out with the wheel either on or off the vehicle. If balanced on the vehicle, ensure that the wheel-to-hub relationship is marked in some way prior to subsequent wheel removal so that it may be refitted in its original position (see also Section 22).

General tyre wear is influenced to a large degree by driving style – harsh braking and acceleration or fast cornering will all produce more rapid tyre wear. Interchanging of tyres may result in more even wear, but this should only be carried out where there is no mix of tyre types on the vehicle. However, it is worth bearing in mind that if this is completely effective, the added expense of replacing a complete set of tyres simultaneously is incurred, which may prove financially restrictive for many owners.

Front tyres may wear unevenly as a result of wheel misalignment. The front wheels should always be correctly aligned according to the settings specified by the vehicle manufacturer.

Legal restrictions apply to the mixing of tyre types on a vehicle. Basically this means that a vehicle must not have tyres of differing construction on the same axle. Although it is not recommended to mix tyre types between front axle and rear axle, the only legally permissible combination is crossply at the front and radial at the rear. When mixing radial ply tyres, textile braced radials must always go on the front axle, with steel braced radials at the rear. An obvious disadvantage of such mixing is the necessity to carry two spare tyres to avoid contravening the law in the event of a puncture.

In the UK, the Motor Vehicles Construction and Use Regulations apply to many aspects of tyre fitting and usage. It is suggested that a copy of these regulations is obtained from your local police if in doubt as to the current legal requirements with regard to tyre condition, minimum tread depth, etc.

25 Fault diagnosis – suspension and steering

Symptom	Reason(s)
Heavy steering	Defective steering joints Tyre pressures incorrect Incorrect suspension and/or steering geometry
Lost motion in steering	Defective steering joints Wear in the rack assembly Steering column universal joint or flexible coupling worn or loose Steering wheel loose
Wheel wobble or vibration, steering pulling to one side, swerving, excessive tyre wear	Hub bearings worn or loose Tyre pressures incorrect Suspension bushes worn Wheels out of balance, or nuts loose Braking system defect Steering box height incorrect Incorrect steering geometry Variations in underbody height Steering box or balljoint wear Incorrect tyre fitment Driveshafts worn or damaged Shock absorbers defective, or mountings worn or loose Stub axles out of alignment (rear axle)
Poor roadholding and cornering	Dampers unserviceable Tyre pressures incorrect
Excessive sensitivity to road camber	Wear in the suspension balljoints Hub bearings worn

Power-assisted steering

Lack of power assistance	Pump drivebelt slack or broken Fluid level too low Pump or regulator valve faulty Internal leak in steering rack

Chapter 12 Bodywork and fittings

For modifications, and information applicable to later models, see Supplement at end of manual

Contents

1 General description

The bodywork is of all-steel monocoque construction, the integral components being spot welded together. In addition to the normal hinged body panels, the front wings are removable, being bolted into position.

Apart from the normal cleaning, maintenance and minor body repairs, there is little that the DIY owner can do in the event of structural defects caused by collision damage or possibly rust. This Chapter is therefore devoted to the normal maintenance, removal and refitting of those parts of the vehicle body and associate components that are readily dismantled.

Although the underbody is given a protective coating when new, it is still likely to suffer from corrosion in certain exposed areas or where road dirt deposits can congeal. Light corrosion can be treated as described in Section 4 but severe rusting of a structural area in the underbody must be repaired by your Renault dealer or competent vehicle body repair shop.

2 Maintenance – bodywork and underframe

The general condition of a vehicle's bodywork is the one thing that significantly affects its value. Maintenance is easy but needs to be regular. Neglect, particularly after minor damage, can lead quickly to further deterioration and costly repair bills. It is important also to keep watch on those parts of the vehicle not immediately visible, for instance the underside, inside all the wheel arches and the lower part of the engine compartment.

The basic maintenance routine for the bodywork is washing – preferably with a lot of water, from a hose. This will remove all the loose solids which may have stuck to the vehicle. It is important to flush these off in such a way as to prevent grit from scratching the finish. The wheel arches and underframe need washing in the same way to remove any accumulated mud which will retain moisture and tend to encourage rust. Paradoxically enough, the best time to clean the underframe and wheel arches is in wet weather when the mud is thoroughly wet and soft. In very wet weather the underframe is usually cleaned of large accumulations automatically and this is a good time for inspection.

Periodically, except on vehicles with a wax-based underbody protective coating, it is a good idea to have the whole of the underframe of the vehicle steam cleaned, engine compartment included, so that a thorough inspection can be carried out to see what minor repairs and renovations are necessary. Steam cleaning is available at many garages and is necessary for removal of the accumulation of oily grime which sometimes is allowed to become thick in certain areas. If steam cleaning facilities are not available, there are one or two excellent grease solvents available, such as Holts Engine Cleaner or Holts Foambrite, which can be brush applied. The dirt can then be simply hosed off. Note that these methods should not be used on vehicles with wax-based underbody protective coating or the coating will be

removed. Such vehicles should be inspected annually, preferably just prior to winter, when the underbody should be washed down and any damage to the wax coating repaired using Holts Undershield. Ideally, a completely fresh coat should be applied. It would also be worth considering the use of such wax-based protection for injection into door panels, sills, box sections, etc, as an additional safeguard against rust damage where such protection is not provided by the vehicle manufacturer.

After washing paintwork, wipe off with a chamois leather to give an unspotted clear finish. A coat of clear protective wax polish, like the many excellent Turtle Wax polishes, will give added protection against chemical pollutants in the air. If the paintwork sheen has dulled or oxidised, use a cleaner/polisher combination such as Turtle Extra to restore the brilliance of the shine. This requires a little effort, but such dulling is usually caused because regular washing has been neglected. Care needs to be taken with metallic paintwork, as special non-abrasive cleaner/polisher is required to avoid damage to the finish. Always check that the door and ventilator opening drain holes and pipes are completely clear so that water can be drained out. Bright work should be treated in the same way as paint work. Windscreens and windows can be kept clear of the smeary film which often appears by the use of a proprietary glass cleaner like Holts Mixra. Never use any form of wax or other body or chromium polish on glass.

3 Maintenance – upholstery and carpets

Mats and carpets should be brushed or vacuum cleaned regularly to keep them free of grit. If they are badly stained remove them from the vehicle for scrubbing or sponging and make quite sure they are dry before refitting. Seats and interior trim panels can be kept clean by wiping with a damp cloth and Turtle Wax Carisma. If they do become stained (which can be more apparent on light coloured upholstery) use a little liquid detergent and a soft nail brush to scour the grime out of the grain of the material. Do not forget to keep the headlining clean in the same way as the upholstery. When using liquid cleaners inside the vehicle do not over-wet the surfaces being cleaned. Excessive damp could get into the seams and padded interior causing stains, offensive odours or even rot. If the inside of the vehicle gets wet accidentally it is worthwhile taking some trouble to dry it out properly, particularly where carpets are involved. *Do not leave oil or electric heaters inside the vehicle for this purpose.*

4 Minor body damage – repair

The colour bodywork repair photographic sequences between pages 32 and 33 illustrate the operations detailed in the following sub-sections.
Note: *For more detailed information about bodywork repair, the Haynes Publishing Group publish a book by Lindsay Porter called The Car Bodywork Repair Manual. This incorporates information on such aspects as rust treatment, painting and glass fibre repairs, as well as details on more ambitious repairs involving welding and panel beating.*

Repair of minor scratches in bodywork

If the scratch is very superficial, and does not penetrate to the metal of the bodywork, repair is very simple. Lightly rub the area of the scratch with a paintwork renovator like Turtle Wax New Color Back, or a very fine cutting paste like Holts Body + Plus Rubbing Compound to remove loose paint from the scratch and to clear the surrounding bodywork of wax polish. Rinse the area with clean water.

Apply touch-up paint, such as Holts Dupli-Color Color Touch or a paint film like Holts Autofilm, to the scratch using a fine paint brush; continue to apply fine layers of paint until the surface of the paint in the scratch is level with the surrounding paintwork. Allow the new paint at least two weeks to harden; then blend it into the surrounding paintwork by rubbing the scratch area with a paintwork renovator or a very fine cutting paste, such as Holts Body + Plus Rubbing Compound or Turtle Wax New Color Back. Finally, apply wax polish from one of the Turtle Wax range of wax polishes.

Where the scratch has penetrated right through to the metal of the bodywork, causing the metal to rust, a different repair technique is required. Remove any loose rust from the bottom of the scratch with a penknife, then apply rust inhibiting paint, such as Turtle Wax Rust Master, to prevent the formation of rust in the future. Using a rubber or nylon applicator fill the scratch with bodystopper paste like Holts Body + Plus Knifing Putty. If required, this paste can be mixed with cellulose thinners, such as Holts Body + Plus Cellulose Thinners, to provide a very thin paste which is ideal for filling narrow scratches. Before the stopper-paste in the scratch hardens, wrap a piece of smooth cotton rag around the top of a finger. Dip the finger in cellulose thinners, such as Holts Body + Plus Cellulose Thinners, and then quickly sweep it across the surface of the stopper-paste in the scratch; this will ensure that the surface of the stopper-paste is slightly hollowed. The scratch can now be painted over as described earlier in this Section.

Repair of dents in bodywork

When deep denting of the vehicle's bodywork has taken place, the first task is to pull the dent out, until the affected bodywork almost attains its original shape. There is little point in trying to restore the original shape completely, as the metal in the damaged area will have stretched on impact and cannot be reshaped fully to its original contour. It is better to bring the level of the dent up to a point which is about $\frac{1}{8}$ in (3 mm) below the level of the surrounding bodywork. In cases where the dent is very shallow anyway, it is not worth trying to pull it out at all. If the underside of the dent is accessible, it can be hammered out gently from behind, using a mallet with a wooden or plastic head. Whilst doing this, hold a suitable block of wood firmly against the outside of the panel to absorb the impact from the hammer blows and thus prevent a large area of the bodywork from being 'belled-out'.

Should the dent be in a section of the bodywork which has a double skin or some other factor making it inaccessible from behind, a different technique is called for. Drill several small holes through the metal inside the area – particulary in the deeper section. Then screw long self-tapping screws into the holes just sufficiently for them to gain a good purchase in the metal. Now the dent can be pulled out by pulling on the protruding heads of the screws with a pair of pliers.

The next stage of the repair is the removal of the paint from the damaged area, and from an inch or so of the surrounding 'sound' bodywork. This is accomplished most easily by using a wire brush or abrasive pad on a power drill, although it can be done just as effectively by hand using sheets of abrasive paper. To complete the preparation for filling, score the surface of the bare metal with a screwdriver or the tang of a file, or alternatively, drill small holes in the affected area. This will provide a really good 'key' for the filler paste.

To complete the repair see the Section on filling and re-spraying.

Repair of rust holes or gashes in bodywork

Remove all paint from the affected area and from an inch or so of the surrounding 'sound' bodywork, using an abrasive pad or a wire brush on a power drill. If these are not available a few sheets of abrasive paper will do the job just as effectively. With the paint removed you will be able to gauge the severity of the corrosion and therefore decide whether to renew the whole panel (if this is possible) or to repair the affected area. New body panels are not as expensive as most people think and it is often quicker and more satisfactory to fit

a new panel than to attempt to repair large areas of corrosion.

Remove all fittings from the affected area except those which will act as a guide to the original shape of the damaged bodywork (eg headlamp shells etc). Then, using tin snips or a hacksaw blade, remove all loose metal and any other metal badly affected by corrosion. Hammer the edges of the hole inwards in order to create a slight depression for the filler paste.

Wire brush the affected area to remove the powdery rust from the surface of the remaining metal. Paint the affected area with rust inhibiting paint like Turtle Rust Master; if the back of the rusted area is accessible treat this also.

Before filling can take place it will be necessary to block the hole in some way. This can be achieved by the use of aluminium or plastic mesh, or aluminium tape.

Aluminium or plastic mesh or glass fibre matting, such as the Holts Body + Plus Glass Fibre Matting, is probably the best material to use for a large hole. Cut a piece to the approximate size and shape of the hole to be filled, then position it in the hole so that its edges are below the level of the surrounding bodywork. It can be retained in position by several blobs of filler paste around its periphery.

Aluminium tape should be used for small or very narrow holes. Pull a piece off the roll and trim it to the approximate size and shape required, then pull off the backing paper (if used) and stick the tape over the hole; it can be overlapped if the thickness of one piece is insufficient. Burnish down the edges of the tape with the handle of a screwdriver or similar, to ensure that the tape is securely attached to the metal underneath.

Bodywork repairs – filling and re-spraying

Before using this Section, see the Sections on dent, deep scratch, rust holes and gash repairs.

Many types of bodyfiller are available, but generally speaking those proprietary kits which contain a tin of filler paste and a tube of resin hardener are best for this type of repair, like Holts Body + Plus or Holts No Mix which can be used directly from the tube. A wide, flexible plastic or nylon applicator will be found invaluable for imparting a smooth and well contoured finish to the surface of the filler.

Mix up a little filler on a clean piece of card or board – measure the hardener carefully (follow the maker's instructions on the pack) otherwise the filler will set too rapidly or too slowly. Alternatively, Holts No Mix can be used straight from the tube without mixing, but daylight is required to cure it. Using the applicator apply the filler paste to the prepared area; draw the applicator across the surface of the filler to achieve the correct contour and to level the filler surface. As soon as a contour that approximates to the correct one is achieved, stop working the paste – if you carry on too long the paste will become sticky and begin to 'pick up' on the applicator. Continue to add thin layers of filler paste at twenty-minute intervals until the level of the filler is just proud of the surrounding bodywork.

Once the filler has hardened, excess can be removed using a metal plane or file. From then on, progressively finer grades of abrasive paper should be used, starting with a 40 grade production paper and finishing with 400 grade wet-and-dry paper. Always wrap the abrasive paper around a flat rubber, cork, or wooden block – otherwise the surface of the filler will not be completely flat. During the smoothing of the filler surface the wet-and-dry paper should be periodically rinsed in water. This will ensure that a very smooth finish is imparted to the filler at the final stage.

At this stage the 'dent' should be surrounded by a ring of bare metal, which in turn should be encircled by the finely 'feathered' edge of the good paintwork. Rinse the repair area with clean water, until all of the dust produced by the rubbing-down operation has gone.

Spray the whole repair area with a light coat of primer, either Holts Body + Plus Grey or Red Oxide Primer – this will show up any imperfections in the surface of the filler. Repair these imperfections with fresh filler paste or bodystopper, and once more smooth the surface with abrasive paper. If bodystopper is used, it can be mixed with cellulose thinners to form a really thin paste which is ideal for filling small holes. Repeat this spray and repair procedure until you are satisfied that the surface of the filler, and the feathered edge of the paintwork are perfect. Clean the repair area with clean water and allow to dry fully.

The repair area is now ready for final spraying. Paint spraying must be carried out in a warm, dry, windless and dust free atmosphere. This condition can be created artificially if you have access to a large indoor working area, but if you are forced to work in the open, you will have to pick your day very carefully. If you are working indoors, dousing the floor in the work area with water will help to settle the dust which would otherwise be in the atmosphere. If the repair area is confined to one body panel, mask off the surrounding panels; this will help to minimise the effects of a slight mis-match in paint colours. Bodywork fittings (eg chrome strips, door handles etc) will also need to be masked off. Use genuine masking tape and several thicknesses of newspaper for the masking operations.

Before commencing to spray, agitate the aerosol can thoroughly, then spray a test area (an old tin, or similar) until the technique is mastered. Cover the repair area with a thick coat of primer; the thickness should be built up using several thin layers of paint rather than one thick one. Using 400 grade wet-and-dry paper, rub down the surface of the primer until it is really smooth. While doing this, the work area should be thoroughly doused with water, and the wet-and-dry paper periodically rinsed in water. Allow to dry before spraying on more paint.

Spray on the top coat using Holts Dupli-Color Autospray, again building up the thickness by using several thin layers of paint. Start spraying in the centre of the repair area and then, with a single side-to-side motion, work outwards until the whole repair area and about 2 inches of the surrounding original paintwork is covered. Remove all masking material 10 to 15 minutes after spraying on the final coat of paint.

Allow the new paint at least two weeks to harden, then, using a paintwork renovator or a very fine cutting paste such as Turtle Wax New Color Back or Holts Body + Plus Rubbing Compound, blend the edges of the paint into the existing paintwork. Finally, apply wax polish.

5   Major body damage – repair

The principle of construction of these vehicles is such that great care must be taken when making cuts, or when renewing major members, in order to preserve the basic safety characteristics of the structure. In addition, the heating of certain areas is not advisable.

In view of the specialised knowledge necessary for this work, and of the alignment jigs and special tools frequently required, the owner is advised to consult a specialist body repairer.

6   Bonnet – removal, refitting and adjustment

1   Raise the bonnet, and have an assistant hold it whilst removing the bolts from the swan neck brackets. Pencil round the brackets first, to assist refitting (photo).
2   Alternatively, remove the nuts at the lower end of the swan neck brackets.
3   Disconnect the screen washer pipe, and remove the bonnet.
4   Refit in reverse order. Reset the screen washer jet if necessary.

Adjustment
5   Loosen the bolts securing the swan neck brackets, and remove the access plugs, one in each wing.
6   Close the bonnet, and re-align as necessary.
7   Secure two bolts, using a socket and extension through the access holes.

8   Open the bonnet, tighten the remaining bolts, and refit the access plugs.
9   Adjust the flush fit of the bonnet by increasing or reducing the heights of the pads on top of the wings, adjacent to the headlamp top brackets, using a screwdriver.

7   Front wing – removal and refitting

1   Disconnect the battery earth lead.
2   Gently ease off the trim.
3   Remove the light unit complete (see Chapter 10).
4   Disconnect the bracket from the front bumper.
5   Remove the top fixing bolts, and those to the lower front pillar. Free the wing closure panel, and straighten the tags where necessary.
6   Remove the wing.
7   To refit, reverse the removal procedure, taking care that all seams are sealed with an appropriate sealant.
8   Ensure that the foam rubber strips are renewed if in poor condition.
9   Correctly align the wing before fastening.
10 Apply a sound-deadening compound inside the wing, and paint as required.

8   Doors, front and rear – removal, refitting and adjustment

Removal
1   Obtain the services of an assistant to hold the door.
2   Punch out the door check pin, using a $\frac{3}{16}$ in (5 mm) diameter pin punch (photo).
3   Punch out the top and bottom hinge pins with a suitable sized drift (photo).
4   Remove the door.

Refitting
5   Reverse the removal procedure. Use a suitable pin through one hinge to obtain alignment, whilst refitting the other hinge pin.

Adjustment, front doors
6   Remove the trim from the door pillar.
7   Adjust the hinge packing pieces as necessary, to obtain correct alignment of the door and body.
8   Adjust the height by using the slots in the pillar.
9   Tighten the hinge fixings, and refit the trim.

Adjustment, rear doors
10 Proceed as for the front doors, noting that the safety belt retractor casing must be removed to give access to the hinge nuts.

9   Boot lid, counterbalance and lock – removal and refitting

Boot lid
1   Mark round the swan neck brackets to assist refitting, and remove the four bolts to remove the lid.
2   To refit, first lift a portion of the boot rubber weatherstrip to check the snubber condition. Renew as necessary, and refit the weatherstrip.
3   Refit the lid by refitting the bolts.

6.1 The bonnet left-hand support bracket and bolts

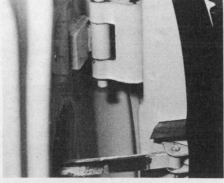
8.2 A door check and upper hinge, showing pivot pins

8.3 A door lower hinge

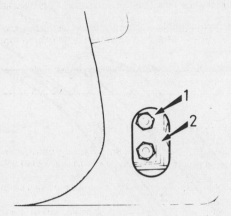

Fig. 12.1 Front door hinge fixings (Sec 8)

1   Hinge fixing nut   2   Packing pieces

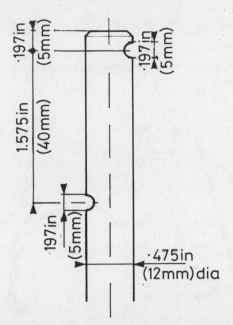

Fig. 12.3 Special bar, to relieve the load on the torsion bars (Sec 9)

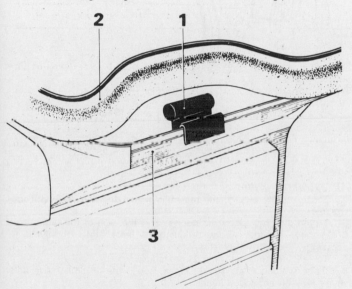

Fig. 12.2 Boot lid snubber arrangement (Sec 9)

1   Lid snubber   3   Boot flange
2   Weatherstop (partially removed)

Counterbalance

4   Remove the lid (see paragraph 1).
5   Make up a bar as shown in Fig. 12.3, and use this to take the load from the torsion bars (Fig. 12.4).
6   Remove the bars and plates by removing the securing nuts.
7   To refit, reverse the removal procedure.

Lock

8   Open the lid, and via an aperture in the lining remove the clip holding the knob to the outer panel.
9   Remove the pin and remove the locking fingers.
10 To refit, reverse the removal procedure.

10 Front grille – removal and refitting

1   Remove the two (or three) screws, depending upon the model, which secure the grille to the crossmember (photo).
2   Lift the grille dowels free at the bottom, and remove the grille.
3   Check that the sleeves which receive the dowels are in position and are serviceable.
4   To refit, insert the dowels and secure the grille with the two (or three) screws.

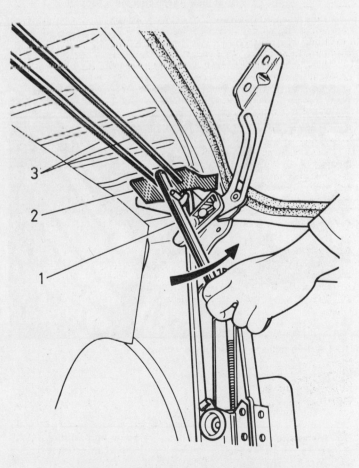

Fig. 12.4 Load-relieving bar in use (Sec 9)

1   Special bar   3   Torsion bars
2   Plates

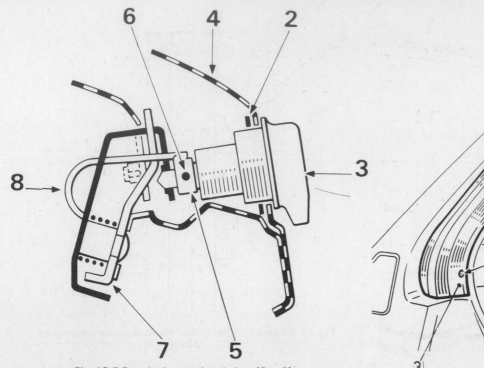

Fig. 12.5 Boot lock – sectional view (Sec 9)

2   Clip   6   Pin
3   Knob   7   Lock mechanism
4   Boot lid   8   Spring
5   Locking finger

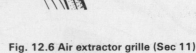

Fig. 12.6 Air extractor grille (Sec 11)

1   Door aperture seal   2   Screws
    (partly removed)   3   Embellisher

Air extractor grilles

4   These are situated to the rear of the rear doors, in line with the windows.
5   To remove, pull away the door seal in the area of the grille.
6   Remove the four screws retaining the grille, and withdraw it. If necessary, take off the plastic nuts.
7   To refit, reverse the removal procedure.
8   If the welded rivets should break, pop rivets of similar size and shape may be employed instead.

12 Rubbing strip, front doors – removal and refitting

1   Pull the rubbing strip from the clips, ensuring that the metal insert in the strip is not damaged.
2   Turn the clips through 90° with pliers, thus disengaging them, and releasing the window lower embellisher.
3   Commence refitting with the embellisher and fit the clips through the door frame slot. Turn them through 90° to secure.
4   Refit the rubbing strip to the clips.

13 Body rubbing strips and plastic window trim (where fitted) – removal and refitting

Body rubbing strips

1   To remove, carefully prise away the mouldings using a screwdriver. Protect the paint as appropriate.
2   Take the clips from the moulding, if necessary.
3   Commence refitting by placing the clips in the moulding slots.
4   Squeeze the clips to reduce their size sufficiently, and progressively introduce them to the holes in the panel.
5   Finally press the clips home.

Plastic window trim

6   To remove, proceed as in paragraph 1.
7   To refit, snap the trim home over the clips.

10.1 A front grille-to-crossmember screw

11 Air intake and extractor grilles – removal and refitting

Air intake grilles

1   These are situated in front of the windscreen, adjacent to the wiper bosses.
2   To remove, protect the paint and use a screwdriver between the panel and grille. Carefully lever the grille out.
3   To refit, enter the lugs in the panel, then tap home smartly with the hand.

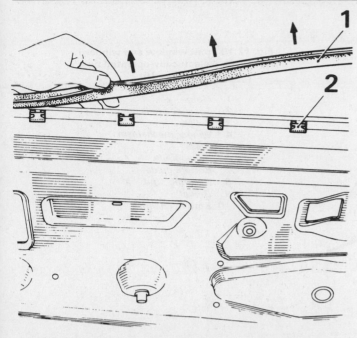

Fig. 12.7 Front door rubbing strip being removed (Sec 12)

1 *Rubbing strip* 2 *Clips*

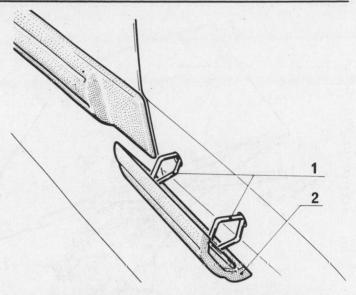

Fig. 12.8 Body rubbing strip, partly removed (Sec 13)

1 *Clips* 2 *Moulding*

14 Roof embellisher – removal and refitting

1   Remove the embellisher from the clips by using a screwdriver, protecting the blade to avoid damage to the paint.
2   Detach the end pieces.
3   To refit, ensure that the clips are in position over the roof flange. Clip the embellisher into place.
4   Slip the end fittings inside the embellisher, and clip them to the flange.

15 Bumper bars – removal and refitting

Front bumper
1   Move the air filter aside to provide access to the left-hand mountings.
2   Remove the 4 bolts (2 per bracket) holding the front brackets, from inside the engine compartment.
3   Pull the bar from the side mountings.
4   If necessary, remove the side mounting bolt and withdraw the component parts, noting arrangement details.
5   The rubber trim on the bumper bar is removed, if necessary, by grinding off the rivet heads at the back.
6   To refit, reverse the removal procedure. Align the bar before finally tightening the bolts

Rear bumper
7   To remove, proceed as in paragraphs 2 to 5, but working inside the boot.
8   Refit as described in paragraph 6.

16 Front windows and window winders – removal and refitting

Winder mechanism – removal
1   Lower the winder $8\frac{11}{16}$ in (22 cm)
2   Remove the trim panel. See Section 22.
3   Disconnect the electrical plug at the winder mechanism, where applicable.
4   Remove the four nuts securing the winder mechanism (photo), and push rearwards to free the three rollers from the runners.
5   Remove the mechanism via the door bottom opening.

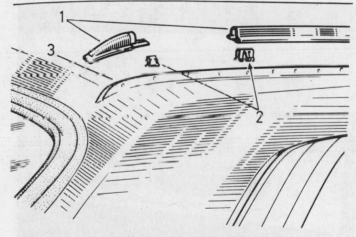

Fig. 12.9 Roof embellisher (Sec 14)

1 *Embellisher* 3 *Roof flange*
2 *Clips*

Windows – removal
6   Proceed as in paragraphs 1 to 5, then tip the front of the glass down.
7   Lift the rear edge, thus releasing the glass from the front channels.
8   Remove the nut securing the lower roller guide and remove the guide, noting the dowel.

Refitting
9   If the bottom glass channel is to be renewed, position it $2\frac{3}{64}$ in (52 mm) from the glass rear edge, using a mallet and a piece of wood.
10   Loosely refit the lower roller guide.
11   Place the winder mechanism in the bottom of the door.
12   Tilt the glass, front end down, insert it in the door, and pivot it to seat in the glass channels.
13   Invert the rollers in their guides.
14   Fit the winder mechanism into place, and loosely fit the securing nuts.
15   Reconnect the wiring plug, where applicable. Alternatively, loosely refit the winder handle.
16   Raise the window, and tighten the winder mechanism nuts and bottom guide nut.
17   Refit the remaining items.

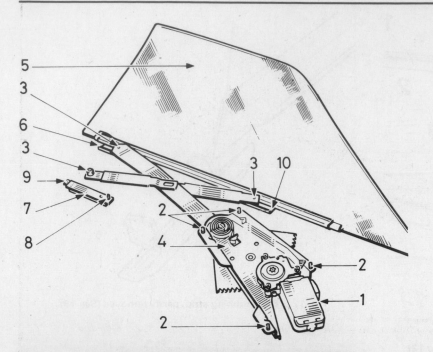

Fig. 12.10 Front window and winder mechanism – electrically-operated type shown (Sec 16)

1   Motor
2   Securing studs
3   Rollers
4   Winder mechanism
5   Glass
6   Roller guide
7   Bottom guide
8   Securing stud
9   Dowel
10 Bottom channel

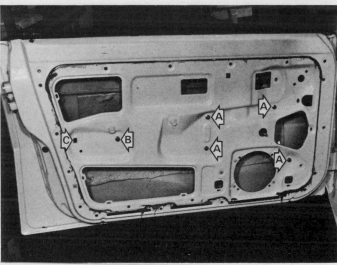

16.4 A door with trim panel removed, showing winder gear securing nuts (A), bottom roller guide nut (B) and bottom roller guide dowel (C)

17 Rear windows and window winders – removal and refitting

Winder mechanism – removal

1   Remove the door trim (see Section 22).
2   Lower the windows to 9¾ in (25 cm) above the door frame.
3   Remove the three nuts which secure the winder, and push the studs clear.
4   Move the mechanism rearwards until the roller leaves the runner.
5   Remove the mechanism via the aperture in the door. Lower the window.

Winder mechanism – refitting

6   To refit, reverse the removal procedure. Lubricate the mechanism before refitting, and do not fully tighten the securing nuts until the window is wound right up.

Windows – removal

7   Proceed as in paragraphs 1 to 5.

8   Remove the fixings at the top and bottom of the vertical window pillar, and withdraw the pillar.
9   Remove the rubber channel around the wind-down window aperture.
10 Remove the fixed window by levering carefully with two screwdrivers.
11 Take out the wind-down window.

Windows – refitting

12 If necessary, fit the bottom channel and rubber profile to the bottom of the glass, 2 in (50 mm) from the back, using a block of wood and a mallet.
13 Reverse the procedure given in paragraphs 8 to 11.
14 Position the window 9¾ in (25 cm) above the door frame.
15 Lubricate the winder, pass it through the door aperture, and invert the rollers in the guide.
16 Refit the winder nuts, but do not tighten.
17 Wind up the window. Tighten the securing nut.
18 Refit the door trim (see Section 22).

18 Door locks – description, removal and refitting

Description

1   A mechanical lock may be fitted. Alternatively an electro-magnetic type may be employed, consisting of the mechanical lock plus a changeover switch and an electro-magnetic actioner. The lock is thus operated either by the front door lock barrels or by the electro-magnetic actioner. Further details are given in Chapter 10. Owners of cars not fitted with electrically-operated locks should disregard those instructions below which are not applicable.

Removal

2   Close the window, and detach the locking pushbutton.
3   Remove the door trim panel, and the plastic sheet (Section 22).
4   Disconnect the electrical connector from the actioner.
5   Pull the changeover switch from the lock barrel.
6   Disconnect the rod at the remote control by carefully prising it from the clip. Free the rod from the door frame bearing.
7   Withdraw the three screws from the door latch.
8   Remove the actioner retaining bolt.
9   Remove the complete assembly via the door frame bottom aperture.

Refitting

10 To refit, reverse the removal procedure. Check that the rod to the

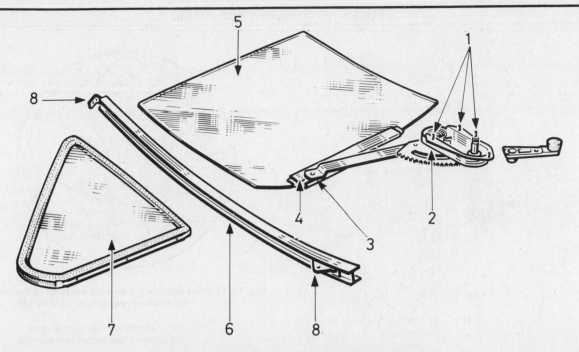

Fig. 12.11 Rear window and winder mechanism (Sec 17)

1	Securing studs	3 Roller	5 Window	7 Fixed window
2	Mechanism	4 Runner	6 Pillar	8 Channel fixings

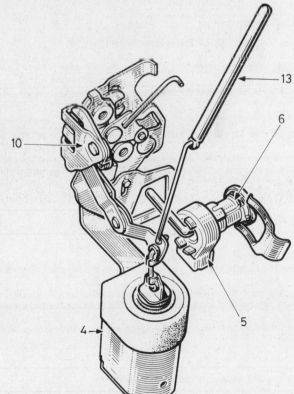

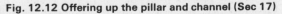

Fig. 12.12 Offering up the pillar and channel (Sec 17)

1 Pillar	3 Attachment points
2 Channel	

Fig. 12.13 Front door lock (electrically-operated type shown)
(Sec 18)

4 Actioner	10 Remote control plate
5 Changeover block	13 Locking tell-tale
6 Barrel	

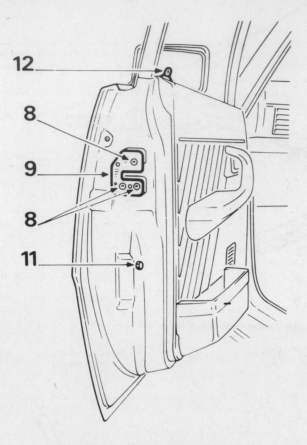

Fig. 12.14 Front door latch (Sec 18)

8   Securing screws   12   Locking tell-tale
9   Latch   (protector fitted)
11 Actioner securing bolt

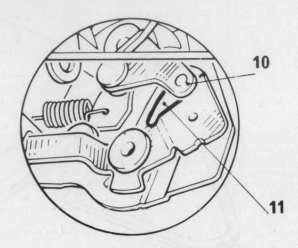

**Fig. 12.15 Rear door lock – correct position of the handle control
lever when reassembling (Sec 18)**

10 External handle control lever
11 Lever to be within this triangle

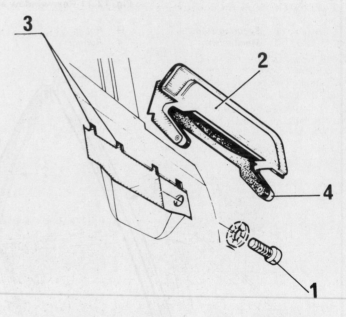

Fig. 12.16 Exterior door handle – exploded view (Sec 19)

1   Securing screw and washer   3   Locating notches
2   Handle   4   Handle lugs

remote control is situated between the window winder and door
frame.
11 In the case of a rear door lock, check that the external handle
control lever is in the position shown in Fig. 12.15 before reassembling
the lock.

19 External door handle – removal and refitting

1   Lock the doors.
2   Take the screw from the door flange.
3   Push the handle forward, releasing it from the notches, tilt it down,
and pull away.
4   To refit, lock the door (if not already done).
5   Insert the handle, lugs leading, and tilt to bring the notches and
lugs into line.
6   Push the handle rearwards to engage the lugs and notches, and fit
the screw.

20 Door lock barrel – removal and refitting

1   Remove the handle (see Section 19).
2   Working through the handle aperture, remove the forked clip on
the barrel holder.
3   Remove the barrel holder.
4   On electro-magnetic locks only, disconnect the changeover assembly.
5   To change the barrel, tap out the pin to disconnect the barrel
locking finger.
6   To refit, reverse the removal procedure.

21 Dashboard – removal and refitting

Removal
1   Remove the accessories plate and bracket (see Chapter 10).
2   Remove the console.
3   Remove two half cowls round the steering column.
4   Remove the two top console bracket bolts.
5   Remove the steering wheel.
6   Disconnect the speedometer cable, and the three wing plugs
behind the instrument panel. Refer to Chapter 10 if necessary.
7   Remove the glove compartment and bracket, ensuring that the
illumination wires are disconnected before removing the bracket.
8   Take out the three dashboard fixing bolts, unclip the end furthest
from the steering column, unclip the other end, and remove the
dashboard.

Refitting

9 To refit, reverse the removal procedure. Clip the dashboard into place at the steering column end first.

22 Door trim panel – removal and refitting

Front door panel, removal

1 Where applicable, prise out the window winder handle cover, take off the centre nut using a suitable socket, and remove the handle (photos).
2 Where applicable, unscrew the ring on the exterior mirror control.
3 Remove the two screws from the armrest and remove it, tilting $\frac{1}{4}$ turn where applicable to free the peg(s) (photo).
4 Depending upon model, remove one or two cross-head screws from the door pocket, lift it over the locating pegs, and remove the pocket (photo).
5 Where applicable, remove the two door pocket locating pegs and screws.
6 Remove the door locking button or tell-tale cover, screwing or levering off depending upon model.
7 Remove the screw from the remote control assembly, and lift it away by turning slightly to free the internal connection (photo).
8 Remove the four screws in the speaker grille, and lift it away (photo).
9 Carefully lever out the panel snap fixings at the bottom and sides, beginning at the bottom.
10 Lift the panel over the door locking button post and door rim, and remove it.
11 Unstick the plastic sealing sheet.

Rear door panel, removal

12 Proceed basically as in paragraphs 1 to 11. Remove the embellisher on the remote control assembly, if applicable.

Front and rear panels, refitting

13 To refit, reverse the removal procedure. Ensure that the plastic sheet is properly refitted, using a suitable adhesive mastic. Renew the sheet if it is damaged.

23 Luggage space internal trim, Estate car – removal and refitting

Rear door and side panel internal trim

1 To remove, work round the panel and ease out the snap fasteners carefully with a screwdriver. Protect the paint as appropriate.
2 To refit, position the panel and tap the fasteners smartly into place.

Rear floor covering

3 Remove the screws from the rear protecting strip. Lift the strip away.
4 Remove the floor covering.
5 To refit, reverse the removal procedure.

24 Seats, front and rear – removal and refitting

Front seats

1 Slide the seat to obtain access to the front slider retaining bolts, and remove them.
2 Under the floor, remove the rear slide retaining nuts.
3 Free the clip on the seat tensioner with a screwdriver blade.
4 To refit, reverse the removal procedure.

Rear seats

5 Remove the cushion by pulling the tongue to unlock. Lift the front, freeing the frame legs from the floor mounting, and remove the cushion.
6 To refit the cushion, push it under the seat back. Tap the base with the hand to position the frame legs in the floor mounting, and lock with the clip.
7 To remove the seat back, lift the lugs at the sides.
8 Turn the knob (0 in Fig. 12.19) one quarter turn.

22.1a Prising out the winder handle cover

22.1b Removing the winder handle centre nut

22.3 Removing the armrest

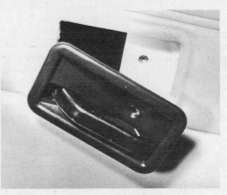
22.4 Removing a door pocket

22.7 The remote control assembly, partially removed

22.8 Removing a screw from the speaker grille

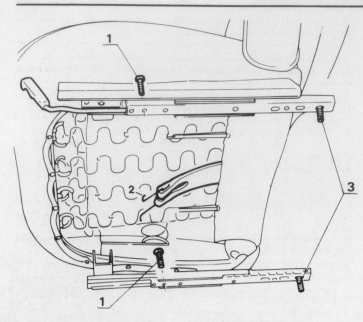

Fig. 12.17 Front seat – underneath view (Sec 24)

1   Securing bolts   3   Securing studs
2   Tensioner clip

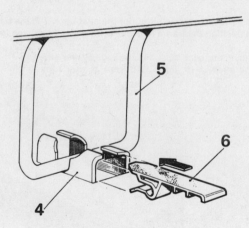

Fig. 12.18 Rear seat cushion (Sec 24)

4   Floor mounting   6   Locking clip
5   Cushion frame legs

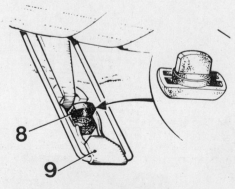

Fig. 12.19 Seat back securing clip (Sec 24)

8   Knob   9   Frame

9   Raise the seat back to free the two mountings above.
10   To refit, reverse the removal procedure.

25 Headrests – adjustment, removal and refitting

Adjustment of the sliding friction
1   Remove the rest (see below), and turn the nuts on the mounting bars to decrease the friction.
2   Reset the height of the mountings, and resecure the nuts.
3   Refit the rest.

Removal and refitting
4   To remove, turn the sleeves in the seat one quarter turn, and withdraw the headrest.
5   Refitting is the reverse of the removal procedure.

Fitting for the first time
6   To fit a rest where none is thus far fitted, prise out each plug in the top of the seat.
7   Insert the collars in the sleeves. Fit the one with an arrow on the RH side. Ensure that the base of the triangle on both sleeves enters the seat first, and that the triangles face to the rear. Once fitted, the sleeves cannot be removed intact.

26 Front and rear screens – removal and refitting

Renewing a broken windscreen is one of the few tasks which the DIY owner is advised to leave to an expert. The fitting charge made by a windscreen specialist is insignificant compared with the cost which will be incurred if a new screen is accidentally broken.
For the owner who wishes to do the job himself, the following guidance is given.

Removal
1   The two types of screen used are either a toughened or a laminated type. These carry identifications of AS1 (laminated) or AS2 (toughened).
2   Where the glass has shattered, stick a self-adhesive paper sheet to the screen on both sides, before removal.
3   Cover the bonnet, and all apertures both inside and outside the car, to prevent scratching and the ingress of broken glass.
4   On rear screens, disconnect the demister wires.
5   Obtain assistance, and use a round tool to lever the rubber seal over the flange of the body. Push the screen out from the inside.

Refitting
6   With the screen laid flat, position the rubber seal round the screen. Fit the embellisher (if applicable), using a silicone lubricant to aid installation.
7   Feed strong string or thin cord round the groove of the seal, allowing the ends to hang down at the centre, and to overlap by about 8 in (200 mm).
8   Locate the lower edge of the screen surround in the screen opening, allow the string ends to hang inside the car, and have an assistant push against the screen whilst the string is pulled out of the groove. Take the string round one side to the top centre, and then the other, pulling it at right angles to the windscreen frame flange.
9   Always use a new seal, to avoid the possibility of water leaks.
10   Reconnect the demister wires (rear screen only).

27 Heated rear screen – renovation

1   A repair paint is available from Renault agents (part No 7701 400 794) for repairing the heating resistance. Whether or not the screen needs to be removed from the vehicle depends upon accessibility of the defect.
2   A voltmeter or ohmmeter can be used to assist with location of defects. Look for an open-circuit or a sudden change in voltage, reading along the resistance wires.
3   To repair, degrease the affected area and wipe dry.
4   Apply tape at each side of the conductor, to mask the adjacent area.

5   Shake the repair paint thoroughly, and apply a thick coat with a fine paint brush. Allow to dry between coats, and do not apply more than three.
6   Allow to dry for at least one hour, before removing the tape.
7   Rough edges may be trimmed with a razor blade if necessary, after a drying time of some hours.

28 Heater control panel – removal and refitting

1   Disconnect the battery.
2   Remove the console, unscrew the heater control panel, and remove the console bracket. Leave the glove compartment in place.
3   Remove the lower steering column half cowl.
4   Remove the accessories plate bracket, leaving the wires connected.
5   Pull off the rheostat and control panel illumination leads.
6   Remove the clips from the five control cables, and disconnect the cables from the panel.
7   Remove the panel.
8   Refitting is the reverse of the removal procedure.

29 Heater cables – adjustment

Heater valve cable
1   Proceed as in Section 28, paragraphs 1 to 5, and draw the panel away.
2   With the cable connected to the heater valve and the sliding knob, close the heater valve (Fig. 12.22).
3   Slide the knob to the right, leaving a clearance of $\frac{1}{8}$ to $\frac{5}{32}$ in (3 to 4 mm), and clamp the cable using clip (19), Fig. 12.21.
4   Check the action of the control, and that the valve closes properly.

Ventilation flap cable
5   With the cable connected as shown in Fig. 12.23, close the ventilation flap.
6   Refer to Fig. 12.21 and remove clip (26). Slide the knob (15) to the right, leaving a $\frac{1}{8}$ to $\frac{5}{32}$ in (3 to 4 mm) clearance. Clamp the cable with clip (26).
7   Check the action of the control, and that the flap closes properly.

Airflow flap cable
8   With the cable connected as shown in Fig. 12.24, close the heater

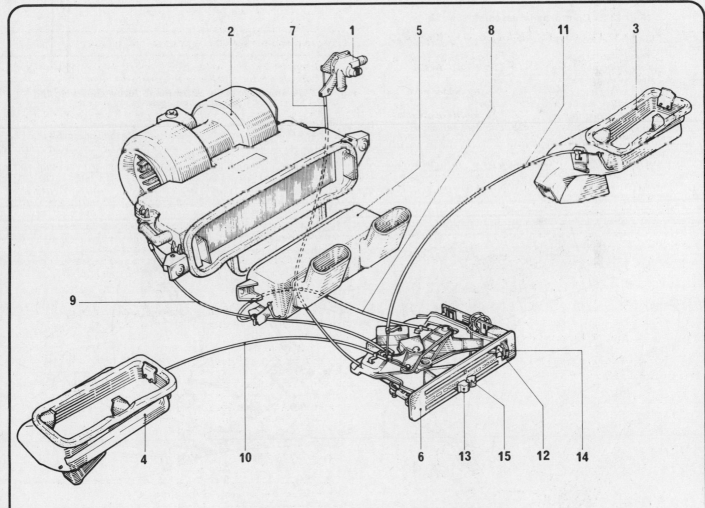

Fig. 12.20 Heating and ventilation system (Secs 28 to 34)

1 Heater valve	5 Ventilation box (fresh air only)	9 Airflow flap cable	12 Distribution knob (head – feet)
2 Heater casing (with fan motor and matrix)	6 Control panel	10 LH side heating duct cable	13 Heater valve knob
3 RH heating duct	7 Heater valve cable	11 RH side heating duct cable	14 Fan motor knob
4 LH heating duct	8 Ventilation flap cable		15 Fresh air entry knob

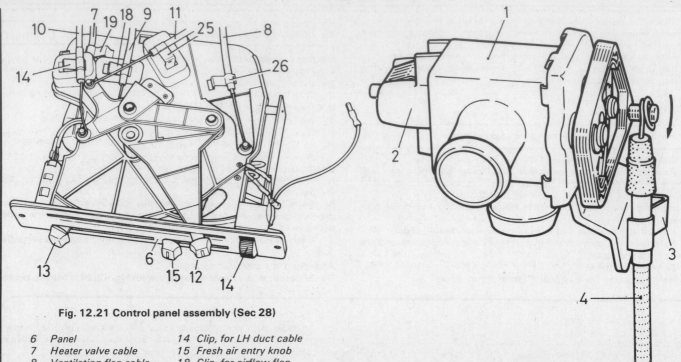

Fig. 12.21 Control panel assembly (Sec 28)

6   Panel
7   Heater valve cable
8   Ventilation flap cable
9   Airflow flap cable
10 LH heating duct cable
11 RH heating duct cable
12 Head/feet distribution
    knob
13 Heater valve knob

14 Clip, for LH duct cable
15 Fresh air entry knob
18 Clip, for airflow flap
    cable
19 Clip, for heater valve
    cable
25 Clip, for RH duct cable
26 Clip, for ventilation
    cable

Fig. 12.22 Heater valve cable attachment. Arrow shows closing direction (Sec 29)

1   Valve body
2   Valve clips

3   Cable clip
4   Heater valve cable

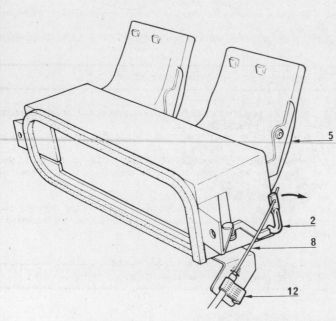

Fig. 12.23 Ventilation box and cable (Sec 29)

2   Flap
5   Box

8   Cable
12 Cable clip

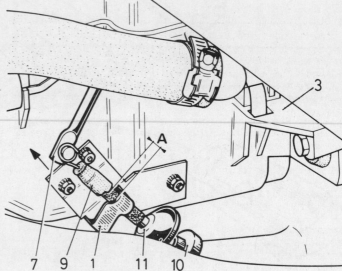

Fig. 12.24 Airflow flap arrangement (Sec 29)

1   Clip
3   Heater casing
7   Airflow flap lever
9   Cable

10 Grommet
11 Washer
A   See text

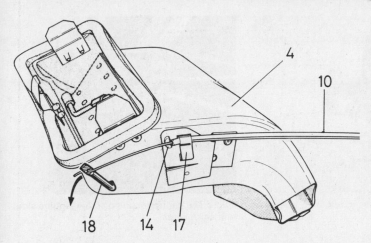

Fig. 12.25 Heater duct, LH. Arrow shows direction to close flap (Sec 29)

4 Duct	17 Cable clip
10 Cable	18 Duct flap
14 Boss for cable	

unit airflow flap.
9   Refer to Fig. 12.21 and remove clip (18). Slide knob (12) to the right, leaving a clearance of $\frac{1}{8}$ to $\frac{5}{32}$ in (3 to 4 mm), and clamp the cable using clip (18).

LH heater duct cable
10 With the cable connected as shown in Fig. 12.25, set the heating duct in the closed position (arrowed). Leave off the clip (17).
11 Referring to Fig. 12.21, move knob (12) to the left, leaving a clearance of $\frac{1}{8}$ to $\frac{5}{32}$ in (3 to 4 mm). Clamp the cable using clip (17), Fig. 12.25.

RH heater duct cable
12 Proceed as in paragraphs 10 and 11, using the relevant items.

30 Heater cables — removal and refitting

Heater valve cable
1   Remove the heater control panel (Section 28).
2   Disconnect cable (7) at the sliding knob (13). See Fig. 12.21.
3   On the valve, free the clip (3) and disconnect the cable. See Fig. 12.22.
4   To refit, renew the scuttle grommet if this is in poor condition, and feed the end of the cable with the rubber seal through the grommet from the passenger side.
5   Fit the cable eye to the lever on the heater valve.
6   Route the cable above the accelerator cable, and attach it to the lever on the knob.
7   Adjust the cable (see Section 29).
8   Refit the heater control panel (Section 28).

Ventilation flap cable
9   Remove the heater control panel (Section 28).
10 Remove clip (26) on the control panel, and disconnect the cable (8) from knob (15) (Fig. 12.21).
11 Remove the clip (12), Fig. 12.23, and release the cable from the ventilation flap. Remove cable.
12 To refit, reverse the removal procedure. Ensure that the outer cable at the flap end butts against the boss.

Airflow flap cable
13 Remove the heater control panel (Section 28).
14 Remove clip (18) on the control panel, and disconnect the cable (9) from knob (12). See Fig. 12.21.
15 Remove clip (1) from the cable, and release the cable from the airflow flap. Remove the cable. See Fig. 12.24.
16 To refit the cable, proceed as described in paragraph 4.

17 Fit the eye on the cable to the airflow flap.
18 Secure the cable with clip (1). Clearance A must be $\frac{5}{64}$ to $\frac{1}{8}$ in (2 to 3 mm) (Fig. 12.24).
19 Push washer (11) over the grommet (10), and fit the cable to the lever (Fig. 12.24).
20 Refit the heater control panel.

LH heater duct cable
21 Proceed as in Section 28, paragraphs 1 to 5.
22 Remove the clip (17), and disconnect cable (10) from the panel sliding knob (12) (Figs. 12.21 and 12.25).
23 Leaving the duct in place, remove the clip (17). See Fig. 12.25.
24 Disconnect the end of the cable (10) from the lever on the flap and remove the cable, noting the hole in the pedal assembly bracket through which it is routed.
25 To refit, feed the cable through the hole previously noted, and attach the bayonet end to the flap lever (18), Fig. 12.25.
26 With the cable outer against the boss, refit the clip (17).
27 Run the cable above the accelerator cable and attach the free end to the sliding lever (12) (Fig. 12.21).
28 Adjust the cable (see Section 29).
29 Check that the mechanism operates freely, and that flap (18) closes fully (Fig. 12.25).

RH heater duct cable
31 Proceed as in paragraphs 20 to 29, but substituting the components which relate to the RH cable and duct. Ignore the remarks concerning the pedal assembly bracket.

31 Heater valve — removal and refitting

1   Clamp off the coolant hoses to the valve.
2   Disconnect the hoses from the valve.
3   Remove the clip on the valve cable, and free the cable from the valve.
4   Pull the valve forward and out, freeing it from the securing clips.
5   To refit, offer up the new valve until secured by the clips.
6   Refit the hoses, and remove the clamps.
7   Refit the cable to the valve, and adjust it as described in Section 29.
8   Bleed the cooling system (see Chapter 2) and top up if necessary.

32 Heating ducts — removal and refitting

1   Disconnect the battery.
2   Remove the console and the glove compartment, or the accessory plate bracket, as appropriate.
3   Remove the glove compartment bracket (if applicable), noting the screw on the ventilation box.
4   Remove the clip securing the outer cable to the duct, then disconnect the inner cable. It may be necessary to disconnect the control panel.
5   Pull the duct down to remove it.
6   Refitting is the reverse of the removal procedure. Tap the duct smartly into position. Adjust the cable on completion as described in Section 29.

33 Heater motor — removal and refitting

Removal
1   Disconnect the battery.
2   Clamp off the hoses to the heater matrix, and disconnect them. (Apply the clamp to the LH hose on the engine side of the bleed screw).
3   Take the clip from the airflow flap cable, and disconnect the cable (photo).
4   Disconnect the electrical lead to the fan motor, and the screen wiper junction block.
5   Remove the screws (3 positions) securing the casing on the engine side (photos).
6   Under the dashboard on the passenger side, remove the nut

33.3 The airflow flap cable and clip

33.5a One of the bulkhead retaining screws (engine side)

33.5b The centre retaining screw (engine side) and earth wires

33.6a A bulkhead retaining screw (passenger side bulkhead)

33.6b A bulkhead retaining screw (driver's side bulkhead)

33.8a A heater case clip

33.8b The heater case centre screw

33.8c The halves of the heater case, separated

securing the motor casing. Similarly remove the nut on the driver's side, after removing the lower half steering column cowl (photos).
7   Lift out the heater assembly complete.
8   Remove the main seal round the edge of the heater assembly, remove the clips and the centre screw, and separate the two halves of the case. (Note that the halves are stuck with a sealant) (photos).
9   Lift out the motor assembly, which cannot be repaired. If defective, obtain a replacement unit. Note the orientation of the fan blades in the case.

Refitting
10 Clean the casing joint faces.
11 Reposition the motor assembly in the case, ensuring that the direction of rotation is correct. Check that the blades do not touch the

case, particularly at the ends.
12 Lightly smear a sealer on the joint face, reposition the half casings, and refit the centre bolt and clips.
13 Check that the fans are still free to revolve.
14 Refit the main seal, and relocate the heater assembly on the bulkhead. Check that when the studs are pushed through the bulkhead, they also locate the ventilation box on the passenger side.
15 Refit the bolts on the engine side, but do not tighten. Remember to secure the earth leads for the heater and screen wiper motors.
16 Fit and tighten the two nuts on the passenger side of the bulkhead.
17 Tighten the two bolts on the engine side of the bulkhead, followed by the remaining front centre bolt.
18 Refit the remaining items in reverse order, adjusting the airflow flap cable as necessary as described in Section 29.
19 Bleed the cooling system (see Chapter 2).

34 Heater matrix – removal and refitting

1 Remove and dismantle the heater unit as described in Section 33, paragraphs 1 to 8.
2 Lift out the heater matrix and disconnect the short hose.
3 To refit, fit the hose to the matrix.
4 Fit the matrix into the lower half case.
5 Continue reassembly as described in Section 33, paragraphs 10 to 19.

35 Bonnet release cable – general

1 If cable breakage occurs, make up a hook from stiff wire about $\frac{5}{32}$ in (4 mm) in diameter, as shown in Fig. 12.26.

2 Operating through the front grille, secure the hook over the opening lever and pull it to unlock the bonnet.
3 No information is at present available covering removal and refitting of the cable or bonnet internal pull lever (photos).

36 Rear view mirrors – removal and refitting

Interior rear view mirror, bonded type

1 To refit (eg after accident or screen renewal) thoroughly degrease and dry the glass.
2 Mark the mirror position on the glass.
3 Employing Locquic NF 312 activator, apply a coat to the surfaces to be joined and leave to dry for 5 minutes.
4 Apply a coat of Loctite 312 sparingly to the mirror base.

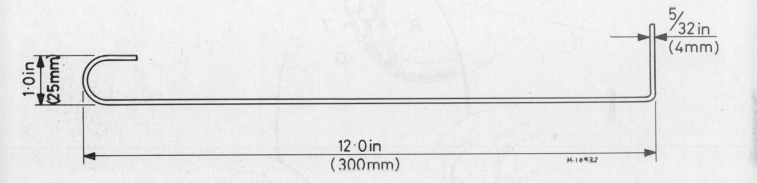

Fig. 12.26 Hook, to assist with bonnet lock operation (Sec 35)

35.3a The bonnet release pull lever

35.3b The bonnet release cable and catch

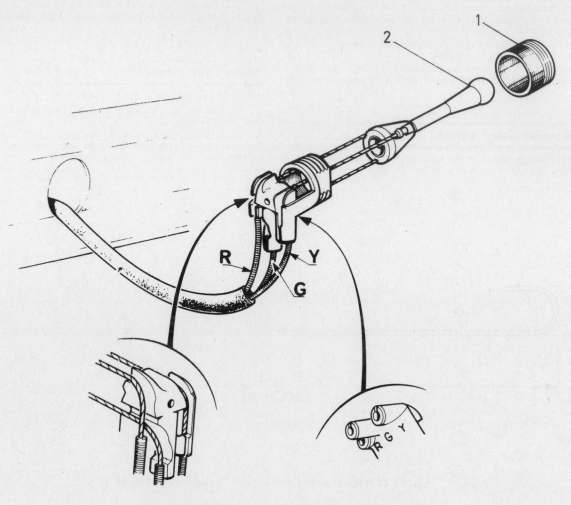

Fig. 12.27 Remote controlled rear view mirror (Sec 36)

1 Support	R Red cable	G Green cable	Y Yellow cable
2 Control lever			

5   Position the base, and press for a minute or so, locating the base 1⅞ in (45 mm) from the top edge of the screen.

6   After a few more minutes, fit the mirror to the base.

Exterior rear view mirror, remote controlled – removal

7   Close the window, and remove the door trim panel as described in Section 22, together with the plastic sheet.

8   Remove the sleeve and unscrew support (1 in Fig. 12.27).

9   Carefully prise away the cover on the mirror base, and take out the screws securing the base to the door.

10   Push the control lever inside the door frame, and pull the mirror to free the cable whilst guiding the lever (2) behind the window channel.

Exterior rear view mirror, remote controlled – refitting

11   Place the control lever and cable through the hole in the door panel, and guide the lever behind the window channel and through the inside hole in the door frame.

12   Refit the mirror base and tighten the two screws.

13   Screw the support (1) home.

14   Check for correct window and mirror functions.

15   Refit the plastic sheet and door trim.

16   Refit the cover to the base of the mirror.

37 Seat belts, front and rear – removal and refitting

Front belts, removal

1   Remove the cover plate from the retractor housing by lifting it up, and off the two studs.

2   Remove the cover from the top anchorage point, where fitted.

3   Remove the upper and lower seat belt anchorage bolts, noting the correct assembly of washers and spacers.

Rear belts, removal

4   Anchorage points are provided at different points depending upon the version.

5   Belts (where fitted) are removed simply by removing the anchorage bolts together with the fittings.

Refitting (all bolts)

6   To refit front or rear belts, reverse the removal procedure.

38 Towing attachment – general

1   A towing attachment kit is available from Renault parts department. The kit comprises all the necessary items, both mechanical and electrical, and the part number is 77 01 401 981. A cover for the ball is available separately, part number 77 01 391 176.

2   Full instructions for fitting the attachment, and for wiring the electrical socket, are contained in the kit. The towing bar should be painted before fitting.

Chapter 13 Supplement:
Revisions and information on later models

Contents

1 Introduction

This Supplement contains specifications, information and maintenance procedures for Renault 18 models manufactured from 1981 to 1986, in particular for Turbo models and North American models. Chapters 1 to 12 contain specifications, information and maintenance procedures for pre 1981 models, but some of the procedures apply to later models. It is suggested that this Supplement is referred to **before** Chapters 1 to 12, particularly if working on models manufactured after 1981.

2 Specifications

The specifications listed here are revisions of, or supplementary to, the main specifications given at the beginning of each Chapter

OHV Engines
Type

R1340 (automatic)	847-B7-21
R1350, R2350	847-A7-20
R1345, R1355	807-A7-27 or A5L
R1341, R1351:	
UK	A2M (manual), 841 (automatic)
US/Canada	841
R1342, R1352	A6M
R1348, R1358	843

Specifications are as for 841 engines in Chapter 1, except for the following. All 847 specifications are as in Chapter 1

Renault 18 Turbo

UK models

Bore:
A5L, 807 .... 3.0315 in (77.0 mm)
Cubic capacity:
A5L, 807 .... 1565 cc
Compression ratio:
A5L, 807 .... 8.6 to 1
Cylinder head (A5L, 807, A6M, 843) height:
Standard .... 3.681 in (93.5 mm)
After machining .... 3.661 in (93.0 mm)
Valve assemblies (A5L, 807, A6M, 843):
Valve seat width:
Inlet .... 0.065 in (1.65 mm)
Exhaust .... 0.073 in (1.85 mm)
Valve seat outside diameter:
Inlet .... 1.654 in (42.0 mm)
Exhaust .... 1.461 in (37.1 mm)
Valve head diameter:
Inlet .... 1.5335 in (38.95 mm)
Exhaust .... 1.3681 in (34.75 mm)
Valve guide outside diameter:
Standard .... 0.5158 in (13.1 mm)
Oversize .... 0.5256 in (13.35 mm)
Valve spring free length .... 1.66 in (42.2 mm)
Valve assemblies (A2M):
Valve head diameter:
Inlet .... 1.4193 in (36.05 mm)
Exhaust .... 1.250 in (31.75 mm)
Valve guide outside diameter:
Standard .... 0.5158 in (13.1 mm)
Oversize .... 0.5256 in (13.35 mm)

Valve timing:

	A5L, 807	A6M, 843	A2M
Inlet opens	10° BTDC	30° BTDC	10° BTDC
Inlet closes	50° ABDC	72° ABDC	54° ABDC
Exhaust opens	50° BBDC	72° BBDC	54° BBDC
Exhaust closes	10° ATDC	30° ATDC	10° ATDC

Pushrods:

	A5L, 807, A6M, 843	A2M
Length:		
Inlet	3.110 in (79.0 mm)	3.484 in (88.5 mm)
Exhaust	4.331 in (110.0 mm)	3.484 in (88.5 mm)

Cylinder liners (A5L, 807):
Bore .... 3.0315 in (77.0 mm)
Gudgeon pin (A5L, 807):
Length .... 2.6142 in (66.4 mm)
Oil filter:
A2M, A6M, 841, 843 (from June '84) .... Champion F103
A5L, 807 (from June '84 on) .... Champion F102
A5L, 807 (up to June '84) .... Champion C102

Torque wrench settings

	lbf ft	Nm
Cylinder head bolts (A5L, 807, A6M, 843):		
Cold	57 to 61	77.5 to 82.5
Hot	63 to 66	85 to 90
Cylinder head bolts (A2M):		
Cold	52 to 55	70 to 75
Hot	57 to 61	77.5 to 82.5
Crankshaft pulley bolt (A5L, 807, A6M, 843)	66	90

US/Canada models

Compression ratio:
841 .... 8.3 to 1
843 .... 8.6 to 1
Cylinder head (843) height:
Standard .... 3.681 in (93.5 mm)
After machining .... 3.661 in (93.0 mm)
Valve assemblies (843):
Valve seat width:
Inlet .... 0.059 to 0.071 in (1.5 to 1.8 mm)
Exhaust .... 0.039 to 0.053 in (1.0 to 1.35 mm)
Valve seat outside diameter:
Inlet .... 1.654 in (42.0 mm)
Exhaust .... 1.457 in (37.0 mm)
Valve head diameter:
Inlet .... 1.524 in (38.7 mm)
Exhaust .... 1.358 in (34.5 mm)
Valve guide outside diameter:
Standard .... 0.5158 in (13.1 mm)
Oversize .... 0.5256 in (13.35 mm)

Valve spring free length:
    Outer .... 2.138 in (54.3 mm)
    Inner .... 1.843 in (46.8 mm)
Valve timing (843):
    Inlet opens .... 21° BTDC
    Inlet closes .... 59° ABDC
    Exhaust opens .... 59° BBDC
    Exhaust closes .... 21° ATDC
Tappets (843):
    Outside diameter .... 0.748 in (19.0 mm)
    Oversize .... 0.756 in (19.2 mm)
Pushrods (843):
    Length:
       Inlet .... 3.071 in (78.0 mm)
       Exhaust .... 4.331 in (110.0 mm)

Torque wrench settings

	lbf ft	Nm
Cylinder head bolts (841, 843):		
Cold	57 to 61	77.5 to 82.5
Hot	63 to 66	85 to 90
Crankshaft pulley bolt (843)	66	90

OHC Engines
Type
1995 cc J6R
2165 cc 851

Number of cylinders 4

Bore .... 3.46 in (88 mm)

Stroke
J6R .... 3.22 in (82 mm)
851 .... 3.50 in (89 mm)

Compression ratio .... 9.2 to 1

Firing order 1–3–4–2

Valve clearances
Inlet .... 0.004 in (0.10 mm)
Exhaust .... 0.010 in (0.25 mm)

Cylinder head
Maximum distortion .... 0.002 in (0.05 mm)

Valves
Seat angles (included):
    Inlet .... 120°
    Exhaust .... 90°
Seat width:
    Inlet .... 0.0709 in (1.8 mm)
    Exhaust .... 0.0630 in (1.6 mm)
Outside diameter:
    Inlet .... 1.7717 in (45 mm)
    Exhaust .... 1.5551 in (39.5 mm)
Stem diameter (inlet and exhaust) .... 0.3150 in (8.0 mm)
Head diameter:
    Inlet .... 1.7323 in (44.0 mm)
    Exhaust .... 1.5157 in (38.5 mm)
Spring free length:
    Early models .... 1.8583 in (47.2 mm)
    Late models .... 1.8110 in (46.0 mm)
Valve guides:
    Bore .... 0.3150 in (8.0 mm)
    Outside diameter:
       Nominal .... 0.5118 in (13.0 mm)
       Repair (1 groove) .... 0.5158 in (13.1 mm)
       Repair (2 grooves) .... 0.5217 in (13.3 mm)

Camshaft
Number of bearings 5

Endfloat:
    J6R .... 0.003 to 0.005 in (0.07 to 0.13 mm)
    851 .... 0.003 to 0.006 in (0.07 to 0.15 mm)

Valve timing

	J6R	851
Inlet opens	17° BTDC	17° BTDC
Inlet closes	63° ABDC	63° ABDC
Exhaust opens	63° BBDC	63° BBDC
Exhaust closes	17° ATDC	17° ATDC

Timing belt deflection
.... 0.217 to 0.276 in (5.5 to 7.0 mm)

Cylinder liners
Bore diameter .... 3.4646 in (88.0 mm)
Base locating diameter .... 3.6850 in (93.6 mm)
Protrusion (less O-ring) .... 0.003 to 0.006 in (0.08 to 0.15 mm)

Pistons
Type .... Alloy with 3 rings
Connecting rods:
    Type .... Press fit in small end, floating in piston

Connecting rods
Endfloat .... 0.012 to 0.022 in (0.31 to 0.57 mm)

Crankshaft
Number of main bearings 5
Endfloat:
    J6R .... 0.003 to 0.010 in (0.07 to 0.25 mm)
    851 .... 0.006 to 0.012 in (0.15 to 0.30 mm)
Main bearing journal diameter:
    Nominal .... 2.4761 in (62.892 mm)
    Regrind .... 2.4662 in (62.642 mm)
Crankpin diameter:
    Nominal:
        J6R .... 2.0589 in (52.296 mm)
        851 .... 2.2164 in (56.296 mm)
    Regrind:
        J6R .... 2.0491 in (52.046 mm)
        851 .... 2.2065 in (56.046 mm)

Lubrication system
Oil type/specification .... Multigrade engine oil, viscosity SAE 15W/40, 20W/40 or 20W/50, to API SE (Duckhams Hypergrade)

Oil capacity:
    Without oil filter .... 8.8 Imp pt; 5.3 US qt; 5.0 litre
    With oil filter .... 9.7 Imp pt; 5.8 US qt; 5.5 litre
Oil filter:
    1995 cc models up to 1984 .... Champion C102
    1995 cc models from 1984 on .... Champion F105
Oil pump pressure at 80°C (176°F):
    Idling .... 11.5 lbf/in² (0.8 bar)
    3000 rpm .... 43.5 lbf/in² (3.0 bar)
Oil pump rotor clearances:
    Rotor tip-to-inner housing .... 0.002 to 0.005 in (0.05 to 0.12 mm)
    Gear endfloat .... 0.0008 to 0.0039 in (0.02 to 0.10 mm)

Torque wrench settings

	lbf ft	Nm
Cylinder head bolts:		
1st stage	37	50
2nd stage	59	80
3rd stage. Loosen ½ turn then	65 to 72	87.5 to 97.6
4th stage (cold after running). Loosen ½ turn then	65 to 72	87.5 to 97.6
Camshaft drive sprocket bolt	37.5	51
Crankshaft pulley bolt	56 to 64	75 to 85
Connecting rod cap bolts	34 to 37.5	45 to 51
Crankshaft main bearing cap bolts	65 to 71	87.5 to 97.5
Flywheel bolts (manual transmission)	41 to 45	55 to 60
Driveplate bolts (automatic transmission)	49 to 52.5	65 to 70
Oil pump bolts	30 to 34	40 to 45
Intermediate shaft sprocket bolt	37.5	51
Rocker shaft plug	15.0	20

OHC engine – USA/Canada

Type/suffix
Manual transmission model J7T/718

Automatic transmission model J7T/719

Capacity .... 2165 cc (132 cu in)

Details as type 851 engine but with following differences:

Compression ratio .... 8.7 to 1

Valve rocker arm clearances
Inlet .... 0.004 to 0.006 in (0.10 to 0.15 mm)
Exhaust .... 0.008 to 0.010 in (0.20 to 0.25 mm)

Valve seat angles
Inlet .... 60°
Exhaust .... 45°

Camshaft endfloat .... 0.002 to 0.005 in (0.05 to 0.13 mm)

Valve timing
Inlet opens .... 12° BTDC
Inlet closes .... 52° ABDC
Exhaust opens .... 52° BBDC
Exhaust closes .... 12° ATDC

Oil pump
Gear end clearance .... 0.001 to 0.004 in (0.02 to 0.10 mm)

Torque wrench settings

	lbf ft	Nm
Crankshaft pulley bolt	88 to 100	120 to 135
Connecting rod cap bolts/nuts	44 to 48	60 to 65

Cooling system
Coolant capacity (including heater)
UK models:
  R1350, R2350 .... 10.6 Imp pt; 6.3 US qt; 6.0 litre
  R1351, R1342, R1352, R1345, R1355 .... 11.1 Imp pt; 6.7 US qt; 6.3 litre
  R1343, R1353 .... 14.1 Imp pt; 8.5 US qt; 8.0 litre
USA models:
  R1348, R1358 .... 11.1 Imp pt; 6.7 US qt; 6.3 litre

Drivebelt tension
Engine type:
  A2M, A6M, 807, A5L .... 0.10 to 0.14 in (2.5 to 3.5 mm)
  J6R .... 0.22 to 0.26 in (5.5 to 6.5 mm)
  843 .... 0.14 to 0.18 in (3.5 to 4.5 mm)

Thermostat (OHC engines)
Opening temperature .... 83°C (181°F) or 88°C (190°F)
Fully open at .... 92° to 96°C (198° to 205°F)

Fuel and exhaust systems
Air cleaner element
1565 cc and 1995 cc models .... Champion W115

Fuel filter
1565 cc and pre '84 1995 cc models .... Champion type not available
1995 cc 1984-on models .... Champion L206

Carburettor specifications and adjustment data
Solex 32 EITA 708:
    Choke tube 24
    Main jet 122.5
    Air compensating jet 170
    Idling jet 45
    Enrichment device 85
    Needle valve 1.5
    Initial throttle opening:
        Extreme cold .... 0.043 in (1.10 mm)
        Medium cold .... 0.035 in (0.9 mm)
    Float level .... 0.5 in (12.4 mm)
    Accelerator pump jet 40
    Pneumatic part opening .... 0.141 in (3.6 mm)
    Choke flap part opening .... 0.394 in (10.0 mm)
    Idling speed .... 675 rpm

Solex 32 DIS:
- Choke tube 24
- Main jet 122.5
- Air compensating jet 135
- Idling jet 41
- Enrichment jet 75
- Boost enrichment jet (E2) 50
- Boost enrichment jet (E1) 80
- Needle valve 1.7
- Accelerator pump jet 50
- Accelerator pump stroke .... 0.275 in (7.0 mm)
- Initial throttle opening (extreme cold) .... 0.029 in (0.75 mm)
- Pneumatic part opening .... 0.216 in (5.5 mm)
- Idling speed .... 650 rpm
- CO% .... 1.0 to 2.0

Solex 32 SEIA 795:
- Choke tube 24
- Main jet 127.5
- Idling jet 45
- Air compensating jet 160
- Needle valve 1.5
- Enrichment device 80
- Accelerator pump jet 35
- Initial throttle opening (extreme cold) .... 0.032 in (0.8 mm)
- Choke flap part opening .... 0.165 in (4.2 mm)
- Defuming valve .... 0.118 to 0.157 in (3.0 to 4.0 mm)
- Idling speed .... 775 rpm
- CO% .... 1.5 to 2.5
- Float level .... 0.461 in (11.7 mm)

Solex 32 MIMSA:

	1st	2nd
Choke tube	23	24
Main jet	105	112.5
Idle jet	41	45
Air compensation jet	155	125
Emulsifier	x3	x2
Enrichment	50	130
Needle valve	1.5 (ball valve)	
Float level	33 mm	
Accelerator pump jet	50	
Initial throttle opening	0.90	
Mechanical part-open setting	2 (gap on high side)	
Defuming valve	1 mm	
Clearance before accelerator pump actuated	1.7 mm	
Idle speed	650 ± 50 rpm	
CO%	1.0 to 2.0	

Zenith 32 IF:
- Choke tube 24
- Main jet 121
- Compensating jet .... 150 (defuming 165)
- Slow running jet 70
- Accelerator pump jet 50
- Accelerator pump stroke .... 0.898 in (22.8 mm)
- Needle valve 1.25
- Float level .... 0.471 in (11.95 mm)
- CO% .... 1.5 to 2.5

Weber 32 DIR 98:

	Primary	Secondary
Choke tube	23	24
Main jet	115	125
Air compensating jet	185	145
Idling jet	57	40
Mixture centralizer	3.5	4.5
Enrichment device	50	90
Needle valve	1.75	
Defuming valve setting	0.014 to 0.026 (0.35 to 0.66 mm)	
Initial throttle opening	0.035 in (0.90 mm)	
Choke flap setting	0.18 to 0.22 in (4.5 to 5.5 mm)	
Float level	0.276 in (7.0 mm)	
Float travel	0.315 in (8.0 mm)	
Emulsifier	F20 (Primary) F6 (Secondary)	
Accelerator pump jet	50	
Pneumatic part-opening	0.157 in (4.0 mm)	
Idling speed	650 rpm	
CO%	1.0 to 2.0	
Secondary barrel locking mechanism setting	0.08 to 0.12 in (2.0 to 3.0 mm)	

Weber 32 DIR:

Barrel specifications:	98-98C 1st	2nd	106 1st	2nd
Choke tube	23	24	23	24
Main jet	115	125	115	125
Air compensating jet	185	145	185	145
Idling jet	47	40	47	40
Mixture centralizer	3.5	4.5	3.5	4.5
Enrichment device	50	90	50	90
Emulsifier	F20	F6	F20	F6

	98-98C	106
Needle valve assembly	1.75	1.75
Defuming valve setting	0.014 to 0.026 in (0.35 to 0.66 mm)	0.014 to 0.026 in (0.35 to 0.66 mm)
Initial throttle opening	0.04 in (0.90 mm)	0.03 in (0.85 mm)
Fuel height under joint face	0.28 in (7.0 mm)	0.28 in (7.0 mm)
Float travel	0.32 in (8.0 mm)	0.32 in (8.0 mm)
Accelerator pump jet	50	50
Accelerator pump stroke	Cam-operated	Cam-operated
Pneumatic part-open setting	0.16 in (4.0 mm)	0.16 in (4.0 mm)
Choke flap setting	0.18 to 0.22 in (4.5 to 5.5 mm)	0.18 to 0.22 in (4.5 to 5.5 mm)
Cold start enrichment device	–	150
Idling speed	600 to 700 rpm	600 to 700 rpm
CO%	1.0 to 2.0	1.0 to 2.0
Secondary barrel locking mechanism setting	0.08 to 0.12 in (2.0 to 3.0 mm)	0.08 to 0.12 in (2.0 to 3.0 mm)

Weber 32 DARA:

	Mark 38		Mark 39		Mark 42	
Idling speed	775 to 825 rpm		625 to 675 rpm		775 to 825 rpm	
CO%	1.0 to 2.0		1.0 to 2.0		1.0 to 2.0	
Suffix	38		39		42	
Barrel specifications:	1st	2nd	1st	2nd	1st	2nd
Choke tube	24	26	24	26	26	26
Main jet	132	150	132	150	132	132
Air compensating jet	180	145	180	145	160	145
Idling jet	47	45	47	45	52	45
Mixture centralizer	3.5	4.5	3.5	4.5	3.5	4.0
Emulsifier	F53	F6	F53	F6	F58	F6

	Mark 38	Mark 39	Mark 42
Initial throttle opening (extreme cold)	0.06 in (1.4 mm)	0.06 in (1.4 mm)	0.05 in (1.3 mm)
Needle valve assembly	2.00	2.00	2.25
Fuel height under joint face	0.28 in (7.0 mm)	0.28 in (7.0 mm)	0.28 in (7.0 mm)
Float travel	0.32 in (8.0 mm)	0.32 in (8.0 mm)	0.32 in (8.0 mm)
Accelerator pump jet	60	60	60
Pneumatic part-open setting:			
Compensator compressed	0.18 in (4.5 mm)	0.18 in (4.5 mm)	0.22 in (5.5 mm)
Compensator released	0.30 in (7.5 mm)	0.30 in (7.5 mm)	0.41 in (10.5 mm)
Deflooding	0.39 in (10.0 mm)	0.39 in (10.0 mm)	0.35 in (9.0 mm)

Weber 32 DARA 41:

Barrel specifications:	1st	2nd
Choke tube	26	26
Main jet	135	140
Air correction jet	155	140
Idling jet	57	42/105
Enricher	60	110
Secondary venturi	3.5	4
Emulsion tube	F58	F6
Initial throttle opening (medium cold)	0.04 in (1.05 mm)	–

	1st
Needle valve	2.25
Fuel level	0.28 in (7.0 mm)
Float travel	0.32 in (8.0 mm)
Accelerator pump injector	60
Pneumatic part-open setting:	
Compensator depressed	0.22 in (5.5 mm)
Compensator not depressed	0.39 in (10.0 mm)
Deflooding	0.35 in (9.0 mm)
Idling speed	850 to 950 rpm
CO%	1.0 to 2.0

Weber 32 DARA (Canada):

Barrel specifications:	27 1st	2nd	28 1st	2nd
Choke tube	23	24	23	24
Main jet	120	120	120	120
Idle jet	50	55	52	55
Air compensating jet	170	145	170	145
Emulsifier	F20	F6	F20	F6
Main nozzle (auxiliary venturi)	4-step	4.5	4-step	4.5

	27	28
Accelerator pump jet	60	60
Needle valve	1.75	1.75
Float level	0.28 in (7.0 mm)	0.28 in (7.0 mm)

Float travel .... 0.32 in (8.0 mm)   0.32 in (8.0 mm)
Initial throttle opening .... 0.05 in (1.20 mm)   0.06 in (1.45 mm)
Air opening:
    Stabilizer depressed .... 0.157 in (4.0 mm)   0.157 in (4.0 mm)
    Stabilizer extended .... 0.276 in (7.0 mm)   0.276 in (7.0 mm)
Deflooding .... 0.315 in (8.0 mm)   0.315 in (8.0 mm)
Idle speed .... 800 to 900 rpm   600 to 700 rpm
CO%:
    Without air injection .... 1.0 to 2.5   1.0 to 2.5
    With air injection .... 1.0 maximum   1.0 maximum

Turbocharger system

Fuel pump output .... 1.32 Imp gal (63 US qt, 6.0 litre) per hour at 36.3 lbf/in^2 (2500 mbar)
Fuel pressure regulator setting .... 3.6 to 4.4 lbf/in^2 (250 to 300 mbar)
Turbocharger type .... Garrett T3
Wastegate release pressure .... 8.0 to 9.0 lbf/in^2 (550 to 620 mbar)
Boost pressure (over 3000 rpm, full load at carburettor input) .... 11.6 to 13.1 lbf/in^2 (800 to 900 mbar)
Idling speed .... 600 to 700 rpm
CO% .... 1.0 to 2.0
Ignition cut-out switch pressure .... 13.1 to 14.5 lbf/in^2 (900 to 1000 mbar)

Fuel injection system (US models)

Type .... Bosch L-Jetronic with oxygen sensor
Thermo-time switch cut-off temperature .... 35°C (95°F)
Coolant temperature sensor operating temperature:
    R1348 .... 20°C (68°F)
    R1358 .... 80°C (176°F)
Fuel pump:
    Flow rate .... 13.2 Imp gal (63 US qt, 6.0 litre) per hour
    Pressure at idle (R1348):
        Vacuum connected .... 28 lbf/in^2 (2000 mbar)
        Vacuum disconnected .... 36 lbf/in^2 (2500 mbar)
Idling speed:
    Manual gearbox .... 750 to 850 rpm
    Automatic transmission .... 600 to 700 rpm

Ignition system
Type (later models) .... Electronic, either transistorized or integral

Spark plugs
Type:
    807, A5L, A7L .... Champion N3G
    A2M, 843 .... Champion RN9YCC or RN9YC
    A6M .... Champion RN7YCC or RN7YC
    J6R .... Champion S7YCC or S279YC
    JT7, 851 (US/Canada) .... Champion RS9YC
Electrode gap:
    Champion N3G, RN7YC, RN9YC .... 0.6 mm (0.024 in)
    Champion RN7YCC, RN9YCC, S7YCC, S279YC .... 0.8 mm (0.031 in)
    Champion RS9YC .... 0.9 mm (0.035 in)

HT leads
J6R .... Champion CLS 3, boxed set

Distributor
Transistorized system sensor-to-reluctor air gap .... 0.012 to 0.024 in (0.3 to 0.6 mm)
Ignition timing (idling with vacuum disconnected):
    R1350 (847) .... 9 to 11° BTDC
    R1351 (841-C7-25) .... 5 to 7° BTDC
    R1351 (841-D7-26) .... 9 to 11° BTDC
    R1345 (807) .... 11 to 13° BTDC
    R1341, R1351 (A2M) .... 7 to 9° BTDC
    R1342, R1352 (A6M) .... 9 to 11° BTDC
    R1343, R1353 (J6R) .... 9 to 11° BTDC
    R1340, R1350 (C1J) .... 9 to 11° BTDC
    R1345, R1355 (A5L) .... 13 to 15° BTDC
    R1341, R1351 (841 US/Canada):
        Manual .... 3 to 5° BTDC
        Automatic .... 7 to 9° BTDC
    R1348, R1358 (843 US/Canada) .... 9 to 11° BTDC
    R135B (J7T US/Canada) .... 9 to 11° BTDC

Clutch
Type
R1350 and R2350 (pre 1984)	180 DBR 335
R1351, R1342, R1352, R1348 and R1358	200 DBR 350
R1345	215 DBR 410, 215 CP 410 or 215 CP 450
R1343 and R1353	215 CP 410
R1340, R1350, R1359 and R2350 (later)	180 CP 310
R1340, R1350, R1359 and R2350 (1984 on)	180 CP 335
R1341 and R1351 (1984 on)	200 DBR 375

Disc (driven plate)
215 DBR 410, 215 CP 410, 215 CP 450:
Number of splines	21
Spring colour	Pink, Blue
Outside diameter	8.465 in (215 mm)
Friction face thickness	0.303 in (7.7 mm)

200 DBR 375:
Number of splines	21
Spring colour	Yellow, Green, Blue
Outside diameter	7.874 in (200 mm)
Friction face thickness	0.303 in (7.7 mm)

Cable operating clearance
R1348 and R1358	0.118 to 0.158 in (3.0 to 4.0 mm)

Manual transmission
Type
NG0	Four-speed
NG1 and NG3	Five-speed

Ratios
	NG0	NG1	NG3
1st	3.82 or 4.09	3.82 or 4.09	3.82 or 4.09
2nd	2.18	2.18	2.18
3rd	1.41	1.41	1.41
4th	0.97 or 1.03	0.97 or 1.03	0.94 or 1.03
5th	–	0.86 or 0.78	0.86 or 0.78
Final drive	3.78 or 3.56	3.78, 3.56 or 3.44	3.78 or 3.44

Setting-up limits
Crown wheel and pinion backlash	0.0047 to 0.010 in (0.12 to 0.25 mm)
Preload, differential bearings:	
Re-used bearings	No play, but free to turn
New bearings	9 to 13.5 lbf (40 to 60 N)
Primary shaft bearing endfloat (NG0 type)	0.0008 to 0.005 in (0.02 to 0.12 mm)

Lubrication
Oil type/specification	Hypoid gear oil, viscosity SAE 80EP to API GL5 (Duckhams Hypoid 80S)
Oil capacity (Types 352, 395, NG0, NG1 and NG3)	3.5 Imp pts (2.1 US qts, 2.0 litres)

Torque wrench settings
	lbf ft	Nm
Primary shaft nut (NG1, NG3)	96	130
Secondary shaft nut (NG0, NG1, NG3)	110	150
Reverse selector lever (NG0, NG1, NG3)	18	25
Casing bolts (NG0, NG1, NG3)	18	25

Braking system
Front brakes
UK models – R1345 and R1355 (1983 on):
Discs:	
Diameter	10.236 in (260.0 mm)
Minimum thickness	0.697 in (17.7 mm)
Pad thickness (including backing):	
New	0.728 in (18.5 mm)
Minimum	0.354 in (9.0 mm)

US/Canada models – R1341, R1348, R1351 and R1358:
Discs:	
Diameter	9.370 in (238.0 mm)
Minimum thickness	0.433 in (11.0 mm)
Pad thickness (including backing):	
New	0.709 in (18.0 mm)
Minimum	0.276 in (7.0 mm)

Rear brakes
R1350 and R1351 UK models and all US/Canada models:
  Drum internal diameter:
    New .... 8.996 in (228.5 mm)
    Maximum .... 9.035 in (229.5 mm)
  Minimum lining thickness .... 0.020 in (0.5 mm) above rivet heads or web
R1345 and R1355 (1983 on):
  Disc diameter .... 9.284 in (235.8 mm)
  Minimum disc thickness .... 0.413 in (10.5 mm)
  Pad thickness (including backing):
    New .... 0.551 in (14.0 mm)
    Minimum .... 0.276 in (7.0 mm)

Handbrake
R1345 and R1355 (1983 on) UK models and all
US/Canada models:
  Adjustment setting .... 12 notches minimum

Brake pressure limiter cut-off pressure
With driver and full tank of fuel
UK models:
  R1350 and R1351 .... 406 to 464 lbf/in^2 (28 to 32 bar)
  R1345 (to 1983) .... 377 to 435 lbf/in^2 (26 to 30 bar)
  R1342 (to 1983) .... 595 to 653 lbf/in^2 (41 to 45 bar)
  R1352 and R1353 .... 638 to 696 lbf/in^2 (44 to 48 bar)
  R1343 .... 377 to 435 lbf/in^2 (26 to 30 bar)
  R1345 (1983 on) .... 305 to 363 lbf/in^2 (21 to 25 bar)
  R1355 .... 348 to 406 lbf/in^2 (24 to 28 bar)
  R1342 (1983 on) .... 522 to 580 lbf/in^2 (36 to 40 bar)
US/Canada models:
  R1341, R1348, R1351, R1358 .... 435 to 493 lbf/in^2 (30 to 34 bar)

Brake servo unit
Pushrod clevis adjustment dimension L:
  R1342, R1343, R1352 and R1353 .... 5.000 in (120.0 mm)

Torque wrench settings

	lbf ft	Nm
Rear caliper mounting bolts	48	65
Hydraulic hose union:		
With copper washer	15	20
Tapered	9.5	13

Electrical system
Starter motor
Make (R1345) .... Paris-Rhone, Ducellier or Bosch

Fuse identification

Fuse number	Rating (amp)	Circuit
UK models (later):		
1	8	Flasher unit
2	–	Not used
3	5	Stop-light switch on Cruise control
4	5	Windscreen wiper Park
5	5	Radio (driving aid)
6	8	Interior lights
7	–	Not used
8	16	Windscreen wiper/washer switch
9	–	Not used
10	5	Left side and rear lights
11	10	Left front door window
12	5	Right side and rear lights
13	10	Right front door window
14	5	Instrument panel feed
15	5	Reversing light switch
16	1.5	Automatic transmission
17	16	Rear window demister
18	16	Heater fan air conditioning
19	5	Rear foglight switch
US/Canada models:		
1	8	Flasher unit, stop-light switch
2	5	Fixed stop windscreen wiper
3	–	Not used
4	–	Not used

Fuse number	Rating (amp)	Circuit
5	–	Not used
6	8	Cigarette lighter, interior lights
7	–	Not used
8	16	Windscreen wiper/washer
9	–	Not used
10	5	Console lights, left side and rear lights
11	10	Left front door window
12	5	Right side and rear lights
13	10	Right front door window
14	5	Instrument panel
15	16	Reverse light, windscreen wiper timer
16	1.5	Automatic transmission
17	16	Heater fan, air conditioning
18	16	Rear window demister

Suspension and steering
Front wheel geometry
Castor angle (unladen):
　UK:
　　1981 models (H5 – H2 = 2.165 in/55 mm) .... 2° 30' to 4° 00'
　　R1345 1981 (H5 – H2 = 2.165 in/55 mm) .... 2° 00' to 3° 00'
　　1982 models (H5 – H2 = 2.165 in/55 mm) .... 2° 00' to 3° 00'
　US/Canada. All models (H5 – H2 = 2.165 in/55 mm):
　　Power steering .... 2° 00' to 3° 00'
　　Manual steering .... 1° 00' to 2° 00'
Toe-out (unladen):
　All models .... 0 to 0.08 in (0 to 2.0 mm) equivalent to 0° to 0° 20'
Kingpin inclination (unladen):
　Negative offset suspension .... 12° 30' to 13° 30' equal on both sides to within 1°

Rear wheel geometry
Toe-in/out (unladen):
　Except R1351 .... Toe-out 0 to 0.06 in (0 to 1.5 mm) equivalent to 0° to 0° 15'
　R1351 .... Toe-in 0 to 0.12 in (0 to 3.0 mm) equivalent to 0° to 0° 30'

Wheels and tyres
UK models:
　Wheel size:
　　R1340 (1983 on) 5B13
　　R1342, R1343, R1352, R1353 .... 5½B13
　　R1350, R1351, R2350 .... 5B13
　Tyre size:
　　R1340 (1983 on) .... 145 SR 13 or 155 SR 13
　　R1342, R1343 .... 165 SR 13
　　R1350, R1351, R2350 .... 155 SR 13 or 165 SR 13
　　R1352, R1353 .... 165 SR 13

Tyre pressures in lbf/in² (bar):	Front	Rear
R1340 (1983 on)	25 (1.7)	26 (1.8)
R1342, R1343	28 (1.9)	31 (2.1)
R1352, R1353	28 (1.9)	35 (2.4)
R1350, R1351, R2350	25 (1.7)	35 (2.4)

US/Canada models:
　Tyre size:
　　R1348, R1358 .... 155x13 or 175/70x13

Tyre pressures in lbf/in² (bar)	Front	Rear
R1348	28 (1.9)	28 (1.9)
R1358	28 (1.9)	34 (2.3)

Torque wrench settings	lbf ft	Nm
Bearing retaining bolts (negative offset)	11	15
Driveshaft nut (negative offset)	185	250
Suspension balljoint nuts (negative offset)	48	65

3 Engine

Part A – OHV engines

General description
1　Several additional OHV engines have been introduced, as listed in the Specifications. The main difference is in the design of the cylinder head. The 841 engine described in Chapter 1 is of side venting design, but with one exception the engines described in this Supplement are of crossflow design with the inlet manifold on the right-hand side and the exhaust manifold on the left-hand side.

2　Procedures are as given in Chapter 1 for the 841 engine except for the information given in this Section.

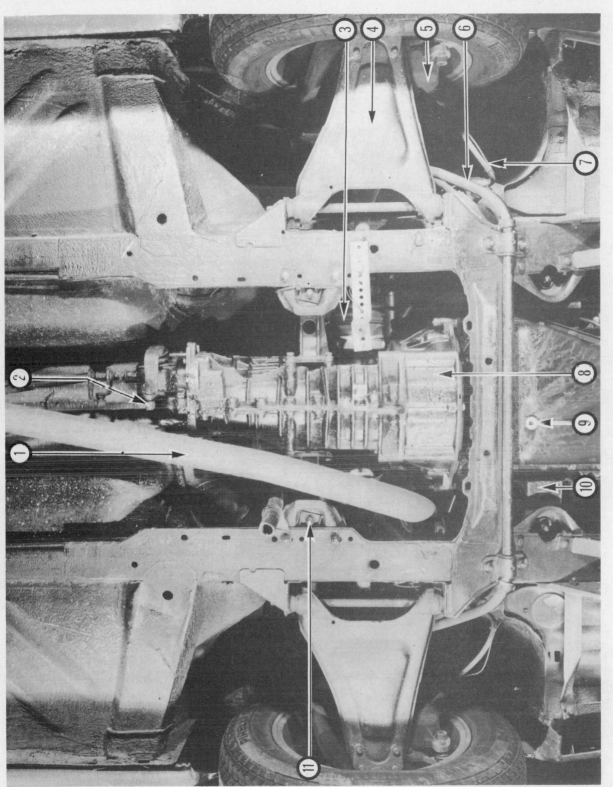

Front under-body view (Turbo model)

1 Exhaust pipe
2 Gearchange linkage
3 Driveshaft
4 Lower suspension arm
5 Brake caliper
6 Anti-roll bar
7 Brake hose
8 Gearbox (manual)
9 Engine oil drain plug
10 Engine mounting
11 Gearbox mounting

Rear under-body view (Turbo model)

1 Rear exhaust silencer
2 Rear axle beam
3 Rear suspension centre arm
4 Rear brake caliper
5 Rear suspension side arm
6 Brake pressure limiter
7 Fuel pump
8 Fuel filter
9 Exhaust mounting
10 Intermediate exhaust silencer
11 Anti-roll bar
12 Handbrake cable
13 Rear shock absorber lower mounting

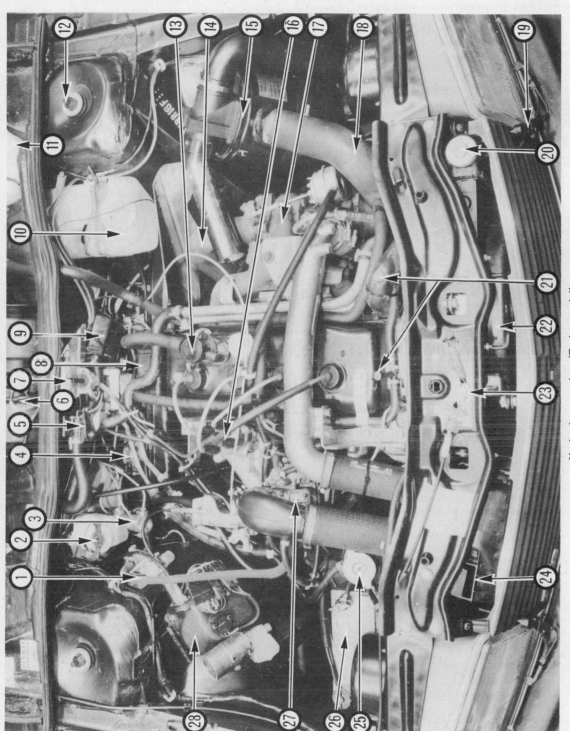

Under-bonnet view (Turbo model)

1 Fuel pipe from computer
  flowmeter
2 Brake fluid reservoir
3 Brake master cylinder
4 Steering gear
5 Heater valve
6 Heater motor

7 Turbo pressure cut-out switch
8 Oil separator
9 Engine wiring harness connector
10 Headlight/windscreen washer
   bottle
11 Coolant expansion tank
12 Front suspension upper mounting

13 Oil filler cap/crankcase
   ventilation unit
14 Exhaust downpipe
15 Air cleaner
16 Distributor
17 Turbocharger unit

18 Air inlet hose
19 Headlight wiper
20 Radiator cap
21 Coolant bleed screws
22 Radiator cooling fan unit
23 Bonnet lock and cable

24 Turbo intercooler
25 Power steering fluid reservoir
   cap
26 Battery
27 Carburettor
28 Cruise control unit

Engine (807 and A5L Turbo) – removal and refitting
3   The procedure is similar to that described in Part B of this Section for the OHC engine, but with the following exceptions.

   (a)   *Remove the oil separator from the rear of the cylinder head (photo).*
   (b)   *To prevent the engine tilting to the left, attach an additional lifting chain to the exhaust manifold*
   (c)   *When connecting the exhaust downpipe tighten the nuts until the springs are coil bound, then back them off one and a half turns before locking with the locknuts.*

14   To remove the housing, first remove the drivebelt and unbolt the pulley.
15   Fit the Renault locating tool Mot 258 to the rear of the camshaft then unbolt and remove the housing. Remove the gasket and the lubrication channel O-ring.
16   Drive out the oil seal and press in the new one.
17   Fit the housing in reverse order using a new O-ring and gasket, then tighten the bolts.
18   Refit the pulley and drivebelt and tension the belt so that its deflection is between 0.10 and 0.12 in (2.5 and 3.0 mm).

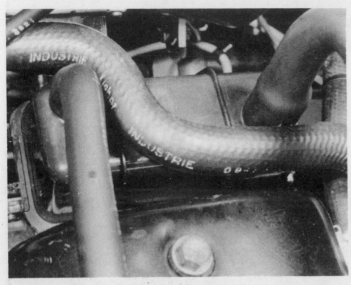

3A.3 Oil separator on Turbo models

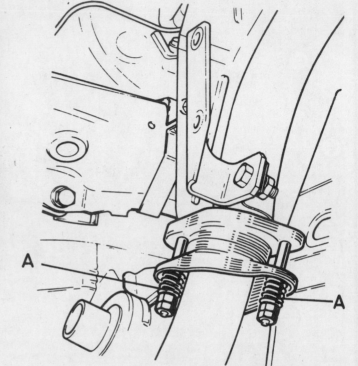

Fig. 13.1 Exhaust downpipe clamp on 807/A5L Turbo engines (Sec 3A)

A   Springs

Engine (US/Canada models) – removal and refitting
4   The procedure is similar to that described in Chapter 1, except for the disconnection of the fuel injection and emission control components, described in Section 5.

Cylinder head (crossflow engines) – removal and refitting
5   The procedure is similar to that described in Chapter 1, Sections 7 and 30, except for the information given in the following paragraphs.
6   With the rocker cover removed unscrew the rocker arm adjusting screws and withdraw the pushrods, keeping them identified for location.
7   Loosen the cylinder head bolts in progressive manner and remove all but the four corner bolts.
8   Remove the rubber rings from the spark plug tubes.
9   Locate a large rubber band over the four corner bolts and lift off the complete rocker assembly.
10   Before refitting the cylinder head, check that the distributor drive pinion is correctly located, as shown in Fig. 13.4, with No 1 cylinder on TDC compression.
11   Refer to Chapter 1, Section 45, and if available use the Renault tools Mot 451 and Mot 446.

Rocker assembly (crossflow engines) – dismantling and reassembly
12   The procedure is straightforward, but **do not** attempt to remove the end plugs from the shafts. Drive out the roll pins using a suitable punch. Note that the inlet and exhaust rocker arms are different, and that numbers 1 and 4 pedestals incorporate lubrication holes. The two shafts are identical.

Camshaft – removal and refitting
13   On models with power-assisted steering and air conditioning the power steering pump is driven by a pulley on the flywheel end of the camshaft. The rear camshaft bearing is incorporated in a housing bolted to the cylinder block.

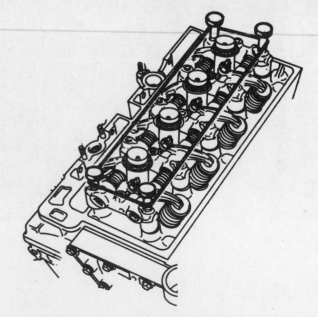

Fig 13.2 Removing the rocker assembly (Sec 3A)

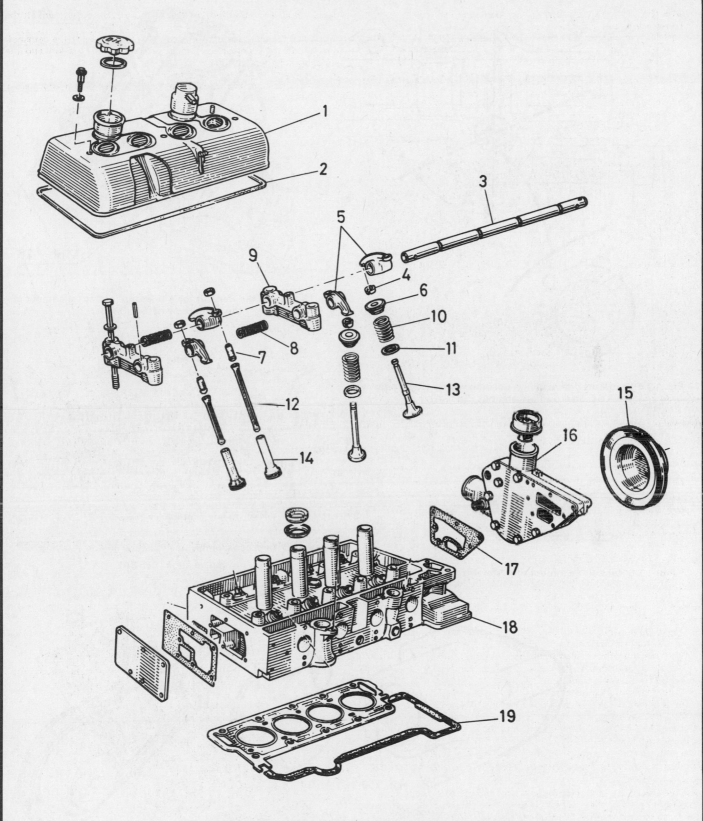

Fig. 13.3 Crossflow type cylinder head components (Sec 3A)

1	Rocker cover	6	Retainer	11	Seat	16	Water pump
2	Gasket	7	Adjuster	12	Pushrod	17	Gasket
3	Rocker shaft	8	Spring	13	Valve	18	Cylinder head
4	Collets	9	Pedestal	14	Tappet	19	Gasket
5	Rockers	10	Spring	15	Pulley		

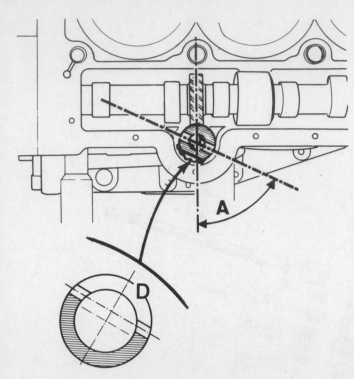

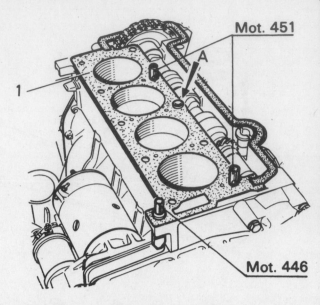

Mot. 451

Mot. 446

Fig. 13.5 Using the Renault tools to ensure correct alignment of the cylinder head (Sec 3A)

1   *Gasket*   A   *Dowel*

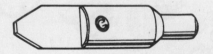

Fig. 13.4 Distributor drive pinion position with No 1 cylinder at TDC on crossflow engines (Sec 3A)

A   *53°*   D   *Offset slot position*

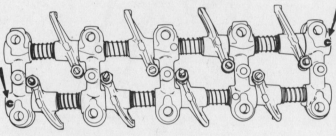

Exhaust

Intake

Fig. 13.7 Rocker assembly on crossflow engine showing roll pin location (Sec 3A)

Fig. 13.6 Renault alignment tool Mot 446 (Sec 3A)

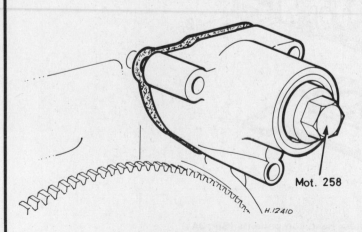

Mot. 258

H.12410

Fig. 13.8 Using the special Renault tool when removing the camshaft rear bearing housing on models with power steering and air conditioning (Sec 3A)

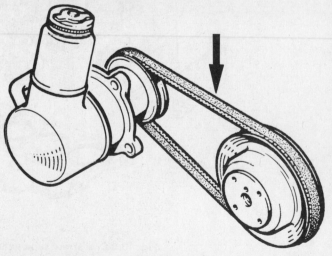

Fig. 13.9 Power steering pump drivebelt tension checking point (Sec 3A)

Part B – OHC engines

General description

1   The ohc engine type is a four-cylinder in-line unit mounted at the front of the vehicle. It incorporates a crossflow design head, having the inlet valves and manifold on the left-hand side of the engine and the exhaust on the right. The inclined valves are operated by a single rocker shaft assembly which is mounted directly above the camshaft. The rocker arms have a stud and locknut type of adjuster for the valve clearances, providing easy adjustment. No special tools are required to set the clearances. The camshaft is driven via its sprocket pulley from the timing belt, which in turn is driven by the crankshaft sprocket pulley. This belt also drives an intermediate shaft which actuates the petrol pump (where applicable) and oil pump shafts, the actuation being cam and gear respectively. The camshaft drives the distributor direct from its opposing end to the pulley sprocket, except on some models when the distributor is driven by the intermediate shaft. A spring-loaded jockey wheel assembly provides the timing belt tension adjustment. A single, twin or triple pulley is mounted on the front of the crankshaft and this drives the alternator/water pump drivebelt, the power steering pump drivebelt and the air conditioning compressor drivebelt, as applicable. The crankshaft runs in the main bearings which are shell type aluminium/tin material. The crankshaft endfloat is taken up by side thrust washers. The connecting rods also have aluminium/tin shell bearing type big-ends.

2   Aluminium pistons are employed, the gudgeon pins being a press fit in the connecting rod small-ends and a sliding fit in the pistons. The No 1 piston is located at the flywheel end of the engine (at the rear). As is typical with Renault engines, removable wet cylinder liners are employed, each being sealed in the crankcase by a flange and O-ring. The liner protrusion above the top surface of the crankcase is crucial; when the cylinder head and gasket are tightened down they compress the liners to provide the upper and lower seal of the engine coolant circuit within the engine. The cylinder head and crankcase are manufactured in aluminium and the engine is inclined at an angle of 15° when installed in the car.

Fig. 13.10 Cutaway view of the overhead camshaft engine (Sec 3B)

Major operations possible with engine fitted

3   The following major operations can be carried out on the engine, whilst still in place in the vehicle:

    (a) Cylinder head removal
    (b) Camshaft removal
    (c) Timing sprockets and belt removal
    (d) Timing belt tensioner removal
    (e) Intermediate shaft removal
    (f) Inlet and exhaust manifolds removal
    (g) Sump removal
    (h) Big end bearings removal
    (j) Oil pump removal
    (k) Pistons and connecting rod assemblies, and cylinder liners removal

Major operations only possible after removing engine

4   The following major operations can be carried out with the engine out of the bodyframe on the bench or floor:

    (a) All items listed in the previous paragraph
    (b) Crankshaft and main bearings – removal, overhaul and refitting

Engine (separate) – removal and refitting

5   Drain the cooling system, as described in Chapter 2. The cylinder block drain plug is located on the right-hand side.
6   Remove the front end panel, crossmember and radiator and disconnect the battery negative lead.
7   Disconnect the accelerator cable at the engine end.
8   Disconnect the engine wiring, noting the location of each wire.
9   Disconnect the coolant hoses and the exhaust system.

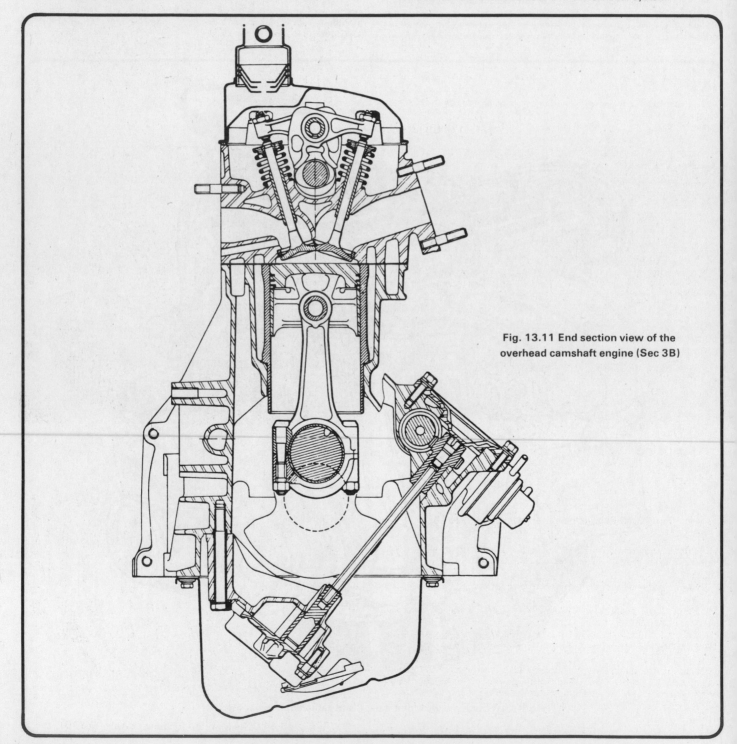

Fig. 13.11 End section view of the overhead camshaft engine (Sec 3B)

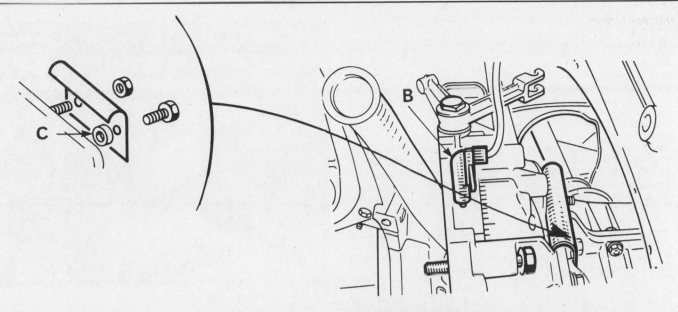

Fig. 13.12 TDC sensor (B) and wiring clip spacer (C)
(Sec 3B)

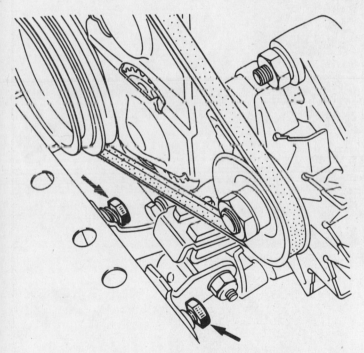

Fig. 13.13 Engine front mounting bolts (Sec 3B)

3B.17 Removing the engine

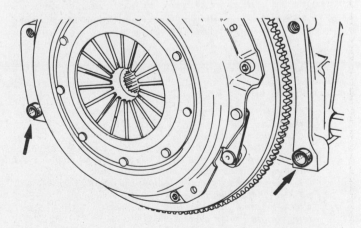

Fig. 13.14 Cylinder block locating dowels (Sec 3B)

10 Unbolt and remove the TDC sensor and the clip supporting the battery positive cable and reverse light wire. Note the location of the spacer.
11 Remove the distributor cap and rotor cam.
12 Disconnect and plug the fuel hoses.
13 Remove the power steering pump (Chapter 11) leaving the hoses connected, and place it on one side.
14 Attach a suitable lifting hoist to the engine and support its weight.
15 Remove the engine mounting nuts and bolts, including the front mounting bolts.
16 Support the gearbox with a trolley jack then unscrew the engine-to-gearbox mounting bolts. Note that the upper left-hand bolt cannot be removed due to the location of the crossmember.
17 Raise the engine slightly and withdraw it from the gearbox then lift it from the engine compartment (photo).
18 Before refitting the engine, check that the locating dowels are

positioned in the cylinder block and that the upper left-hand bolt is located in the gearbox.

19 Refitting is a reversal of removal, but lightly grease the clutch shaft splines, bleed the cooling system, and adjust the clutch cable.

Engine and gearbox – removal and refitting

20 Proceed as described for the engine, however do not remove the TDC sensor, gearbox-to-engine mounting bolts or the reverse light wire and clip.

21 Remove the ignition module (Section 6) to prevent it being damaged.

22 Disconnect the driveshafts, as described in Chapter 8.

23 Disconnect the gearchange and speedometer cable with the front of the car supported on axle stands. Also disconnect the clutch cable.

24 Lower the exhaust system by releasing the rubber mounting rings.

25 Support the gearbox with a trolley jack then unbolt the gearbox mounting.

26 Attach a suitable lifting hoist to the engine and support its weight.

27 Remove the engine mounting nuts and bolts, including the front mounting bolts, then lift the assembly from the engine compartment.

28 Refitting is a reversal of removal, but bleed the cooling system and adjust the clutch cable.

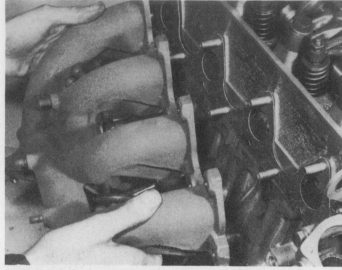
3B.29A Remove the exhaust manifold

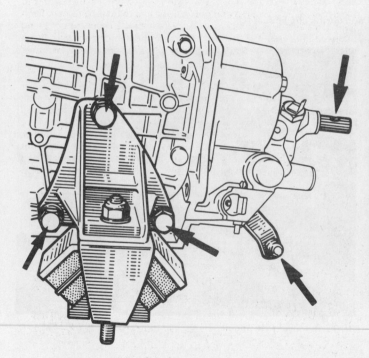

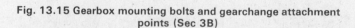

Fig. 13.15 Gearbox mounting bolts and gearchange attachment points (Sec 3B)

3B.29B Remove the inlet manifold

Engine dismantling – ancillaries

29 Remove the following items, taking note of how they are secured and located (photos):

    (a) *The coolant feed pipe to heater from water pump*
    (b) *Exhaust manifold*
    (c) *Water pump and thermostat housing*
    (d) *Inlet manifold and carburettor, noting in particular the inter-connecting hot spot pipe layout from the water pump, to the automatic choke manifold and cylinder head*
    (e) *If not already disconnected, remove the petrol pump unit with gaskets and spacer plate, if applicable*
    (f) *Unscrew and remove the oil pressure and coolant temperature sender switches*
    (g) *Unscrew and remove the oil filter, using a strap wrench if necessary*
    (h) *Remove the clutch unit (see Chapter 5)*

3B.29C Remove the oil pressure switch

30 Before removing the distributor, turn the engine to the top dead centre (TDC) position. This is indicated by the marks on the camshaft and intermediate shaft sprockets being aligned with the timing indicators in the respective apertures in the timing cover. When both marks are in alignment with their pointers, and the No 1 cylinder piston on TDC (firing stroke), scribe a mark on the end of the cylinder block opposite that of the timing mark on the flywheel.

31 Note the position of the distributor rotor arm and make an alignment mark between the distributor flange and the cylinder head. This will act as a timing guide when refitting.

32 Unscrew and remove the three distributor retaining bolts and withdraw the distributor (photo).

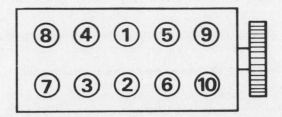

Fig. 13.16 Cylinder head bolts loosening and tightening sequence (Sec 3B)

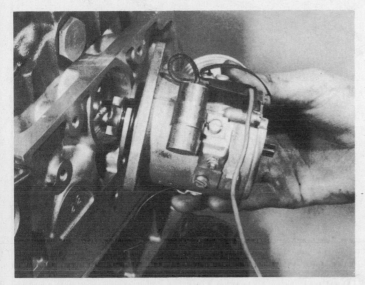

3B.32 Removing the distributor

bolt at the front. You will probably have to tap the head round on its rear corner, using a soft-headed mallet or wood block to break the seal between the head, gasket and block. Once the head is pivoted round and the seal broken, fully release the remaining cylinder bolt and lift the head clear. This action is necessary because if the head were to be lifted directly from the block, the seal between the head gasket and block would be unbroken and the liners would thus be disturbed. Each liner has a seal around its lower flange; where the liners are not being removed, it is essential that this seal is not broken or disturbed. If the liners are accidentally disturbed they must be removed and new lower seals fitted.

40 The rocker shaft assembly and final bolt can be removed from the cylinder head once it is withdrawn from the engine.

41 To prevent the possibility of movement by the respective cylinder liners whilst the head is removed, it is advisable to place a clamp plate over the top edges of the liners. A suitable plate (or bar) can be fastened temporarily in position using the existing head bolt holes, using shorter bolts of the desired thread and diameter with large, flat washers under the heads. Alternatively, use a block of wood with two bolts and spacers, clamping it in position in diagonal fashion (photo).

Cylinder head – removal

33 If the engine is still in the vehicle, drain, disconnect or remove (as applicable) the following items:

(a) Disconnect the battery earth lead
(b) Drain the cooling system and remove the radiator
(c) Disconnect the wires, pipes and leads to the cylinder head, noting the respective connections
(d) Remove the air filter, carburettor (if applicable) and inlet manifold
(e) Remove the exhaust manifold, disconnecting it from the cylinder head and exhaust downpipe flange
(f) Remove the undertray beneath the engine
(g) Where fitted detach the power steering pump drivebelt
(h) Detach the alternator drivebelt
(i) Remove the diagnostic socket
(j) If the cylinder head is to be dismantled and the distributor removed, refer to paragraphs 30 to 32 and check TDC before unbolting the distributor

34 The procedure for cylinder head removal with engine fitted or removed is now the same.

35 Unbolt and remove the timing cover.

36 Remove the rocker cover.

37 With the engine set at TDC, the timing belt tensioner bolts can be loosened and the tensioner pivoted to slacken the belt tension. Remove the timing belt. Note that on some models the tensioner is on the water pump, as shown in Section 4.

38 The cylinder head retaining bolts can now be progressively loosened, and with the exception of the front right-hand bolt, removed. Loosen off the bolts in the sequence shown in Fig. 13.16.

39 The cylinder head is now ready for removal from the block, and it is essential that it is detached in the correct manner. **Do not** lift the head directly upwards from the block. The head must be swivelled from the rear in a horizontal manner, pivoting on the remaining head

3B.41 Clamp the cylinder liners in position

Rocker assembly – dismantling

42 Unscrew the end plug from the shaft and extract the plug and filter. This filter must be renewed at 36 000 mile (60 000 km) intervals, or whenever it is removed.

43 Number the rocker arms and pedestals/bearings. Note that bearing No 5 has two threaded holes to retain the shim which adjusts the camshaft endfloat, and a hole for the roll pin which locates the

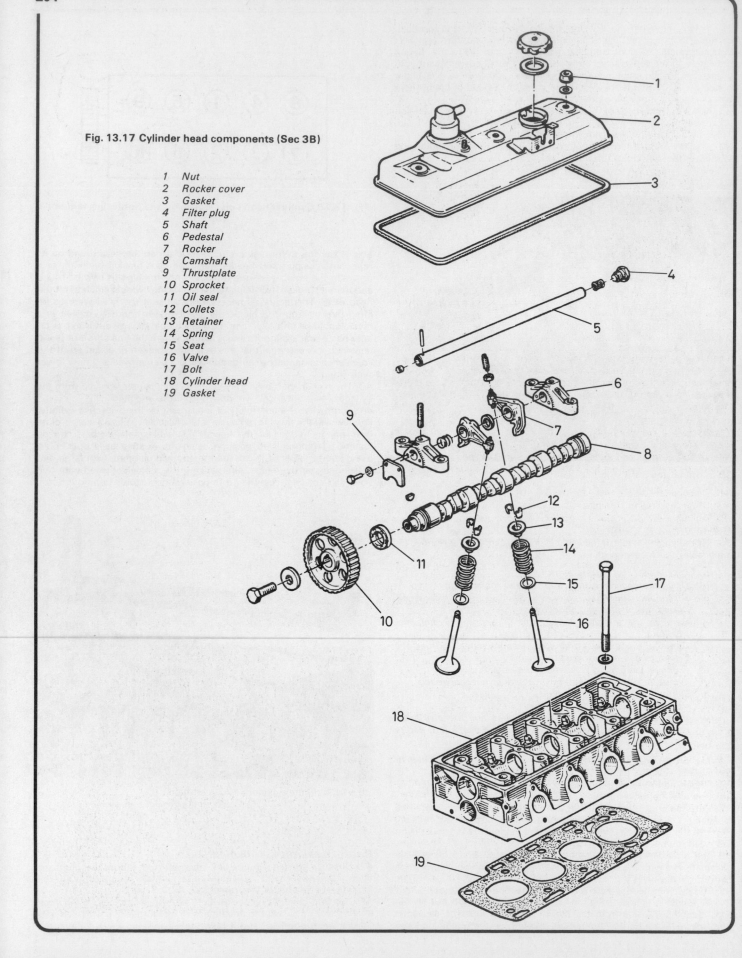

Fig. 13.17 Cylinder head components (Sec 3B)

1   Nut
2   Rocker cover
3   Gasket
4   Filter plug
5   Shaft
6   Pedestal
7   Rocker
8   Camshaft
9   Thrustplate
10 Sprocket
11 Oil seal
12 Collets
13 Retainer
14 Spring
15 Seat
16 Valve
17 Bolt
18 Cylinder head
19 Gasket

3B.43 Removing the No 5 rocker pedestal – note the roll pin hole

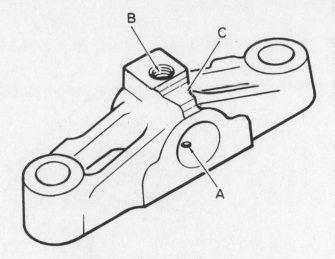

Fig. 13.18 Rocker shaft pedestals 1 to 4 (Sec 3B)

A   Oil hole   C   Timing flat
B   Stud hole

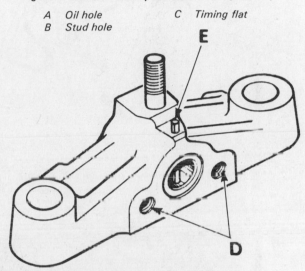

Fig. 13.19 Rocker shaft pedestal 5 (Sec 3B)

D   Thrustplate bolt holes   E   Roll pin (must be solid)

shaft and pedestal (photo). Renew the rollpin if it is not the solid type.
44 Keep the respective parts in order as they are removed from the shaft and note their respective locations. Note also that the machined flat section on top of pedestals 1 to 4 all face towards the camshaft sprocket.

Camshaft – removal

45 The camshaft can be removed with the engine fitted in the car. Where this is the case, proceed as follows in paragraphs 46 to 52. Where the engine/cylinder head are removed, proceed from paragraph 53.
46 Disconnect the battery earth lead.
47 Drain the engine coolant (see Chapter 2).
48 Detach the throttle cable from the carburettor and bracket on the rocker cover.
49 Remove the radiator and, where fitted, the electric cooling fan (see Chapter 2).
50 Unscrew and withdraw the front grille panel.
51 Remove the engine undertray.
52 Detach the warm air trunking.
53 Remove the drivebelts to the alternator and, where fitted, the power steering pump and air conditioning pump.
54 Unbolt and remove the timing cover.
55 Disconnect the respective HT leads and remove the sparking plugs. Next, rotate the crankshaft to locate the No 1 piston (flywheel end) at TDC on firing stroke. This can be checked by ensuring the timing mark on the flywheel is in line with the 'O' graduation marked on the aperture in the clutch housing. Rotate the crankshaft by means of a spanner applied to the crankshaft pulley retaining bolt.
56 Next, turn the crankshaft from the TDC a further quarter of a turn clockwise (viewed from the front).
57 Loosen the timing belt tensioner, release the tension and withdraw the belt.
58 To remove the camshaft sprocket, pass a suitable hardwood dowel rod or screwdriver through a hole in the sprocket and jam against the top surface of the cylinder head to prevent the camshaft from turning. Unscrew the retaining bolt and remove the sprocket. Take great care as the sprocket is manufactured in sintered metal and is therefore relatively fragile (photo). Also, take care not to damage or distort the cylinder head. Remove the Woodruff key.
59 Remove the rocker cover and unscrew the cylinder head retaining bolts in a progressive manner in the sequence shown in Fig. 13.16. With the bolts extracted, remove the rocker assembly and relocate some of the bolts with spacers to the same depth as the rocker pedestals fitted under the bolt heads, to ensure that the cylinder head is not disturbed during subsequent operations. If for any reason the head is disturbed, then the head gasket and respective liner seals will have to be renewed. Remove the distributor (see Section 6).

3B.58 Removing the camshaft sprocket

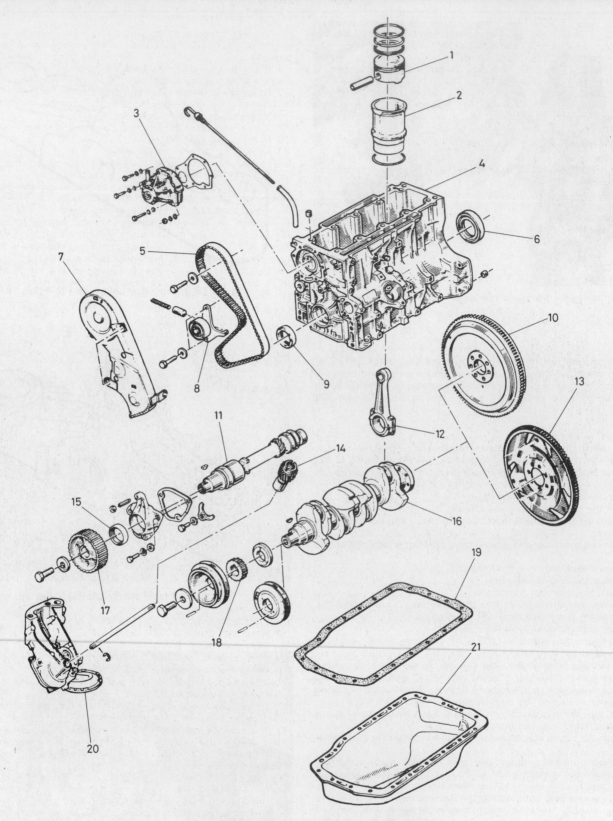

Fig. 13.20 Cylinder block components (Sec 3B)

1	Piston	7	Timing cover	13	Driveplate (automatic	17	Sprocket
2	Liner	8	Tensioner		transmission)	18	Sprocket
3	Water pump	9	Crankshaft front oil seal	14	Drivegear	19	Gasket
4	Cylinder block	10	Flywheel	15	Oil seal	20	Oil pump
5	Timing belt	11	Intermediate shaft	16	Crankshaft	21	Sump
6	Crankshaft rear oil seal	12	Connecting rod				

60 Unscrew and remove the two bolts retaining the camshaft location thrust plate. Remove the plate from the groove in the camshaft.

61 The camshaft front oil seal can now be extracted from its housing. Carefully prise out from the front or preferably drift out from the rear. Take great care not to damage the seal location bore in the head.

62 The camshaft can now be withdrawn carefully through the seal aperture in the front of the cylinder head. Take care during its removal not to snag any of the lobe corners on the bearings as they are passed through the cylinder head.

Crankshaft pulley (sprocket) and front seal – removal

63 The crankshaft pulley and front oil seal can be removed with the engine fitted, as given below. When removing these items with the engine removed, proceed from paragraph 67.

64 Remove the engine undertray panel.

65 Remove the power steering pump drivebelt (where fitted) and the alternator drivebelt.

66 Remove the clutch shield plate.

67 Turn the engine over to the TDC position, with the mark on the flywheel and 'O' mark on the clutch housing in alignment.

68 Remove the timing cover.

69 Loosen the timing belt tensioner and remove the belt.

3B.70 Removing the crankshaft pulley

70 Now remove the crankshaft pulley which is secured by a large bolt and washer (photo). To prevent the shaft turning when measuring the bolt, jam the flywheel starter ring teeth with a large screwdriver blade jammed against a convenient spot on the cylinder block or clutch housing (engine in position).

71 With the pulley bolt removed, the pulley can be withdrawn from the shaft using a puller, or levering free, but take special care. The timing sprocket is located directly onto the pulley rear face and this sprocket, like its counterparts, is relatively fragile, being manufactured of sintered metal. Remove the pulley (and sprocket) from the crankshaft.

72 Prise the key from the keyway in the shaft. Remove the spacer or air conditioning pulley wheel if fitted.

73 Carefully prise out the old seal using a screwdriver blade but do not damage the seal housing or shaft in the process.

74 The timing sprocket is located onto the rear of the pulley by a couple of roll pins, and the two components can be prised apart with care if desired.

75 Seal refitting (engine fitted) is a reversal of the removal process. Lubricate the new seal before fitting with clean engine oil and drive it into position using a suitable pipe drift. The seal must be fitted 'squarely' and accurately or it will leak. If distorted in any way during fitting, renew it.

76 When refitting the pulley retaining bolt, smear the threads with a thread locking compound. Readjust the tension of the timing belt and the respective drivebelts, as applicable. **Note**: *Do not get any oil onto the timing belt during the above operations. If it is oil impregnated it must be renewed before refitting.*

Timing belt/tensioner – removal

77 If the engine is fitted and only the timing belt and tensioner are to

be removed, proceed as given in paragraphs 78 to 89. Where the engine is being dismantled, as given previously, proceed from paragraph 87.

78 Raise the front of the car so that the wheels clear the ground and support it with suitable stands.

79 Unbolt and remove the engine undertray.

80 Remove the power steering pump drivebelt, loosening the pump retaining/adjuster bolts to loosen the tension.

81 Loosen the alternator retaining bolts and remove its drivebelt.

82 Remove the rocker cover.

83 Detach the battery earth terminal.

84 Remove the spark plugs and keep them in order for refitting.

85 Turn the engine over to the TDC position.

86 Drain the cooling system or clamp the inlet manifold heater hose close to the manifold and detach the outlet hose from the manifold at the timing case end. Remove the timing cover.

87 Loosen the tensioner retaining bolts and remove the belt.

88 Take care when removing the tensioner as it is operated by a spring and piston in the crankcase side and the tensioner is under considerable pressure. As the tensioner is unbolted, take care that the spring and piston (photo) don't fly out of their aperture at high speed.

89 If the engine is installed, ensure that the timing is not disturbed whilst the belt and tensioner are removed.

3B.88 Timing belt tensioner spring and piston

Intermediate shaft and oil seal – removal

90 Where the engine is fitted, refer to the previous paragraphs 79 to 82 then continue as follows in paragraphs 91 and 92. Where the engine is removed, continue from paragraph 93.

91 Clamp the inlet manifold heat hose to prevent circulation. Clamp it as close as possible to the inlet manifold and then detach the outlet hose on the manifold.

92 Loosen the timing belt tensioner and remove the timing belt, taking care not to get oil or grease on it.

93 Unscrew the intermediate shaft sprocket bolt, passing a screwdriver shaft or rod through one of the holes in the sprocket to lock against the front of the block and retain the sprocket in position whilst it is being unscrewed.

94 Withdraw the sprocket and remove the Woodruff key from the shaft.

95 Unbolt and remove the intermediate shaft housing.

96 The oil seal can now be extracted. Prise it free with a screwdriver, but take care not to damage the housing.

97 Before the intermediate shaft can be withdrawn, the oil pump drive pinion must first be extracted. To do this, unscrew the two side cover retaining bolts and remove the cover. The oil pump drive pinion is now accessible and can be extracted (photos). Note that on later models the oil pump drive gear is not threaded and, when removing it with the engine *in situ*, a metal brazing rod or similar should be inserted through the gear centre and into the driving rod. This will keep the driving rod located and avoid the possibility of it dropping into the sump.

3B.97A Unscrew the bolts and remove the side cover ...

3B.97B ... then extract the oil pump drive pinion

105 Unscrew the sump bolts and prise the sump free.
106 The oil pump is now accessible for removal. First remove the driveshaft.
107 Unscrew the two pump unit retaining bolts and withdraw the pump. Note that the pump is also located by dowels.
108 Unscrew the four pump cover bolts to inspect the rotors.
109 Extract the relief valve split pin and withdraw the cup, spring, guide and piston. Refer to paragraphs 128 to 135 for overhaul.

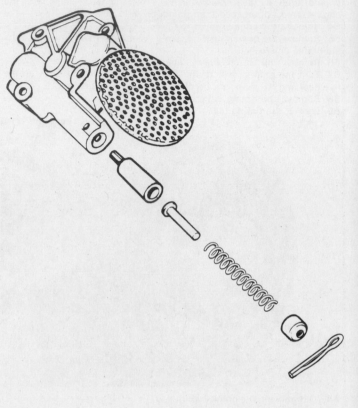

Fig. 13.21 The oil pump and relief valve components (Sec 3B)

98 The intermediate shaft can now be removed from the front of the block.

Flywheel converter driveplate – removal

99 The flywheel/converter driveplate is retained in position by seven bolts on the end of the crankshaft.
100 To loosen these bolts, lock the starter ring with a screwdriver blade or similar implement inserted between the teeth of the starter ring and against a portion of the block.
101 With the flywheel 'jammed' to prevent it turning, release the lockplates (if fitted) and unscrew the retaining bolts. A thread locking compound may have been applied during assembly and they will therefore be tight to unscrew.
102 The bolt holes are not equidistant and so there is no need to mark the flywheel and crankshaft flange alignment positions prior to removal.
103 If the flywheel converter driveplate is to be removed or inspected and the engine is in the car, the engine and clutch will have to be removed for access. It is not possible to withdraw the gearbox.

Sump and oil pump – removal

104 If the engine is still fitted, drain the engine oil, remove the engine undertray and the flywheel coverplate with TDC pick-up unit. Also unscrew and remove the oil level indicator sensor (where applicable) to prevent damaging it when the sump is removed.

Piston/liner assemblies, connecting rods and big-end bearings – removal

110 The procedure is the same as for the OHV engines given in Chapter 1, however, note the following additional points:

(a) If necessary, mark the connecting rods numerically from No 1 (flywheel end) to No 4 on the intermediate shaft side
(b) If removing the big-end bearings with the engine in situ, it is essential to remove the oil pump for better accessibility

Crankshaft and main bearings – removal

111 The procedure is the same as for the OHV engines given in Chapter 1.

Oil pump and lubrication system

112 The engine lubrication system and circuit is shown in Fig. 13.22. Oil from the sump is sucked into the circuit via the pump to the oil filter. From there the oil is passed into the main oil gallery, where it is distributed under pressure to the crankshaft main bearings, big-end bearings and at the front end is circulated via the intermediate shaft. Oil to the camshaft and rocker shaft assemblies is fed up from the rear end of the main gallery. The rocker shaft has a small gauze filter at its oilway entrance. The oil passes along the oilway in the rocker shaft and is dispersed from various interspaced holes to the lobes and bearings of the camshaft.
113 The oil pressure is regulated by a valve unit within the oil pump

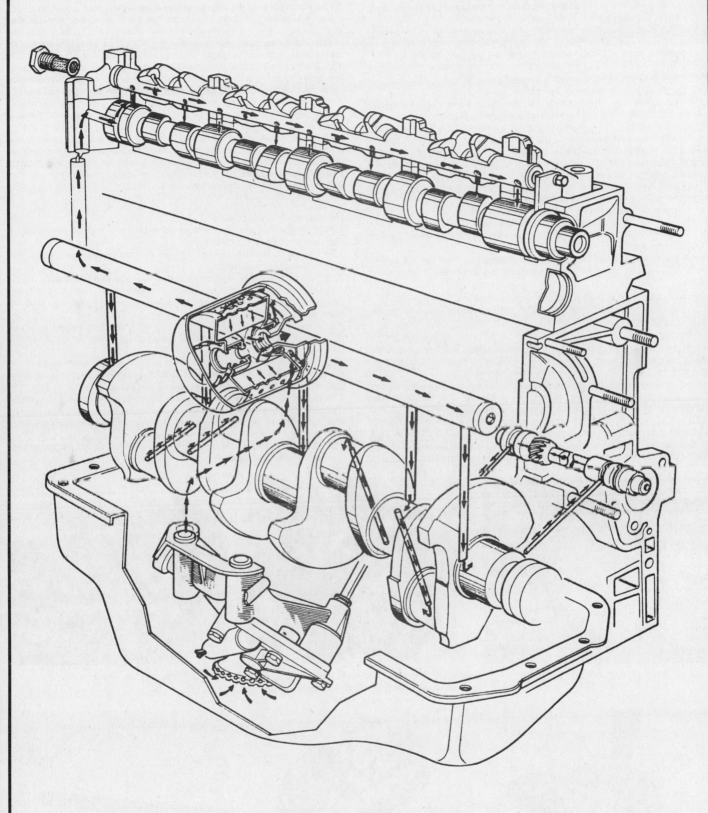

Fig. 13.22 Engine lubrication system (Sec 3B)

unit. The main oil filter is of the disposable cartridge type and is easily accessible for renewal at the specified intervals.

114 The oil pump can be removed with the engine fitted, but due to its location in the underside of the crankcase, the sump must first be removed for access. Once removed, overhaul the pump as described in paragraphs 128 to 135, renewing any defective components.

Oil filter – removal and refitting

115 The oil filter is a disposable canister type, horizontally located in the cylinder block on the right-hand side (photo).

116 If the old filter proves too tight to be removed by hand pressure, use a strap wrench to unscrew it. Alternatively, a worm drive clip (or two joined together to make up the required diameter) strapped around the canister will provide a better grip point. If necessary, the canister can be tapped round using a small hammer.

117 Once the old filter is removed, wipe clean the joint area in the cylinder block and lubricate the new seal ring.

118 Screw the filter carefully into position by hand, ensuring that the seal does not twist or distort as it is tightened.

119 Do not overtighten the filter; it need only be tightened by firm hand pressure.

120 Check and top up the engine oil as required, then run the engine to check for any signs of leakage around the filter seal.

Examination and renovation – general

121 Refer to Chapter 1, Section 17 – the procedure basically applies to the OHC engines.

Timing case components – examination and renovation

122 The principal items of concern in the timing case components are the timing belt, sprockets and belt tensioner unit.

123 If the timing belt is obviously worn, shows any signs of cracking or is coated in oil or grease, it must be renewed. The manufacturers recommend that the belt be renewed every 36 000 miles (60 000 km).

124 Clean and inspect the belt sprockets. They are manufactured in sintered metal and as such are relatively fragile – so handle with care. If they show any signs of wear or damage they must be renewed. Any slight burrs made by the puller during removal must be removed, using

a fine file. Note that the timing belt and sprockets on models produced from 1984 have a different tooth profile than the earlier models, and the timing cover was also modified. Take care when ordering replacements as in each case they are not interchangeable with the earlier type and *vice versa*.

125 The crankshaft sprocket is attached to the pulley wheel by two roll pins and the two components can therefore be separated if necessary. Take care not to damage either component (photo).

126 Examine the timing belt tensioner unit, in particular the jockey wheel tensioner spring and piston. Renew any components showing signs of defects.

127 Where the engine has been dismantled, the front oil seals of the crankshaft, camshaft and intermediate shaft must automatically be renewed on assembly. Failure to do this could cause leakage through one or possibly more of them within a relatively short time and this will, in turn, ruin the timing belt.

Oil pump – overhaul

128 It is essential that all parts of the pump are in good condition for the pump to work effectively.

129 To dismantle the pump, remove the cover retaining bolts and detach the cover (photo).

130 Extract the gears and clean the respective components.

131 Inspect for any signs of damage or excessive wear. Use a feeler gauge and check the clearance between the rotor (gear) tips and the inner housing (photo).

132 Also, check the gear endfloat using a straight-edged rule laid across the body of the pump and feeler gauge inserted between the rule and gears.

133 Compare the clearances with the allowable tolerances given in the Specifications at the start of this Supplement and, if necessary, renew any defective parts, or possibly the pump unit.

134 Do not overlook the relief valve assembly. To extract it, remove the split pin and withdraw the cup, spring, guide and piston. Again, look for signs of excessive wear or damage and renew as applicable (photo).

135 Check the pump driveshaft for signs of wear or distortion and renew if necessary (photo).

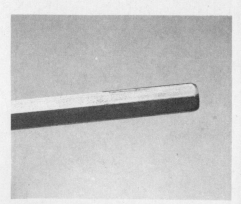

3B.115 Oil filter location

3B.125 The two roll pins connecting the crankshaft pulley to the timing sprocket

3B.129 Removing the oil pump cover

3B.131 Checking the oil pump rotor clearance

3B.134 Oil pump and relief valve components

3B.135 Check the oil pump driveshaft for wear

Intermediate shaft – examination and renovation

136 Clean and inspect the intermediate shaft, the oil pump drive pinion and associate components.

137 Check the front housing for any signs of cracks or distortion.

138 Check the drive pinion teeth for any signs of damage.

139 Renew any defective components

Crankshaft and main bearings – refitting

140 Invert the block and locate the main bearing upper shells into position, engaging the lock tabs into the cut-outs in the bearing recesses (photo). Note that the bearing shells for bearing Nos 1, 3 and 5 are identical and have two oil holes in them whilst the Nos 2 and 4 bearing shells have three holes and an oil groove in them. However, all new shells incorporate grooves.

141 Lubricate the shells with clean engine oil and carefully lower the crankshaft into position (photos).

142 Locate the shells in the main bearing caps in a similar manner to that of the block and lubricate (photo).

143 Fit the bearing caps into position (without the side seals at this stage) and torque tighten the retaining bolts.

144 Now check the crankshaft endfloat using a clock gauge or feeler gauge. Note the total endfloat, and then, referring to the Specifications subtract the specified endfloat. Select a thrust washer of suitable thickness to provide the correct endfloat.

145 Before removing the crankshaft bearing caps to insert the selected thrust washers check the clearance between the block and the bottom of the seal groove on the No 1 and No 5 bearing caps. This clearance can be checked using twist drills to gauge the clearance (photo). Where the clearance is 0.197 in (5 mm) or less use a seal of 0.20 in (5.10 mm) thickness. Where the clearance is in excess of 0.197 in (5.0 mm) select a white code seal of 0.212 in (5.40 mm) thickness.

146 Remove the main bearing caps and crankshaft and insert the selected thrust washers, with their slotted faces towards the

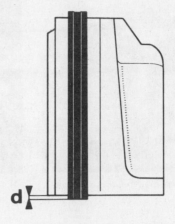

Fig. 13.23 Main bearing cap seal protrusion (Sec 3D)
d = 0.008 in (0.20 mm)

crankshaft. When relocating the bearing caps, fit the side seals to the grooves of the Nos 1 and 5 main bearings. The grooved seal must face out and protrude from the joint face of the cap by approximately 0.008 in (0.20 mm).

147 When fitting the No 1 and No 5 main bearing caps, locate a feeler gauge blade or shim between the crankcase and the seal on each side to ensure that the cap and seals are fitted and do not become dislodged (photo). Take due care when tackling this awkward operation. When fitting the seal into the groove each side, leave a

3B.140 Locate the main bearing shells in the crankcase ...

3B.141A ... and lubricate them

3B.141B Lower the crankshaft into position

3B.142 Fit the shells to the main bearing caps engaging the location tabs

3B.145 Check the main bearing cap seal dimension

3B.147 Using feeler blades to facilitate fitting Nos 1 and 5 main bearing caps

3B.149A Tighten the main bearing cap bolts ...

3B.149B ... and check the endfloat

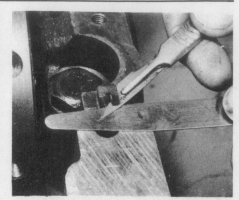

3B.150 Trimming the seals

small amount of seal protruding at the bottom so that it does not slide up the groove as the bearing cap is being fitted.

148 When the end caps are fitted, insert the retaining bolts and check that the crankshaft rotates freely. Tap the bearing housings with a mallet, should the shaft bind. If it continues to bind, a further inspection must be made. Also check that, once in position, there are no gaps between the seals and the crankcase when the shim is removed.

149 Tighten the retaining bolts down to the specified pressure, and recheck the endfloat (photos).

150 Once the shim is removed, the side seals can be trimmed to within 0.020 to 0.028 in (0.5 to 0.7 mm) of the sump joint face. Use a feeler gauge blade of the required thickness to act as a guide when trimming off the seal (photo).

151 Lubricate the new front and rear crankshaft oil seals and carefully locate them into their apertures, tapping them fully into position using a tube drift of a suitable diameter (photo). Ensure that the seals face the correct way round, with the cavity/spring side towards the engine. Should the seal lip accidentally become damaged during fitting, remove and discard it and fit another new seal.

152 If a new clutch spigot bearing is being fitted now is the time to do it. Drive it home using a suitable diameter tube drift.

Note: *When fitting this bearing to the crankshaft, on an engine in which the flywheel bolts are not retained by lockplates, smear the outer bearing surface with a suitable thread locking compound.*

Flywheel converter driveplate – refitting

153 Locate the flywheel or converter driveplate (as applicable) onto the crankshaft and align the bolt holes (which are not equidistant).

154 If the retaining bolts are fitted with a lockplate, use a new plate, and when the bolts have been tightened to the specified torque, bend over the tabs to lock each bolt in position.

155 Where no lockplate is fitted, smear the bolt threads with a thread locking compound and tighten to the specified torque (photo).

156 On manual transmission models, refit the clutch unit, as described in Chapter 5.

Cylinder liners, pistons and connecting rods – refitting

157 Before fitting the piston and connecting rod assemblies into the liners, the liners must be checked in the crankcase for depth of fitting. This is carried out as follows.

158 Although the cylinder liners fit directly onto the crankcase inner flange, O-ring seals are fitted between the chamfered flange and the lower cylinder section, as shown in Fig. 13.24. New O-rings must always be used once the cylinders have been disturbed from the crankcase.

159 First, insert a liner into the crankcase without its O-ring and measure how far it protrudes from the top face of the crankcase. Lay a straight-edge rule across its top face and measure the gap to the top face of the cylinder block with feeler gauges. It should be as given in the Specifications (photo).

160 Now check the height on the other cylinders in the same way and note each reading. Check that the variation in protrusion on adjoining liners does not exceed 0.0016 in (0.04 mm).

161 New liners can be interchanged for position to achieve this if necessary, and when in position should be marked accordingly 1 to 4 from the flywheel end.

162 Remove each liner in turn and position an O-ring seal onto its lower section so that it butts into the corner, taking care not to twist or distort it.

163 Wipe the liners and pistons clean and smear with clean engine oil, prior to their respective fitting.

164 The instructions covering the fitting of the piston rings, pistons

3B.151 Fitting the crankshaft rear oil seal

3B.155 Tightening the flywheel bolts

3B.159 Checking the liner protrusion

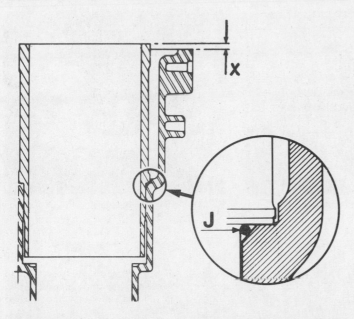

Fig. 13.24 The liner seal location (J) when fitted – measure protrusion X (Sec 3B)

and connecting rods are covered in Chapter 1. Make sure that the following points are adhered to (photos):

(a) *The arrows on top of the pistons must point towards the flywheel*

(b) *The connecting rods and caps must be tightened to the specified torque with the numbered markings in alignment*

(c) *When assembled, reclamp the liners and rotate the crankshaft to ensure it rotates smoothly*

Oil pump – refitting

165 Lubricate the respective parts of the oil pump and reassemble.

166 Insert the rotors (photo) and refit the cover. No gasket is fitted on this face.

167 Tighten the retaining bolts to secure the cover.

168 Insert the oil pressure relief valve assembly, fitting the piston into the spring and the cup over the spring at the opposing end. Compress into the cylinder and insert a new split pin to retain the valve assembly in place.

169 Fit the assembled pump unit into position, together with the driveshaft. Ensure the driveshaft has the C-clip fitted into its groove, and that this is fitted at the pump end of the shaft (photo). Tighten the retaining bolts.

Intermediate shaft – refitting

170 Lubricate the shaft and insert it through the front of the crankcase (photo).

171 Slide the lockplate fork into the protruding shaft location groove and secure the plate with the bolt and washer. Check that the shaft is free to rotate on completion (photo).

172 Fit the new oil seal into the intermediate shaft front cover and lubricate its lips (photo).

173 The timing belt tensioner can also be fitted at this stage (photo). Insert the spring into its housing in the side of the crankcase and locate the plunger over it. Compress the spring and locate the tensioner jockey wheel arm, retaining it with bolts. The spring tension is quite strong and an assistant will probably be required here.

174 Fit the front cover carefully into position with a new gasket, and retain it with the bolts (photo).

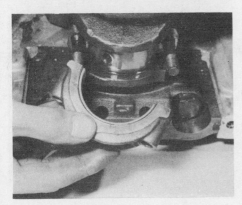

3B.164A Piston crown arrow to flywheel end

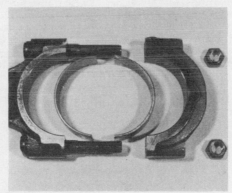

3B.164B The connecting rod bearings and cap laid out ready for assembly

3B.164C Ensure that the rod and cap markings correspond ...

3B.164D ... and fit the caps

3B.164E Tightening the big-end bearing cap nuts

3B.166 Inserting the oil pump rotors

3B.169 Refitting the oil pump – note the C-clip location

3B.170 Inserting the intermediate shaft

3B.171 Fitting the lockplate

3B.172 Fitting the oil seal to the intermediate shaft front cover

3B.173 Fitting the timing belt tensioner

3B.174 Fitting the front cover and gasket

175 Fit the Woodruff key into its groove in the shaft and carefully locate the intermediate shaft drive sprocket into position with its large offset inner face towards the crankcase (photo). Use a suitable diameter drift to tap the sprocket into position over the key.
176 Prevent the sprocket from rotating by inserting a screwdriver blade or similar through a sprocket hole, and tighten the retaining nut (complete with flat washer) to the specified torque.

177 If not already located, the oil pump drive pinion and shaft can now be inserted through the side cover hole in the crankcase. Make sure that the limiting circlip is in position on the oil pump end of the shaft (photo). Once in position, lubricate with engine oil to prevent pinion 'pick-up' on restarting the engine. Where a side-mounted distributor is fitted the pinion must be located as shown in Fig. 13.25 with No 1 cylinder at TDC.

3B.175 Fitting the sprocket

3B.177 Check that the circlip is in position on the oil pump driveshaft

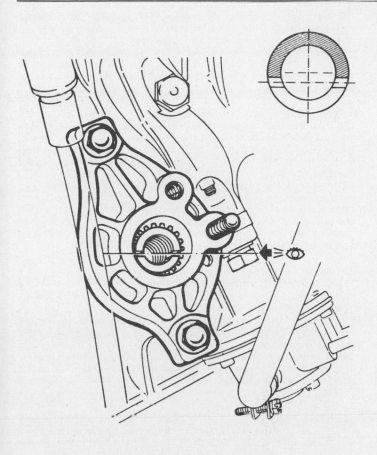

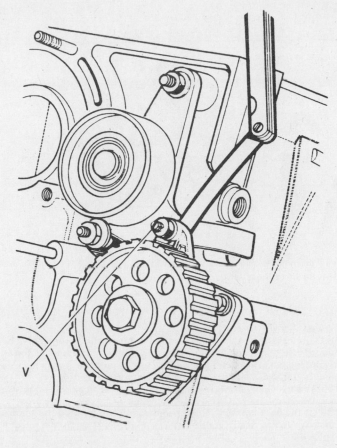

Fig. 13.25 Oil pump drive pinion TDC setting on engines with a side-mounted distributor (Sec 3B)

Fig. 13.26 Tensioner clearance adjustment screw location (V) (Sec 3B)

3B.178 Side cover showing seal location (arrowed)

178 Ensure that the intermediate shaft and oil pump drive rotate freely, then refit the side cover with seal and secure with bolts (photo). Refit the petrol pump.
179 Using a feeler blade, check that the clearance between the lug on the intermediate shaft cover and the tensioner arm is 0.004 in (0.1 mm) (Fig. 13.26). On early models the clearance is not adjustable, but on later models adjust the screw as necessary.

Sump – refitting
180 Check that the mating surfaces of the sump and crankcase are perfectly clean, with no sections of old gasket remaining.
181 Smear an even layer of sealant round the two mating flange surfaces and locate the gasket.
182 Fit the sump carefully into position and locate the retaining bolts (photo). Tighten the bolts progressively by hand in a diagonally opposed sequence.
183 Fit and tighten the sump drain plug, and where applicable refit the oil level indicator sensor.

3B.182 Refitting the sump

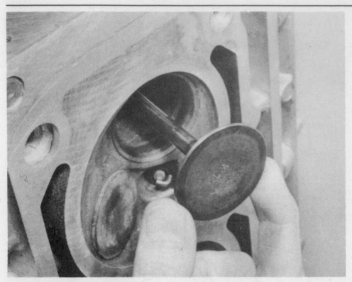

3B.184 Inserting a valve

Valves – refitting

184 Lubricate the valve stems and guide bores with clean engine oil prior to inserting them into their respective positions (photo). All traces of grinding paste around the valve seat and cylinder head seating faces must be removed – check this before each valve is fitted.

185 As each inlet valve is relocated into its guide, fit a new O-ring seal over the stem. These O-rings are not fitted to the exhaust valves.

186 Fit the base washers, springs and spring cap (photo). The springs must be located with the close coil gaps downwards (to the head) if applicable.

187 Using a valve spring compressor, compress the valve springs sufficiently to allow the split collets to be located in the groove of the valve stem, then release the compressor (photo). The use of a little grease will retain the collets in position as the compressor is removed.

188 When all the valves are reassembled, tap the end of each stem in turn with a soft-faced hammer to ensure that the collets are correctly seated.

Rocker shaft – reassembly

189 Lubricate each component as it is assembled with engine oil. Lay the pedestals, spacers, springs and rockers out in order of appearance.

190 Support the rocker shafts in a soft-jawed vice and insert the new filter into the end of it (photo), fit the retaining bolt and tighten it to the specified torque.

191 Assemble the respective pedestals, rocker arms, springs and spacers onto the shaft. When the shaft assembly is complete, compress the last pedestal to align the retaining pin hole in the shaft and pedestal. Drive a new pin into position to secure (photo). Early models fitted with a hollow type roll pin should have the later solid type pin fitted on reassembly.

Camshaft – refitting

192 Check that the respective camshaft location bearings in the cylinder head are perfectly clean and lubricate with some engine oil. Similarly lubricate the camshaft journals and lobes.

193 Insert the camshaft carefully into the cylinder head, guiding the cam sections through the bearing apertures so as not to score the bearing surfaces (photo).

194 With the camshaft in position, the front oil seal can be carefully drifted into position. Lubricate the seal lips with oil and drive into its location using a suitable tube drift.

Cylinder head – refitting

195 Before refitting the cylinder head, check that all the mating surfaces are perfectly clean. Loosen the rocker arm adjuster screws fully back. Also ensure that the cylinder head bolt holes in the

3B.186 Fitting the valve springs

3B.187 Compressing the valve springs

3B.190 Rocker shaft filter

3B.191 Rocker pedestal retaining pin

3B.193 Inserting the camshaft

crankcase are clean and free of oil. Syringe or soak up any oil left in the bolt holes, and in the oil feed hole on the rear left-hand corner of the block. This is most important in order that the correct bolt tightening torque can be applied.

196 Prior to removing the liner clamps, rotate the crankshaft to locate the pistons halfway down the bores. Check that the location dowel is in position at the front right-hand corner.

197 Remove the liner clamps.

198 Fit the cylinder head gasket onto the cylinder block (photo) upper face and ensure that it is exactly located. If possible, screw a couple of guide studs into position. They must be long enough to pass through the cylinder head so that they can be removed when it is in position.

199 Lower the cylinder head into position, engaging with the dowel, and then locate the rocker assembly (photos). Remove the guide studs if they were used.

200 Lubricate the cylinder head bolt threads and washers with engine oil, then screw them into position. Tighten them progressively in the sequence given in Fig. 13.16. Tighten all the bolts to the initial torque specified then further tighten the second torque specified. Now slacken off each bolt one half turn and retighten to the final torque specified.

201 Locate the camshaft retaining plate in the groove of the camshaft and retain it with bolts and washers. Check the camshaft endfloat and, if necessary, adjust by fitting a retainer plate of an alternative thickness (photos).

3B.198 Cylinder head gasket in position

3B.199A Lower the cylinder head onto the gasket ...

3B.199B ... and fit the rocker shaft assembly

3B.201A Fit the camshaft retaining plate ...

3B.201B ... and check the camshaft endfloat

202 Refit the camshaft sprocket and tighten the retaining bolt and flat washer to the specified torque (photo). When fitting the sprocket, ensure that the keyway is in exact alignment with the key. The sprocket inner hub is offset to the camshaft. On later models make sure that the sprocket is fitted the correct way round (Fig. 13.27).
203 The valve rocker clearances can now be adjusted, as described in paragraphs 219 to 223.

compress the piston and spring, then screw in the lower adjustment bolt. Do not tighten it at this stage.
207 Refit the crankshaft pulley unit comprising a spacer washer (or air compressor drive pulley where fitted), the Woodruff key and the timing sprocket and pulley (photo).
208 Rotate the crankshaft to the TDC position with No 1 piston on firing stroke. The bottom sprocket timing mark should be in alignment with the timing case retaining stud, as shown in Fig. 13.28. Check this

3B.202 Tightening the camshaft sprocket bolt

3B.207 Crankshaft spacer washer and Woodruff key

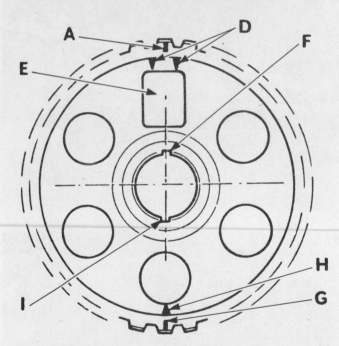

Fig. 13.27 Camshaft sprocket fitted to later models (Sec 3B)

A, D, E and F – 851 engine   G, H and I – J6R engine

Timing sprockets, tensioner and belt – refitting
204 Referring to Section 4 refit the water pump with its pulley and outlet pipe.
205 Refit the thermostat (see Section 4) and relocate the hose between the water pump housing and the thermostat housing, securing with worm drive clips.
206 Refit the timing belt tensioner unit (if not already assembled). Insert the spring and piston into the crankcase aperture. Locate the jockey wheel. Fitting the top bolt first and pushing it inwards,

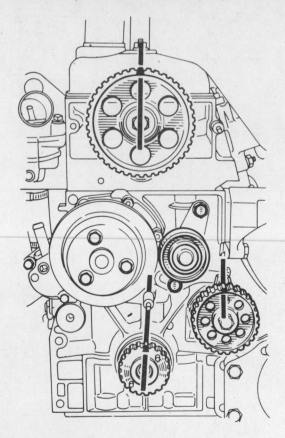

Fig. 13.28 The timing sprocket marks aligned at their respective timing positions (Sec 3B)

Mark on camshaft sprocket in line with rocker cover stud
Mark on crankshaft sprocket in line with central timing case stud
Mark on intermediate shaft sprocket in line with edge of crankcase

at the rear end by the mark on the flywheel, which should be opposite the mark on the rear face of the crankshaft when dismantled.

209 Set the camshaft sprocket with its timing mark aligned with the rocker cover stud.

210 Align the intermediate shaft sprocket vertically in line with the crankcase web.

211 The timing belt can now be relocated over the sprockets and with its outer face bearing on the adjuster jockey wheel. If using the old belt (which is *note* recommended), it should be fitted facing the same way as when removed.

212 To adjust the tensioner, loosen the tensioner bolts about a quarter of a turn to allow the tensioner to automatically take up any belt slack under the spring tension. Tighten the tensioner retaining bolts.

213 Next check the timing belt tension at the mid-point of its longest run (between the camshaft and intermediate shaft pulleys). The correct amount of deflection is about 0.25 in (6.0 mm) (photo).

214 As a double check on the tension, turn the engine through two complete revolutions in a clockwise direction (when facing the sprockets), loosen the tensioner bolts and allow the tensioner to adjust itself, if at all. Then recheck the belt tension after retightening the bolts. **Note:** *Do not turn the engine in an anti-clockwise direction!*

215 Refit the timing cover to complete. Do not forget the spacer tube, which fits onto the stud immediately above the crankshaft pulley and the hose clips on the upper left-hand bolt (photo).

3B.213 Check timing belt tension at point arrowed

3B.215 Timing cover location stud and spacer

Inlet and exhaust manifolds – refitting

216 Check that the mating faces of the manifolds and cylinder head are perfectly clean.

217 Fit the combination gasket onto the exhaust manifold studs and the individual gaskets onto the inlet manifold studs.

218 Refit the manifolds and secure with washers and nuts.

Valve rocker clearances – adjustment

219 The precise adjustment of the valve/rocker clearances is of utmost importance for two main reasons. The first, to enable the valves to be opened and fully closed at the precise moment required by the cycle of the engine. The second, to ensure quiet operation and minimum wear of the valve gear components.

220 Settings made when the engine is on the bench will require rotation of the engine and this may be done by turning the exposed crankshaft pulley bolt. if the engine is in the car and a manual gearbox fitted, select top gear, then jack-up the front so that a front wheel is clear of the ground and can be turned. With automatic transmission, this method is not possible and 'inching' the engine using the starter motor will have to be resorted to.

221 Turn the engine using one of the methods described until the exhaust valve of No 1 cylinder is fully open (flywheel end). The clearances for No 3 cylinder inlet valve and No 4 cylinder exhaust valve can now be checked and adjusted using a feeler blade.

222 If the clearance requires adjustment, loosen the locknut, and with the feeler in position turn the adjuster screw until the feeler blade is nipped and will not move. Now unscrew the adjuster until the feeler blade is a stiff sliding fit. Tighten the locknut and recheck the clearance (photo). Refer to the Specifications for the correct clearances.

3B.222 Adjusting the valve clearances

223 Repeat the adjustment procedure using the following sequence:

Exhaust valve fully open on cylinder:	Inlet valve to adjust on cylinder:	Exhaust valve to adjust on cylinder:
1	3	4
3	4	2
4	2	1
2	1	3

Ancillaries and fittings – refitting

224 The engine ancillary items and fittings are reassembled in the reverse order to removal as given in paragraphs 29 to 32. When refitting the following items, attend to the respective adjustments and special fitting instructions with due care.

225 When the engine is refitted to an automatic transmission, reattach the driveplate bolts before assembling the lower coverplate and TDC pick-up unit. Ensure that the converter and driveplate are realigned as when removed.

226 Reattach the starter motor with the three bolts at the rear and two at the front. Make sure that the wiring connections are correctly made.

227 Fit the power steering pump and bracket, followed by the alternator and respective drivebelts, which must be retensioned.

228 Reconnect the exhaust downpipe, applying some suitable sealant to the pipe joint connections (photo). Use new clamps where the old ones are suspect at the bottom end.

229 Reconnect the radiator and the respective coolant hoses.

230 Reconnect the throttle cable. Any adjustments that may be necessary can be made when the engine is restarted.

231 Top up the engine and gearbox with the relevant oils on completion. Refill the cooling system and bleed it, as described in Chapter 2.

232 Before restarting the engine, check that all fittings and ancillaries have been securely located. Do not leave tools lying around the engine area, and wipe up any spillages.

233 After running the engine for twenty minutes switch it off and allow it to cool for $2\frac{1}{2}$ hours. Then retighten the cylinder head bolts using the following method. Refer to Fig. 13.16 and loosen bolt No 1 half a turn then retighten to the specified torque. Repeat the procedure on each bolt in turn using the correct sequence.

4.1 Additional bleed screw for the cooling system

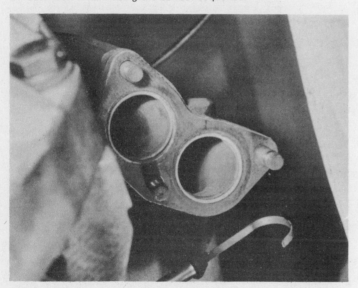

3B.228 The exhaust downpipe and flange gasket

4 Cooling system

Cooling system – draining and filling

1 On later models the cooling system expansion bottle is located on the bulkhead and there is an additional bleed screw next to the valve cover (photo). The coolant level on the new bottle is 1.5 in (35 mm) from the bottom.

Radiator (US/Canada models) – removal and refitting

2 The procedure is similar to that given in Chapter 2. However, on models with air conditioning, the following additional work must be carried out.

3 Remove the radiator cooling fans and disconnect the condenser, leaving the refrigerant hoses connected, and placing the unit on the front bumper. Reverse the procedure for refitting.

Water pump (OHV engines) – removal and refitting

4 On later models the water pump pulley is a press fit on the shaft. When removing the pump the centre bolt can be loosened, but cannot

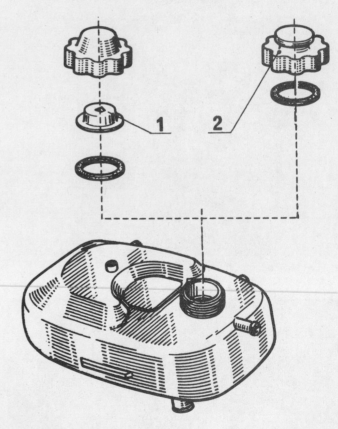

Fig. 13.29 Bulkhead mounted expansion bottle (Sec 4)

1 Valve   2 Cap

be removed because of the pulley location. With this one exception the removal and refitting procedure is as given in Chapter 2.

5 New water pumps are supplied without a pulley and the old one must therefore be transferred. To do this use a two-legged puller acting on a bolt screwed into the shaft. Make sure that the central bolt is located in the pump before pressing the pulley on the shaft with a bolt and spacer. Note that the outer faces of the pulley and shaft must be flush.

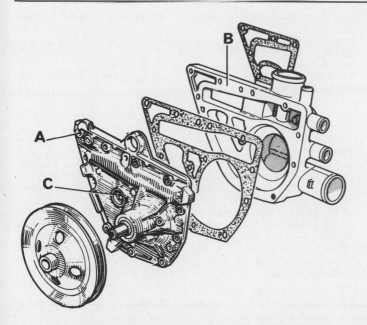

Fig. 13.30 Water pump fitted to later OHV engines (Sec 4)

A   Outer casing   C   Captive bolt location
B   Inner casing

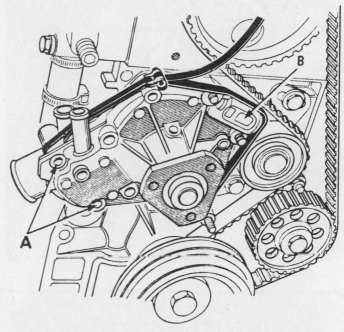

Fig. 13.32 Using a strap to retain the timing belt tensioner on certain OHC engines (Sec 4)

A   Return elbow bolt holes   R   Tensioner

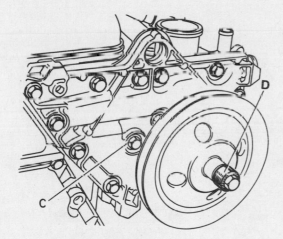

Fig. 13.31 Pressing the pulley onto the water pump on later OHV engines (Sec 4)

C   Captive bolt location   D   Spacer and bolt

Water pump (OHC engines) – removal and refitting

6 Remove the radiator and water pump drivebelt, as described in Chapter 2.
7 Disconnect the hoses and remove the timing cover.
8 Unbolt the pulley from the drive flange (photo).
9 Unscrew the nuts and withdraw the coolant return elbow (photo).
10 Where applicable retain the timing belt tensioner (Fig. 13.32) then unscrew the mounting bolts and tap the water pump free. Remove the gasket (photo).
11 Clean the mating surfaces.
12 Refitting is a reversal of removal, but fit new gaskets and bleed the system as described in Chapter 2.

Thermostat (OHC engine) – removal, testing and refitting

13 The thermostat is located in the cylinder head outlet on the right-hand side of the engine.
14 To remove the thermostat, drain sufficient coolant to bring the

4.8 Water pump and pulley on the OHC engines

4.9 Removing the coolant return elbow on the OHC engines

4.10 Removing the water pump on the OHC engines

4.15 Removing the thermostat on the OHC engines

level below the thermostat, then unbolt the cover; leaving the hose still attached.
15 Lift out the thermostat and remove the O-ring from the housing (photo).
16 Test the thermostat with reference to Chapter 2.
17 Refitting is a reversal of removal, but fit a new O-ring and finally top up and bleed the system.

---

5   Fuel and exhaust systems

Turbocharger – description and precautions
1   The turbocharger uses the exhaust gas pressure to drive a turbine which pressurises the inlet system, thus providing increased efficiency and power. The turbine rotates at a very high speed and has its own oil supply direct from the engine.
2   To prevent possible seizure of the turbine spindle it is **important** to let the engine idle for approximately 30 seconds prior to switching it off, otherwise the turbine may still be rotating without an oil supply.
3   The turbocharger components are shown in Fig. 13.33 and include an intercooler, special carburettor, electric fuel pump and the turbocharger itself.

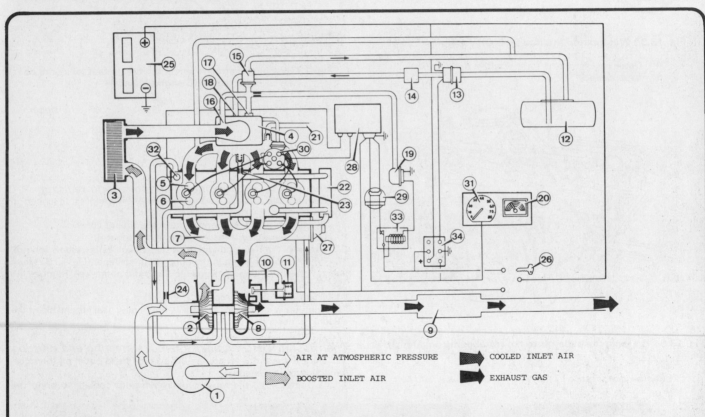

Fig. 13.33 Diagram of turbocharger components and controls (Sec 5)

1   Air filter
2   Inlet air compressor
3   Intercooler
4   Pressure-tight carburettor
5   Inlet valve
6   Exhaust valve
7   Exhaust manifold
8   Exhaust driven turbine
9   Expansion chamber
10 Wastegate control
11 Wastegate valve
12 Fuel tank
13 Fuel pump

14 Fuel filter
15 Fuel pressure regulator
16 Boosted air pressure sensor
17 Boosted air feed to accelerator pump
18 Fuel inlet
19 Pressure cut-out (activated in event of excess boost pressure)
20 Pressure gauge (showing absolute intake pressure)
21 Vacuum take-off to Master-Vac (with check valve)

22 Oil separator
23 Crankcase vapour rebreathing (non-boosted air)
24 Crankcase vapour rebreathing (boosted air)
25 Battery
26 Ignition switch
27 Starter
28 Ignition computer
29 Ignition coil
30 Distributor
31 Rev counter

32 Pinking detector (if pinking is sensed, the detector retards the ignition 4° and re-establishes it after 15 seconds)
33 Cut-out relay (activated by pressure switch 19. Ignition circuit is opened if air pressure is excessive)
34 Tachometer relay (current feed to fuel pump cut off when ignition is on and engine not running)

Turbocharger – removal and refitting

4 Note that the turbocharger wastegate valve is set at the factory and therefore no attempt should be made to adjust it (photo).
5 Remove the air filter-to-turbocharger hose.
6 Remove the air filter and exhaust downpipe.
7 Disconnect the intercooler hose.
8 Unbolt the oil feed pipe flange and oil return pipe flange (photo).
9 Unbolt and remove the support strut.
10 Unscrew the mounting nuts and withdraw the unit. Do not, however, lift it by the wastegate rod (photo).
11 Refitting is a reversal of removal, but before connecting the oil feed pipe turn the engine on the starter with the coil HT lead or ignition module disconnected until oil flows from the pipe. Fit new gaskets and mounting nuts. Allow the engine to idle for several minutes after starting.

Intercooler (Turbocharger system) – removal, checking and refitting

12 Remove the radiator grille and crossmember (photo).
13 Disconnect the hoses and lift out the intercooler (photo).
14 A control flap in the top of the intercooler is operated by a thermostat in order to maintain the inlet air at a pre-determined temperature. To check the thermostat, first immerse the capsule in water at a temperature of 41 to 45°C (105 to 113°F) for five minutes.

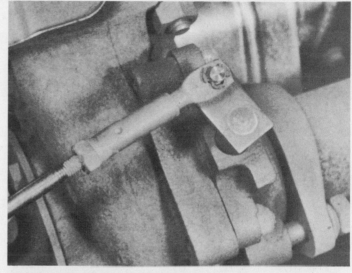

5.10 The turbocharger wastegate rod

5.4 The turbocharger unit

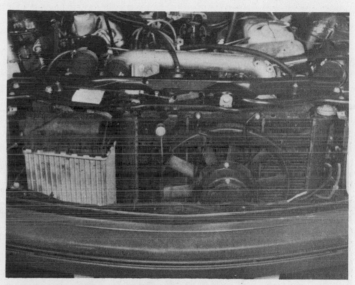

5.12 The intercooler and radiator

5.8 The turbocharger oil return pipe

5.13 Turbocharger intercooler

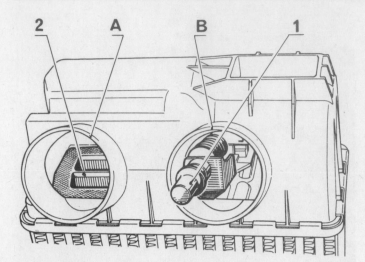

Fig. 13.34 Intercooler control thermostat (Sec 5)

A	Inlet	1	Thermostat capsule
B	Outlet	2	Flap

The internal flap should close, thus preventing air passing through the intercooler. Using the same procedure with the water at a temperature of 45 to 49°C (113 to 121°F) the flap should open.
15 Refitting is a reversal of removal.

Fuel filter (Turbocharger system) – removal and refitting
16 The fuel filter is located next to the fuel pump beneath the rear of the car (photo).

17 Clamp the hoses, unbolt the clip, and remove the filter. Note which way round it is fitted.
18 Refitting is a reversal of removal.

Fuel pump (Turbocharger system) – removal and refitting
19 The fuel pump is also located beneath the rear of the car (photo).
20 Clamp the hoses, then disconnect the wiring , unbolt the clip and release the pump from the hoses.
21 Refitting is a reversal of removal. The pump terminals are marked to ensure correct fitting.

Pressure regulator (Turbocharger system) – removal and refitting
22 The pressure regulator is located on the right-hand side of the engine compartment (photo).
23 Identify the hoses for location then disconnect them.
24 Unscrew the bolts from the bottom and withdraw the regulator from the bracket.
25 Refitting is a reversal of removal.

Carburettor (Turbocharger system) – description and idling adjustment
26 A Solex 32 DIS carburettor is fitted to the Turbocharger system downstream of the turbocharger. All the internal channels are designed to operate under boost pressure (photo).
27 Accurate adjustment of the idling speed is only possible using an exhaust gas analyser and engine tachometer. The air filter and all air hoses must be securely fitted.
28 The procedure is identical to that described in Chapter 3, Section 9 (photos).

Carburettor (Solex 32 DIS – Turbocharger system) – adjustments
29 Adjustment of the initial throttle opening must be made with the carburettor removed. First turn the choke lever and close the choke flap, then use a twist drill to check that the gap between the throttle

5.16 Fuel filter

5.19 Fuel pump

5.22 Pressure regulator

5.26 The Solex 32 DIS carburettor fitted to Turbo models

5.28A Idling speed adjusting screw on the Solex 32 DIS

5.28B Idling mixture adjusting screw on the Solex 32 DIS

valve and barrel is as given in the Specifications. If not, turn the adjusting screw as necessary.

30 To adjust the accelerator pump stroke, insert a twist drill the same diameter as the pump stroke between the throttle valve and the barrel (Fig. 13.36) and check that the accelerator pump lever is just at the end of its travel. If not, turn the adjusting nut as necessary.

Carburettor (Weber 32 DIR) – description and idling adjustment

31 GTL models are fitted with the Weber 32 DIR twin choke carburettor. The idling adjustment procedure is identical to that described in Chapter 3, Section 9.

Carburettor (Weber 32 DIR) – adjustments

32 To adjust the float level, remove the float chamber cover and hold the carburettor vertical with the needle valve shut, but without depressing the spring-tensioned ball. Refer to Fig. 13.37 and check that dimension A is as given in the Specifications. If not, bend the float arm. Similarly check dimension B with the cover held horizontally and, if necessary, bend the float arm end tab. Refit the cover together with a new gasket.

33 To adjust the initial throttle opening, turn the lever to fully close the choke flap then use a twist drill to check that the gap between the primary throttle valve and barrel is as given in the Specifications. If not, loosen the intermediate locknut, turn the screw as necessary, then retighten the locknut.

34 To adjust the choke flap, operate the lever to close the flap and check that the gap between the flap and barrel is as given in the Specifications (see Fig. 13.38). If not, turn the adjusting screw.

35 To adjust the pneumatic part-opening setting depress the pushrod (Fig. 13.39) then close the choke flap and use a twist drill to check that the gap between the choke flap and barrel is as given in the Specifications. If not, extract the plug and turn the adjustment screw as necessary. Refit the plug after making the adjustment.

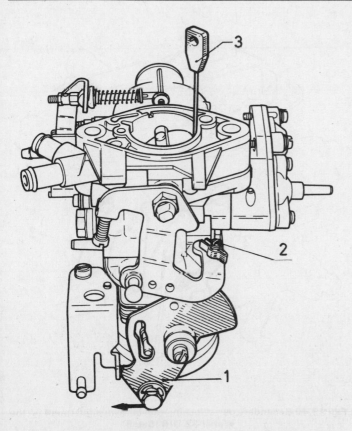

Fig. 13.35 Initial throttle opening adjustment on the Solex 32 DIS (Sec 5)

1 Choke lever   3 Gauge/twist drill
2 Adjusting screw

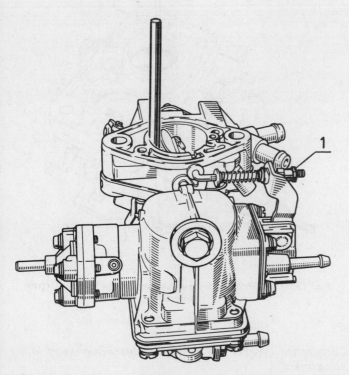

Fig. 13.36 Accelerator pump stroke adjustment on the Solex 32 DIS (Sec 5)

1 Adjusting nut

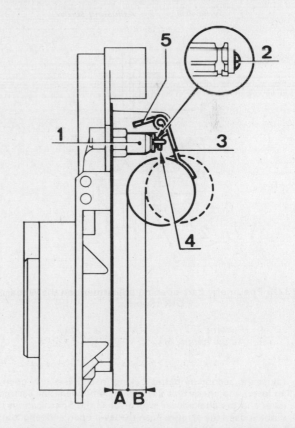

Fig. 13.37 Float level adjustment on the Weber 32 DIR (Sec 5)

1 Needle valve   3 Float arm   5 End tab
2 Spring-tensioned ball 4 Arm guide   A Float level
                                           B Float travel

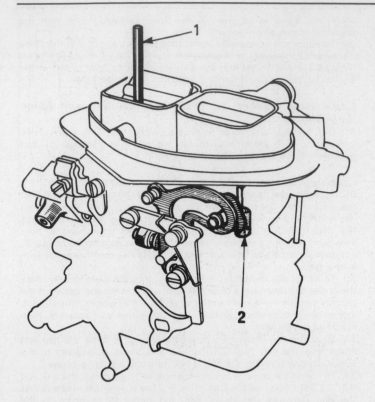

Fig. 13.38 Choke flap adjustment on the Weber 32 DIR (Sec 5)

    1   Twist drill   2   Adjusting screw

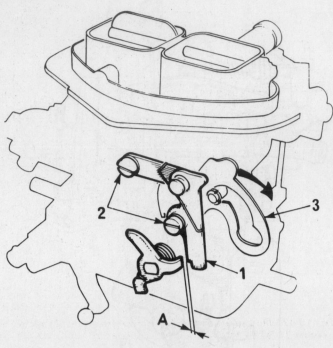

Fig. 13.40 Secondary barrel locking mechanism adjustment on the Weber 32 DIR (Sec 5)

    1   Lever   3   Choke lever
    2   Screws   A   Clearance

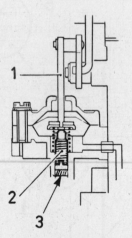

Fig. 13.39 Pneumatic part-opening adjustment on the Weber 32 DIR (Sec 5)

    1   Pushrod   3   Plug
    2   Adjustment screw

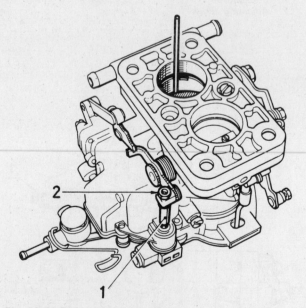

Fig. 13.41 Defuming valve adjustment on the Weber 32 DIR (Sec 5)

    1   Valve spindle   2   Adjusting nut

36 To adjust the secondary barrel locking mechanism fully open the choke flap lever, and check that the clearance between the secondary throttle lever and the choke lever is as given in the Specifications (Fig. 13.40). If not, loosen the screws, push the lever open and retighten the screws.

37 To adjust the defuming valve fitted to some carburettors, fully open the choke flap then push in the defuming valve spindle and use a twist drill to check that the clearance between the primary throttle valve and barrel is as given in the Specifications. If not, turn the adjusting nut on the spindle.

Carburettor (Weber 32 DARA) – description and idling adjustment

38 The Weber 32 DARA carburettor is fitted to models with the A6M and J6R engines. It is of twin choke type and incorporates an automatic choke. The idling adjustment procedure is identical to that described in Chapter 3, Section 9.

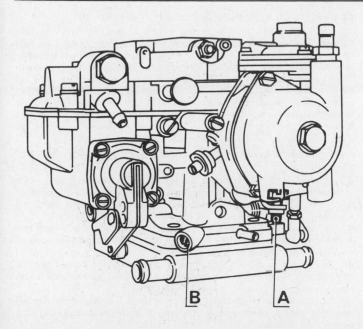

Fig. 13.42 Idling speed screw (A) and mixture screw (B) on the
Weber 32 DARA fitted on models R1342 and R1352 (Sec 5)

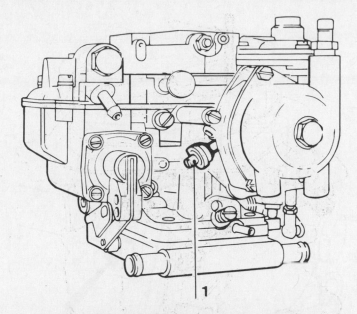

Fig. 13.44 Initial throttle opening adjusting screw (1) on the
Weber 32 DARA (Sec 5)

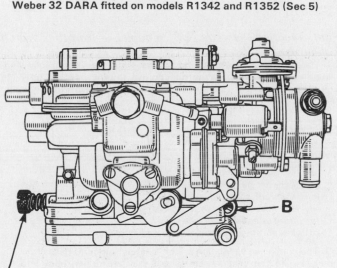

Fig. 13.43 Idling speed screw (A) and mixture screw (B) on the
Weber 32 DARA fitted on models R1343 and R1353 (Sec 5)

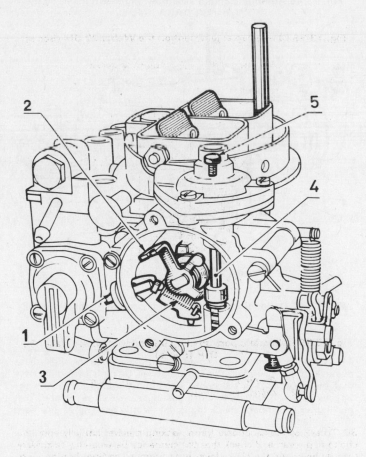

Fig. 13.45 Automatic choke adjustment on the Weber 32 DARA
(Sec 5)

1   Initial throttle opening screw   4   Pushrod
2   Operating lever   5   Adjusting screw
3   Cam

Carburettor (Weber 32 DARA) – adjustments

39 To adjust the initial throttle opening, fully close the choke flap
manually and position the adjusting screw on the top cam flat (except
32 DARA 41) or second from top cam flat (32 DARA 41). Using a
twist drill, check that the gap between the primary throttle valve and
barrel is as given in the Specifications. If not, turn the screw as
necessary.

40 To adjust the automatic choke (pneumatic part-open setting)
remove the cover and bi-metallic spring then manually close the choke
flaps. Fully raise the pneumatic capsule pushrod and turn the choke
operating lever against it. Using a twist drill check that the gap
between the choke flap and barrel is as given in the Specifications. If
not, turn the adjustment screw inside the top of the pneumatic capsule
as necessary. Refit the bi-metallic spring and cover, making sure that
the spring engages the lever correctly and the assembly marks on the
cover and body are in alignment.

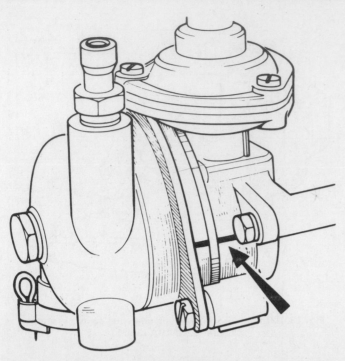

Fig. 13.46 Automatic choke assembly alignment marks on the Weber 32 DARA (Sec 5)

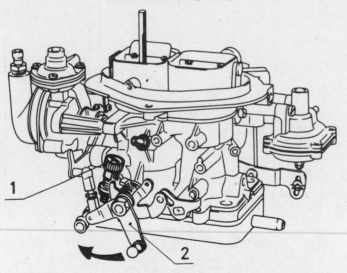

Fig. 13.47 Deflooding mechanism adjustment on the Weber 32 DARA (Sec 5)

1 Adjusting screw	2 Throttle lever

41 To adjust the deflooding mechanism, fully close the choke flaps manually then fully open the throttle lever. Using a twist drill check that the gap between the choke flaps and barrel is as given in the Specifications. If not, turn the adjusting screw as necessary. After making an adjustment check the initial throttle opening, as described in paragraph 39.

42 To adjust the float level proceed as described in paragraph 32.

Carburettor (Canada models) – idling adjustment

43 The procedure is as for UK models except that the air filter intake valves must be plugged during the procedure in order to by-pass the air injection system.

Carburettor (Solex 32 MIMSA) – general

44 At the time of writing, no dismantling or overhaul information was available for this carburettor.

Fuel injection system (US models) – description

45 US models are fitted with the Bosch L-Jetronic fuel injection system. The system components are shown in Fig. 13.48. The fuel mixture adjusting screw is sealed at manufacture and will not normally require adjustment. The idling speed screw, however, is not sealed. The fuel injectors are electronically-controlled for duration and all operate simultaneously. The control unit determines the injection period which proportions the quantity of fuel injected. The airflow meter senses the volume of air entering the engine and sends this information to the control unit.

Fuel injection system (US models) – component removal and refitting

46 The fuel pump and filter are located beneath the rear of the car. To remove either unit, clamp the hoses and disconnect them, together with the wiring, where applicable. Remove the clamp bolt and withdraw the unit. Refitting is a reversal of removal.

47 To remove the auxiliary air regulator, disconnect the wiring and hoses and unbolt the unit. Plug the bolt holes to prevent loss of coolant. Refitting is a reversal of removal.

48 To remove an injector, disconnect the hose by cutting and pull out the injector. Refitting is a reversal of removal, but fit new seals and moisten the new hose with fuel before fitting it.

49 To remove the temperature sensor or thermo-time switch, disconnect the wiring, unscrew the unit and plug the hole. Refitting is a reversal of removal.

50 To remove the airflow meter, disconnect the wiring and hoses then release the clip and withdraw the unit. Refitting is a reversal of removal.

Fuel injection system (US models) – throttle switch adjustment

51 Loosen the switch mounting screws. With the throttle fully shut, slowly turn the switch in the direction of throttle opening until the contacts are heard to operate. Tighten the screws with the switch in this position.

Emission control systems – description

52 All engines incorporate one or more of the following emission control systems:

(a) *Crankcase ventilation system*
(b) *Exhaust gas recirculation (EGR)*
(c) *Vacuum spark advance*
(d) *Air injection system*
(e) *Inlet preheating device*
(f) *Evaporative emission control system*

53 The crankcase ventilation system channels unburnt gases from the crankcase to the inlet system where they are drawn back into the engine for combustion. Maintenance consists of checking the hoses for security (photo).

54 The exhaust gas recirculation system channels some of the inert exhaust gases back into the engine in order to reduce the emission of nitrogen oxides. The system is operated by vacuum from the inlet manifold in conjunction with coolant temperature. On 5-speed models the system is inoperative in 5th gear. The EGR valve operation may be checked visually by accelerating the **hot** engine and observing the valve movement through the special opening.

55 The vacuum spark advance system is vacuum-operated by a thermovalve at coolant temperatures of less than 45°C (113°F). On manual gearbox models it operates only in 5th gear.

56 The air injection system supplies air to the exhaust manifold under certain conditions in order to assist unburnt gases to continue burning. The system functions as a result of vacuum pulses in the exhaust manifold, which draw air from the air filter through two one-way valves. The one-way valves prevent backfiring when the flow of exhaust gases is low.

57 The inlet preheating device maintains the inlet air temperature stable by using a thermostat-controlled flap to divert cool air from the atmosphere or hot air from the exhaust manifold shroud as necessary (photo). With an inlet air temperature of 17.5°C (63°F) the flap closes the cold air supply, but with an inlet air temperature of 26°C (79°F) the hot air supply is closed.

58 The evaporative emission control system prevents the escape of

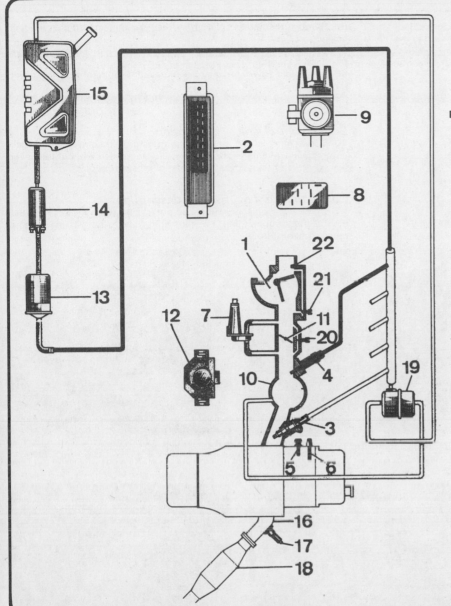

Fig. 13.48 Diagram of the Bosch L-Jetronic fuel injection system (Sec 5)

1 Airflow meter
2 Electronic control unit
3 Injector
4 Cold start injector
5 Coolant temperature sensor
6 Thermo-coolant time switch
7 Auxiliary air regulator
8 Relay
9 Ignition distributor
10 Intake manifold chamber
11 Throttle plate assembly
12 Throttle position switch
13 Fuel filter
14 Fuel pump
15 Fuel tank
16 Exhaust manifold
17 Oxygen sensor
18 Tri-functional catalytic converter
19 Fuel pressure regulator
20 Throttle plate bypass screw
21 Airflow meter bypass screw
22 Air temperature sensor

5.53 Crankcase ventilation unit/oil filler cap on the rocker cover of a Turbo model

5.57 Inlet preheating thermostat and flap

vapour in the form of hydrocarbons from the fuel tank. The fuel tank filler cap is of the sealed type and fuel vapour is channelled to the inlet system. When the engine is stopped the vapours are retained by a charcoal canister, but when the engine is started the vapours are drawn into the inlet system.

Fuel tank gauge unit (Estate models) – removal and refitting

59 Disconnect the battery negative lead.
60 Remove the rear carpet and trim panel.
61 Disconnect the wiring.
62 Use a screwdriver to turn the retaining ring anti-clockwise then withdraw the unit.
63 Check the seal and seating faces.
64 Refitting is a reversal of removal.

Fuel pump (OHC engines) – removal and refitting

65 The fuel pump is located on the left-hand side of the engine.

66 Remove the undertray, if fitted, then disconnect and plug the fuel hoses.
67 Unscrew the mounting nuts and withdraw the fuel pump. Remove the gasket (photo).
68 If required, the filter may be removed after removing the cover (photos).
69 Refitting is a reversal of removal, but fit a new gasket.

Cruise control unit – description

70 Some models are equipped with a cruise control unit which automatically maintains the selected cruising speed without the driver having to keep the accelerator pedal depressed (photos).

Computer fuel flowmeter – description

71 On models equipped with a computer, a flowmeter is incorporated in the fuel supply line on the right-hand side of the engine compartment (photo).

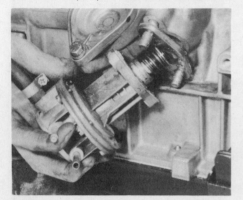

5.67 Removing the fuel pump on the OHC engine

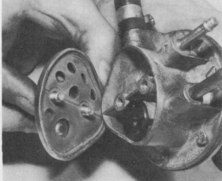

5.68A Removing the fuel pump cover on the OHC engine ...

5.68B ... followed by the filter

5.70A Cruise control switch on the steering wheel

5.70B Cruise control unit incorporated in the throttle cable

5.70C Control motor on the cruise control unit

5.70D Cruise control speed sensor on the right-hand side driveshaft

5.71 Computer flowmeter location

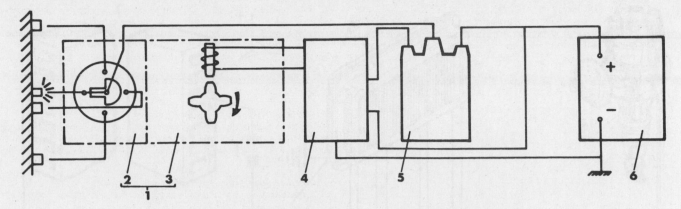

Fig. 13.49 Diagram of the transistorized ignition system (Sec 6)

1	Distributor	3	Pulse generator	5	Ignition coil
2	Rotor arm	4	Electronic module	6	Battery

6 Ignition system

Transistorized ignition system – general description

1   Certain early models are fitted with a transistorized ignition system in place of the conventional system. The system components are shown in Fig. 13.49.

2   The distributor contact points are replaced by a pulse generator; comprising a reluctor and magnetic sensor. The voltage signals from the generator are sent to the electronic module which produces control current for the ignition coil. HT current distribution is via the distributor cap and rotor arm, as in the conventional system. Ignition advance and retard is also controlled as in the conventional system by centrifugal weights and a vacuum capsule on the distributor.

3   On some models, two magnetic sensors are fitted in the distributor, positioned diagonally opposite each other.

4   Early Turbo models are fitted with an uprated version of the transistorized ignition system. The electronic module is incorporated within a computer and the system includes a knock sensor and pressure operated ignition cut-out. The knock sensor detects the onset of pinking and the computer automatically retards the ignition timing between 6 and 9°. The pressure-operated ignition cut-out switches off the ignition if the induction pressure exceeds 11.2 lbf/in^2 (775 mbar).

Transistorized ignition system (non Turbo) – checking and adjustment

5   First check that all wiring and connectors are in good condition and fitted correctly.

6   Remove the distributor cap and disconnect the central (ignition coil) HT lead.

7   With the ignition on and the end of the HT lead held 0.125 to 0.190 in (3.2 to 4.8 mm) from the cylinder head with well-insulated pliers, move a permanent magnet quickly over the magnetic sensor and check that an HT spark occurs. If not, proceed as follows.

8   Switch off the ignition and disconnect the magnetic sensor wiring, then connect an ohmmeter across the sensor terminals. If infinity is registered the sensor coil wiring is open circuited and the coil should be renewed. Now connect the ohmmeter between one of the terminals and the distributor body. If infinity is not registered the coil insulation is faulty and the coil should be renewed. Note that the coil may be damaged if a test lamp is used for these checks.

9   To check the ignition coil and electronic module, connect a voltmeter to the coil as shown in Fig. 13.52 and switch on the ignition. Move a permanent magnet quickly over the magnetic sensor – if the needle remains still the electronic module is proved faulty, and if it moves the coil is proved faulty.

10 To adjust the magnetic sensor-to-reluctor air gap on a single sensor distributor, loosen the two screws shown in Fig. 13.53. Turn

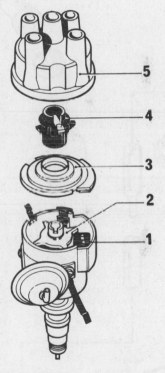

Fig. 13.50 Transistorized ignition distributor components (Sec 6)

1	Sensor coil	4	Rotor arm
2	Reluctor	5	Cap
3	Dust seal		

the engine as necessary to align one of the reluctor arms with the sensor post, then use a non-magnetic feeler blade to check that the air gap is as given in the Specifications. Move the sensor as necessary then tighten the screws. Check the gap for each of the reluctor arms. If, due to wear or damage, it is not possible to obtain the correct gap on all of the arms the distributor must be renewed.

11 Adjustment of the air gap on distributors with two magnetic sensors is identical, but additionally the offset of the sensors must be adjusted. First align one the reluctor arms with the fixed sensor post, then loosen the screws and position the adjustable sensor post as shown in Figs. 13.55 or 13.56. On manual transmission models the offset is 3° and on automatic transmission models 5°. Note that the edge of the sensor post must be aligned with the centre or edge of the arm. Tighten the screws after making the adjustment.

12 Refit the distributor cap and HT lead.

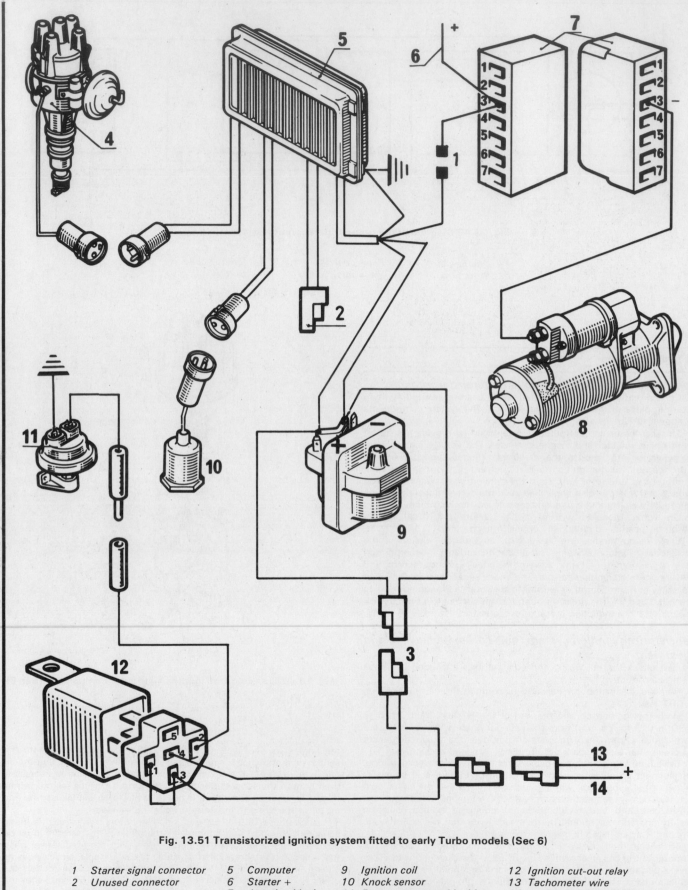

Fig. 13.51 Transistorized ignition system fitted to early Turbo models (Sec 6)

1 Starter signal connector	5 Computer	9 Ignition coil	12 Ignition cut-out relay
2 Unused connector	6 Starter +	10 Knock sensor	13 Tachometer wire
3 Ignition coil +	7 Junction block	11 Pressure-operated ignition	14 + after ignition switch
4 Distributor	8 Starter motor	cut-out	

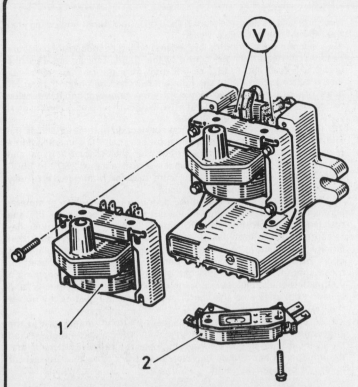

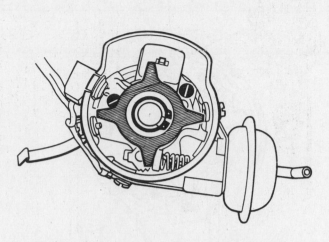

Fig. 12.53 Reluctor air gap adjustment screws (Sec 6)

Fig. 13.52 Testing the transistorized ignition coil (Sec 6)

1   Coil   2   Electronic module

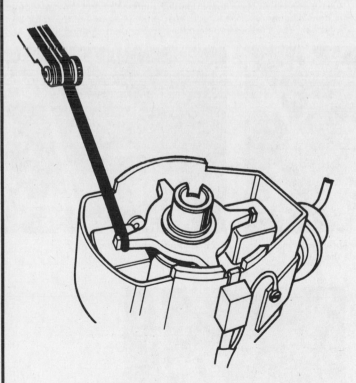

Fig. 13.54 Checking the reluctor air gap (Sec 6)

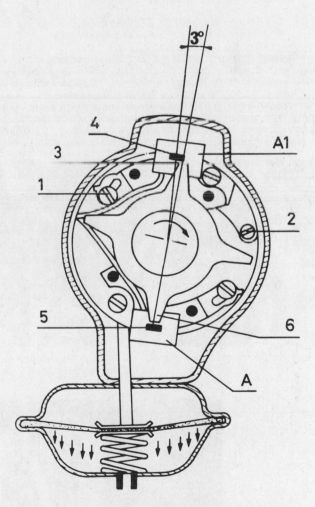

Fig. 13.55 Double sensor offset on manual transmission models (Sec 6)

A	Fixed sensor	3	Reluctor arm edge
A1	Adjustable sensor	4	Sensor post
1	Adjusting screw	5	Sensor post
2	Adjusting screw	6	Centre line

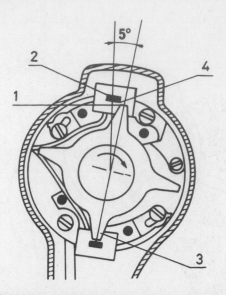

Fig. 13.56 Double sensor offset on automatic transmission models
(Sec 6)

1   Reluctor arm edge   3   Centre line
2   Sensor post   4   Sensor post edge

Transistorized ignition system (Turbo) – checking

13 First check that all wiring and connectors are in good condition and fitted correctly. Note that the ignition coil may be damaged if either of its low tension terminals is inadvertently earthed.

14 Disconnect the wiring from the pressure-operated ignition cut-out and connect an ohmmeter across the terminals – infinite resistance should be registered at an induction pressure of less than 11.2 lbf/in² (775 mbar). If a means of increasing the induction pressure above this level is available zero resistance should be registered, indicating that the internal contacts have closed.

15 Separate the distributor low tension wiring connector then disconnect the HT lead from the centre of the distributor cap and hold the end 0.125 to 0.190 in (3.2 to 4.8 mm) from the cylinder head with well-insulated pliers. With the ignition switched on check that an HT spark occurs when the two terminals in the computer end low tension wiring connector are bridged either with a jump wire or screwdriver. If no spark occurs check the ignition coil.

16 Disconnect the coil wiring, then connect an ohmmeter across the low tension terminals – a resistance of less than 10 ohms should be registered. Connect the ohmmeter between the HT lead terminal and the negative (-) low tension terminal – a resistance of 2500 to 5500 ohms should be registered. Make sure that the ohmmeter lead fully contacts the HT terminal for the test.

17 To check the sensor coil in the distributor, connect an ohmmeter across the low tension wiring connector terminals and check that the resistance is approximately 600 ohms. Connect the ohmmeter between one of the terminals and the distributor body and check that infinity resistance is registered, indicating that the coil insulation is good. Note that the coil may be damaged if a test lamp is used for these checks. The sensor coil can also be checked by connecting a voltmeter across the low tension wiring connector terminals, and spinning the engine on the starter – if the needle registers the pulses the coil is serviceable.

18 To check the knock sensor, connect a strobe timing light to the engine and run the engine at idling speed. Using a soft metal drift tap the cylinder head in quick succession near the sensor and check that the ignition timing retards between 6 and 9°. Note that the sensor must not be tapped directly.

Integral electronic ignition system – general description

19 As from 1982, an integral electronic ignition system is fitted. The components are shown in Fig. 13.57.

20 The system is fully computerized and the main ignition functions take place within the integral ignition module (photo). The distributor is considerably reduced in size as it only incorporates the rotor arm (photos). The computer determines the correct ignition timing and

6.20A Integral electronic ignition module

6.20B Distributor for integral electronic ignition system – remove the screws ...

6.20C ... and withdraw the cap ...

6.20D ... rotor arm ...

6.20E ... and cover

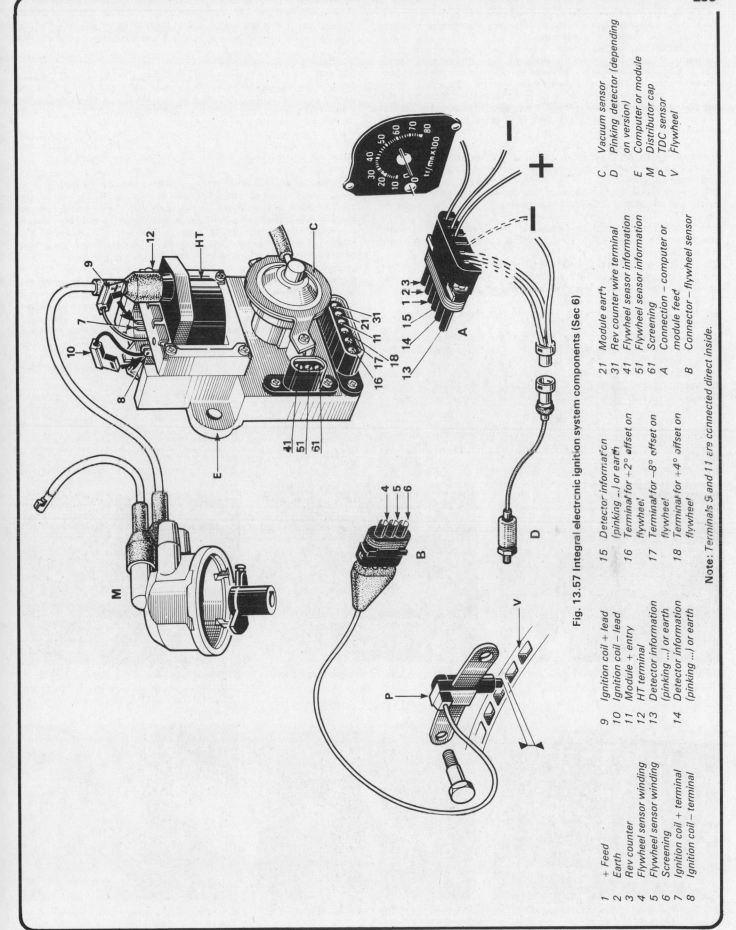

Fig. 13.57 Integral electronic ignition system components (Sec 6)

1 + Feed
2 Earth
3 Rev counter
4 Flywheel sensor winding
5 Flywheel sensor winding
6 Screening
7 Ignition coil + terminal
8 Ignition coil − terminal

9 Ignition coil + lead
10 Ignition coil − lead
11 Module + entry
12 HT terminal
13 Detector information
   (pinking ...) or earth
14 Detector information
   (pinking ...) or earth

15 Detector information
   (pinking ...) or earth
16 Terminal for +2° offset on
   flywheel
17 Terminal for −8° offset on
   flywheel
18 Terminal for +4° offset on
   flywheel

21 Module earth
31 Rev counter wire terminal
41 Flywheel sensor information
51 Flywheel sensor information
61 Screening
A Connection − computer or
   module feed
B Connector − flywheel sensor

C Vacuum sensor
D Pinking detector (depending
   on version)
E Computer or module
M Distributor cap
P TDC sensor
V Flywheel

Note: Terminals 5 and 11 are connected direct inside.

dwell by processing signals from the TDC sensor and vacuum sensor, and, on Turbo models, a knock sensor.
21 Although the vacuum capsule is similar in appearance to the conventional unit, the internal components are different and it **must not** be removed, otherwise a thin wire to the computer will be broken. However, the coil does not incorporate internal connections to the computer and it can therefore be removed separately. Make sure that the coil wires are fitted correctly: the red wire (9) goes to the positive (+) terminal (7), and the black wire (10) goes to the negative (-) terminal (8).

Integral electronic ignition system – checking
22 First check that all wiring and connectors are in good condition and fitted correctly.
23 Connect a voltmeter between the ignition coil terminal 7 and earth, and check that 9.5 volt is registered with the ignition switched on.
24 Disconnect the computer supply multiplug and, using a voltmeter, check that 9.5 volt is registered on terminal 1 with the ignition on and the starter operating.
25 Switch off the ignition and, using an ohmmeter, check that zero ohms is registered between the multiplug terminal 2 and earth. Also check that there is zero ohms between the module terminals 9 and 11 – if not, the computer is faulty.
26 Refit the multiplug and using the voltmeter check that there is 9.5 volt at terminal 9 with the ignition on.
27 Disconnect the TDC sensor plug from the module and, using an ohmmeter between terminals 4 and 5, check that the resistance is 100 to 200 ohm. If not, renew the sensor.
28 Check that the resistance between terminals 5 and 6, and also between terminals 4 and 6, is infinity. Using a feeler blade, check that the clearance between the TDC sensor and the flywheel is 0.02 to 0.06 in (0.5 to 1.5 mm).
29 Reconnect the TDC sensor plug to the module then disconnect the coil wires (9 and 10) and connect a test bulb between them. Spin the engine on the starter and check that the bulb flashes. Switch off the ignition.
30 Connect an ohmmeter between the ignition coil terminals 7 and

12, and check that the resistance is 2500 to 5500 ohm. Also check that the resistance between terminals 7 and 8 is between 0.4 and 0.8 ohm.
31 Reconnect the coil wires.
32 Start the engine and run it at 3000 rpm, then disconnect the pipe from the vacuum capsule on the module and check that the engine speed decreases. If no decrease occurs check the condition of the vacuum pipe, but if this is good a fault is indicated in either the vacuum capsule or the computer.
33 Disconnect the computer supply multiplug and, using an ohmmeter, check that the resistance between the tachometer terminals 2 and 3 is 20 000 ohm. Reconnect the multiplug.

Distributor (OHC engines) – removal and refitting
34 On OHC engines the distributor is located on the rear of the cylinder head and is driven by the camshaft. On models with air conditioning and power steering it is located on the left-hand side of the engine and driven by the intermediate shaft. The removal and refitting procedure is basically as given in Chapter 4, but the unit is secured with bolts and small spacers (photo).

Contact breaker points (OHC engines) – adjustment
35 Access to the contact points is limited where the distributor is located on the rear of the cylinder head, and it is therefore suggested that the distributor be removed if the points are to be inspected. However, an external adjustment nut is provided for simply adjusting the contact points gap or dwell angle (photos).

Contact breaker points (OHC engines) – removal and refitting
36 With the distributor removed, remove the cap, rotor arm and dust shield.
37 Remove the screws and withdraw the bearing plate (photo).
38 The remaining procedure is as given in Chapter 4.

Spark plugs – removal and refitting
39 On crossflow head OHV engines access to the spark plugs is

6.34 Removing the distributor rotor arm on the OHC engine

6.35A Checking the contact points gap on the OHC engine

6.35B Internal view of the distributor on the OHC engine showing (1) moving contact pivot, (2) contact points, (3) fixed contact screw, and (4) external adjuster

6.37 OHC engine distributor bearing plate screw (arrowed)

6.39 Removing the spark plug caps (crossflow head OHV engines)

6.40 Removing the spark plugs (crossflow head OHV engines)

gained through tubes in the valve cover. First pull off the caps then unscrew the extension rods (photo).

40   Unscrew and remove the spark plugs using a long box spanner or socket, preferably with a rubber insert to grip the spark plug (photo).

41   Before refitting the spark plugs clean the threads. Take extra care to ensure that the plugs enter the cylinder head threads correctly, then tighten them and fit the extension rods and caps. The HT lead locations are shown in Fig. 13.58.

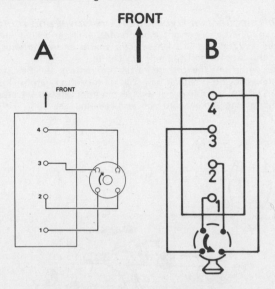

Fig. 13.58 HT lead layout (Sec 6)

A   843 engine   B   OHC J6R and 851 engines

7   Clutch

Clutch shaft and spigot bearing – description

1   Early models are fitted with a long clutch shaft supported at the front by a spigot bearing in the crankshaft. Later models have a shorter shaft without the spigot bearing, but with an additional roller bearing in the clutch housing.

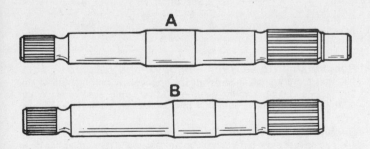

Fig. 13.59 Two types of clutch shaft (Sec 7)

A   With spigot bearing   B   Without spigot bearing

Clutch cable – adjustment

2   The cable clearance at the release lever is increased for North American R1348 and R1358 models. Refer to the Specifications at the beginning of this Supplement.

Clutch cable – lubrication

3   It is important that the clutch cable guide and trunnion are lubricated with grease to ensure that the trunnion operates in a satisfactory manner and the cable aligns with the guide (Fig. 13.60).

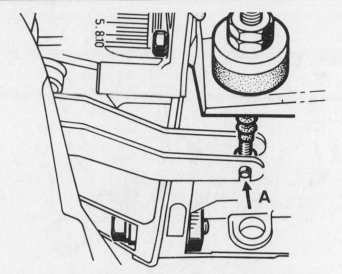

Fig. 13.60 Lubricate the clutch cable and trunnion at A (Sec 7)

8   Manual transmission

Gearbox, type 395 – modifications

1   The type 395 gearbox fitted to 1981-on models incorporates modifications to the rear cover and speedometer drive, the selector mechanism, the primary shaft bearing, and the secondary shaft bearing. Procedures on the gearbox are as given in Chapter 6 except as described in the following paragraphs.

Rear cover (type 395) – removal, dismantling, reassembly and refitting

2   Unscrew the 5th speed shaft detent plug and extract the spring and ball.

3   Select neutral and unscrew the cover bolts, then remove the cover whilst tilting the selector finger.

4   Using a suitable punch, drive the roll pin from the selector finger.

5   Remove the end cap from the cover using a chisel. The end cap will be damaged, so a new one must be obtained.

6   Extract the circlip and pull out the shaft, followed by the spacers, spring and selector finger.

7   To remove the speedometer drivegear prise up the short arms and

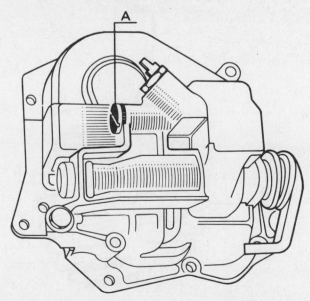

Fig. 13.61 5th speed shaft detent plug (A) – 1981-on type 395 gearbox (Sec 8)

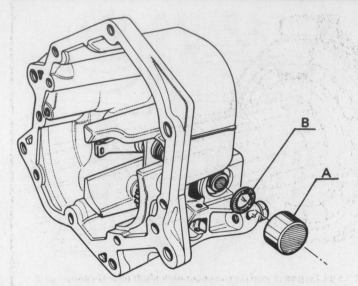

Fig. 13.62 Rear cover endcap (A) and circlip (B) – 1981-on type 395 gearbox (Sec 8)

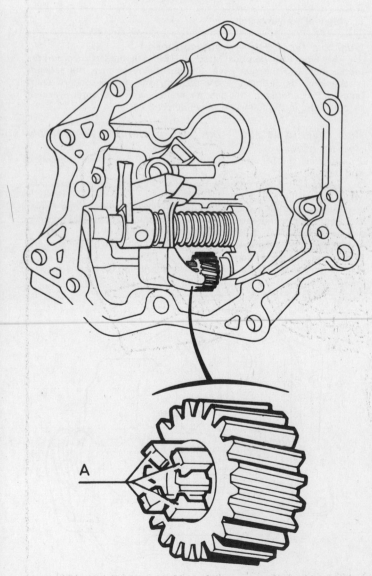

Fig. 13.63 The speedometer drivegear retaining arms (A) – 1981-on type 395 gearbox (Sec 8)

pull out the shaft, followed by the gear. Extract the seals from the cover.

8 Fit new seals in the cover and obtain a new speedometer drivegear.

9 Reassemble in reverse order, making sure that the plastic spacer goes on the selector finger side and that the arms on the speedometer drivegear fully enter the shaft groove.

10 Refitting is a reversal of removal, with reference also to Chapter 6.

Selector mechanism (type 395) – removal and refitting

11 The modified mechanism incorporates an interlocking disc between the 1st/2nd and 3rd/4th selector shafts, and an interlock ball between the 3rd/4th and 5th selector shafts.

12 When removing the 5th gears do not remove the 5th selector shaft, otherwise the interlock ball will fall into the gearbox. Instead, select 3rd or 4th, then drive out the roll pin from the 5th selector fork, and withdraw the 5th synchro unit and selector fork.

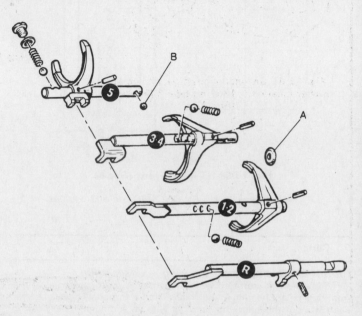

Fig. 13.64 Selector mechanism on 1981-on type 395 gearbox (Sec 8)

A *Interlocking disc*   B *Interlock ball*

Primary shaft (type 395) – removal and refitting

13 The primary shaft incorporates a double row ball-bearing at the 5th gear end and a roller bearing at the clutch shaft end. When refitting the shaft make sure that the circlips fully enter their grooves.

Secondary shaft (type 395) – removal and refitting

14 The spacer plate between the main casing and rear cover has been discontinued and in its place is fitted a thrust washer. Note that the shoulder on the washer faces the taper roller bearing and the cut-out locates over the 5th selector shaft.

15 When removing the secondary shaft position a clip over the pinion end bearing (Fig. 13.66) to prevent the rollers falling out.

NG0, NG1 and NG3 gearbox – dismantling

16 Unbolt the clutch housing.

17 On NG1 and NG3 gearboxes select 3rd or 4th, unscrew the 5th detent plug, and extract the spring and ball.

18 Unscrew the bolts and remove the rear cover whilst tilting the selector finger. On NG0 gearboxes recover the spacer and primary shaft bearing preload shims, noting their location.

19 Support the gearbox on the right-hand side casing then lift off the left-hand side casing.

20 Lift out the differential unit, followed by the secondary shaft and primary shaft.

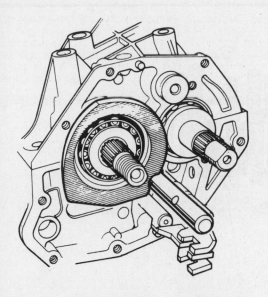

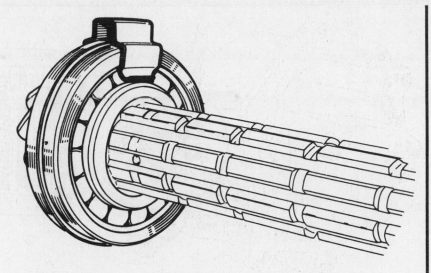

Fig. 13.65 Secondary shaft bearing thrust washer location – 1981-on type 395 gearbox (Sec 8)

Fig. 13.66 Clip positioned on the secondary shaft roller bearing (Sec 8)

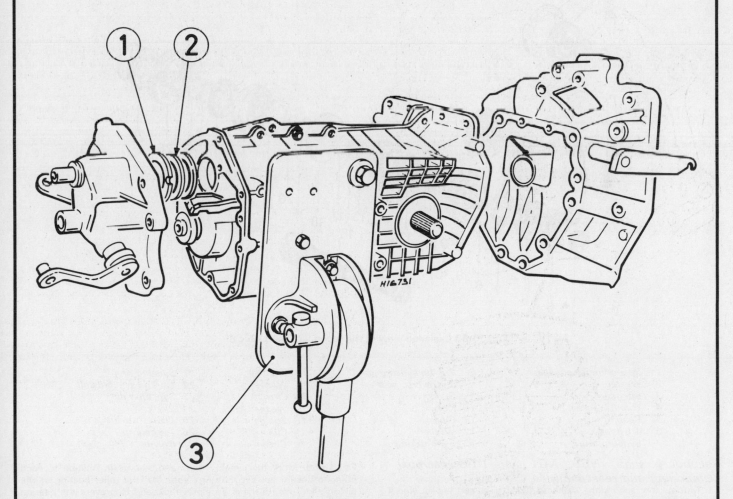

Fig. 13.67 Rear cover removal on NGO gearbox, showing spacer (1) and preload shims (2) – gearbox shown mounted on special Renault dismantling stand (3) (Sec 8)

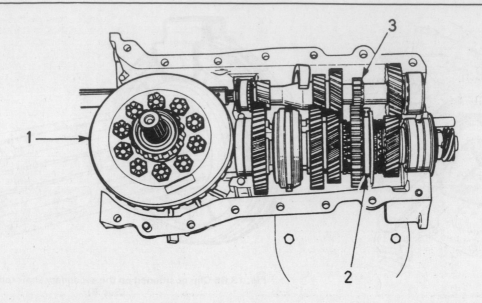

Fig. 13.68 Right-hand casing with shafts and differential unit on NGO gearbox (Sec 8)

1   Differential unit   3   Primary shaft
2   Secondary shaft

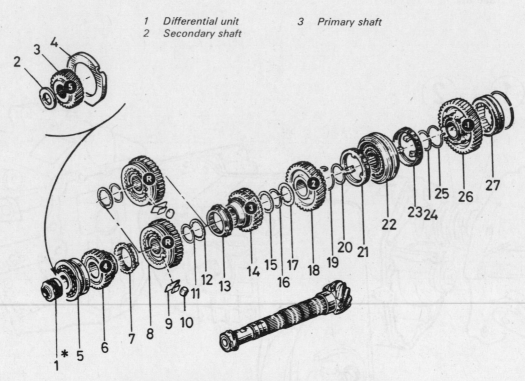

Fig. 13.69 Exploded view of the secondary shaft (Sec 8)

1	Speedometer drivegear	8	3rd/4th synchro unit	14	3rd gear
2	Washer		and reverse gear	15	Washer
3	5th gear	9	Spring	16	Circlip
4	Thrust washer	10	Key	17	Washer
5	Bearing	11	Circlip	18	2nd gear
6	4th gear	12	Washer	19	Clip
7	Synchro ring	13	Synchro ring	20	Circlip

21	Synchro ring
22	1st/2nd synchro unit
23	Synchro ring
24	Circlip
25	Clip
26	1st gear
27	Bearing

Secondary shaft (NGO, NG1 and NG3 gearbox) — dismantling and reassembly

21 Mount the secondary shaft vertical in a soft-jawed vice by clamping the 1st gear.

22 Select 1st then unscrew the speedo drive nut.

23 On NG1 and NG3 gearboxes use a puller to remove the 5th gear.

24 Remove the secondary shaft components in the order shown in

Fig. 13.69. Mark the synchro-hubs and sleeves in relation to each other to ensure correct refitting. Note that the roller bearing at the pinion end does not have an inner track, and to prevent the bearing falling apart a clip can be fitted over it.

25 Renew the circlips and the speedo drive nut, then refit the components in the reverse order to that given in Fig. 13.69. The synchro-hubs are a sliding fit on the shaft splines; however, if the hub

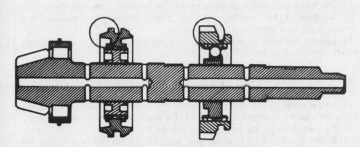

Fig. 13.70 The correct fitted positions of the synchro units on the secondary shaft (Sec 8)

becomes tight, remove it, turn it, and engage it with different splines. Note the correct fitted positions of the synchro units shown in Figs. 13.70 and 13.71.

26 On NG1 and NG3 gearboxes two possible types of 5th gear are fitted. With the first type having continuous inner splines apply a small amount of locking fluid to the splines before fitting the gear. The second type gear is free turning for three-quarters of its splines and the final splines must be pressed into position to give the correct preload to the double taper roller bearing. Position the assembly in the press as shown in Fig. 13.72 with a spring balance and cord around the bearing outer track. Press on the gear until the preload is between 3.4 and 9 lbf (1.5 and 4.0 kgf) and note that the press loading must not be less than 220 lbf (100 kgf) or more than 3307 lbf (1500 kgf).

27 Finally select 1st then fit the speedo drive nut using a little locking fluid, tighten and lock it.

Primary shaft (NG0, NG1 and NG3 gearbox) — dismantling and reassembly

28 Drive out the roll pin and separate the clutch shaft. Remove the special washer.

29 On NG1 and NG3 gearboxes use a puller if necessary to remove the 5th gear and synchro-hub, after unscrewing the nut. Hold the shaft in a soft-jawed vice.

30 Remove the bearings from the shaft using a puller.

31 To reassemble, first press on the bearings or drive them on with a metal tube against the inner tracks.

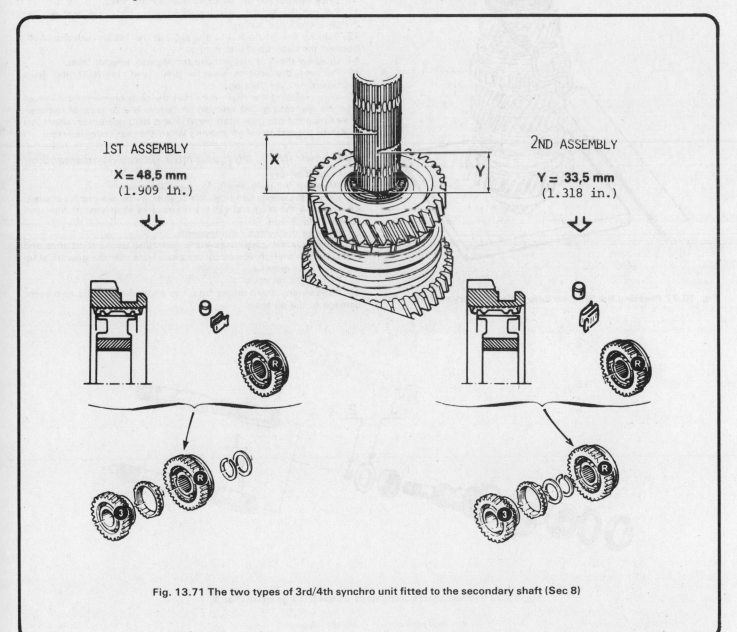

1ST ASSEMBLY

X = 48,5 mm
(1.909 in.)

2ND ASSEMBLY

Y = 33,5 mm
(1.318 in.)

Fig. 13.71 The two types of 3rd/4th synchro unit fitted to the secondary shaft (Sec 8)

32 Two types of 3rd/4th synchro-hub are fitted to NG1 and NG3 gearboxes. On the early type with continuous inner splines apply a little locking fluid to the splines before fitting the unit. On the later type with short splines, support the 4th gear and press on the synchro-hub with a load of not less than 220 lbf (100 kg) or more than 3307 lbf (1500 kg).

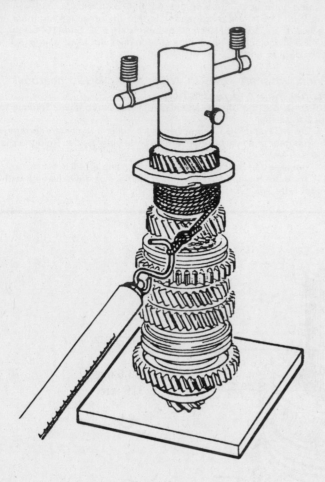

Fig. 13.72 Pressing the 5th gear onto the secondary shaft (Sec 8)

33 On NG1 and NG3 gearboxes fit the washer, then apply a little locking fluid to the threads of the nut before fitting it and tightening it to the specified torque. Lock the nut.
34 Locate the special washer then fit the clutch shaft to the primary shaft and drive in the roll pin to secure.

Reverse shaft (NG0, NG1 and NG3 gearbox) – removal and refitting
35 Extract the circlip then withdraw the shaft followed by the gear, friction washer and guide. Recover the interlocking ball and spring.
36 To refit, locate the spring and ball in the casing.
37 Insert the shaft and locate the gear on it with the hub facing the differential end.
38 Fit the friction washer with the bronze face toward the gear.
39 Locate the guide in the bore then push in the shaft and fit the circlip.

Gear selector mechanism (NG0, NG1 and NG3 gearbox) – removal and refitting
40 On NG1 and NG3 gearboxes, select neutral then pull out the 5th gear shaft and recover the interlock ball.
41 Drive out the roll pin securing the selector forks.
42 Pull out the 3rd/4th selector shaft, remove the fork, and recover the detent ball and spring.
43 Remove the interlock disc and pull out the 1st/2nd selector shaft. Recover the detent ball and spring.
44 Unscrew the bolt and remove the reverse selector lever.
45 Pull out the reverse selector shaft and, on NG0 and NG1 gearboxes, recover the dog.
46 When refitting the shaft, note that the slots on the roll pins must face the rear casing. The refitting procedure is a reversal of removal. After fitting the 5th gear shaft on NG1 and NG3 gearboxes, select 3rd or 4th to prevent the shaft moving when the rear cover is fitted.

Rear cover (NG0, NG1 and NG3 gearbox) – dismantling and reassembly
47 Drive out the roll pin(s) using a suitable punch.
48 Extract the circlip and prise the bush from the selector lever shaft.
49 Unscrew the plug and extract the reverse shaft stop plunger and spring.
50 Remove the selector components.
51 To remove the speedometer drivegear prise up the short arms and withdraw the shaft, followed by the gear. Note that the gear must be renewed after removal.
52 Extract the oil seals.
53 Remove any sharp edges from the ends of the shafts to prevent damage to the oil seals.

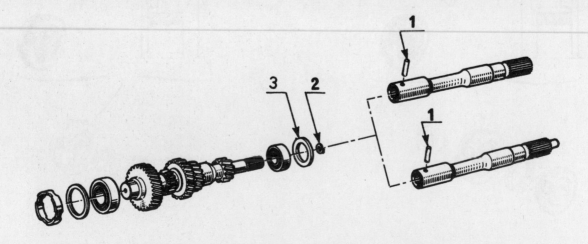

Fig. 13.73 Primary shaft components on the NG0 gearbox (Sec 8)

1 Roll pins	3 Spacer
2 Washer	

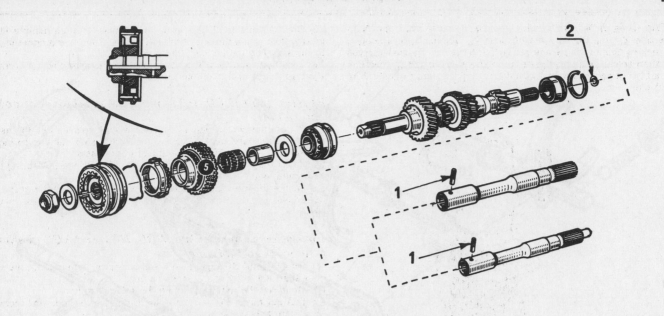

Fig. 13.74 Primary shaft components on the NG1 and NG3 gearboxes (Sec 8)

1   Roll pins   2   Washer

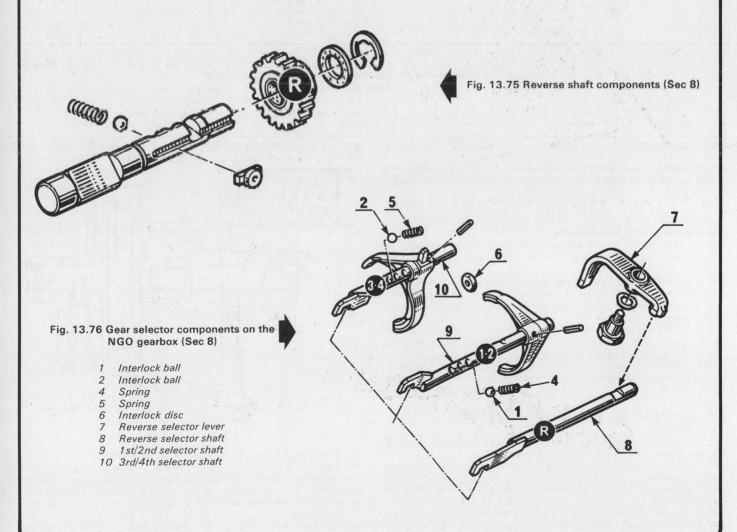

Fig. 13.75 Reverse shaft components (Sec 8)

Fig. 13.76 Gear selector components on the
NGO gearbox (Sec 8)

1   Interlock ball
2   Interlock ball
4   Spring
5   Spring
6   Interlock disc
7   Reverse selector lever
8   Reverse selector shaft
9   1st/2nd selector shaft
10 3rd/4th selector shaft

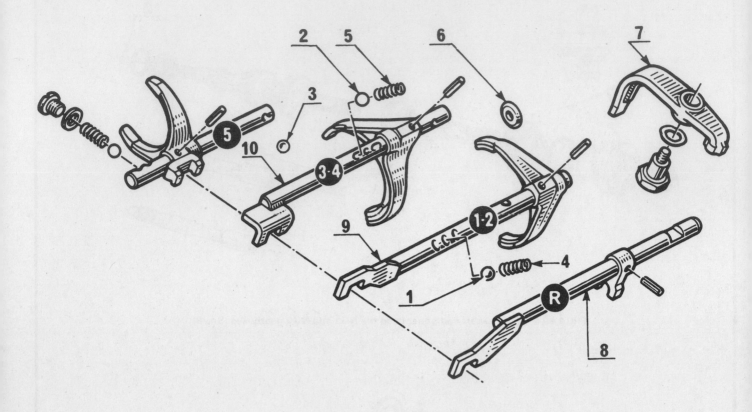

Fig. 13.77 Gear selector components on gearboxes NG1 and NG3
(Sec 8)

1 Interlock ball	4 Spring	7 Reverse selector lever	9 1st/2nd selector shaft
2 Interlock ball	5 Spring	8 Reverse selector shaft	10 3rd/4th selector shaft
3 Interlock ball	6 Interlock disc		

54 Reassembly is a reversal of dismantling, but fit new oil seals and make sure that the speedometer drivegear arms fully enter the groove in the shaft.

Clutch housing (NG0, NG1 and NG3 gearbox) – dismantling and reassembly

55 To remove the clutch release fork, extract the pins using an adapter and slide hammer, withdraw the release arm, then remove the fork.

56 If necessary drive out the release bearing guide, but note that it must be renewed after removal. Extract the oil seal where applicable.

57 Drive a new oil seal into the housing where applicable using a metal tube.

58 Lubricate the guide tube with a little grease and where applicable fit the O-ring, then drive the tube into the housing, making sure that the oil holes line up where applicable.

59 Lubricate the release arm shaft with grease then insert it, together with the release fork. Drive in the pins so that their shoulders protrude 0.04 in (1.0 mm).

5th speed gear cluster (NG1 and NG3 gearbox) – removal and refitting

60 The 5th speed gear cluster can be removed without fully dismantling the gearbox using the following procedure.

61 Select 3rd or 4th, unscrew the 5th detent plug, and extract the spring and ball.

62 Unscrew the bolts and remove the rear cover whilst tilting the selector finger.

63 Select 1st and 5th to lock the gears then unscrew the nuts from the ends of the shafts. Select 3rd or 4th again to hold the 5th selector shaft.

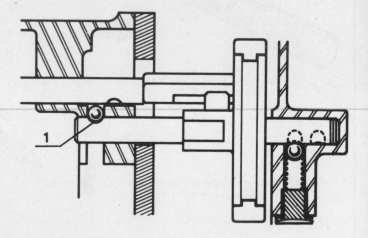

Fig. 13.78 Cross-section showing 5th gear selector shaft and interlock ball (1) (Sec 8)

64 Drive out the roll pin from the 5th selector fork.

65 Mark the 5th synchro-hub and sleeve in relation to each other, then withdraw the synchro unit and both gears using a puller where necessary. Remove the spacer plate from the secondary shaft, noting that the cut-out locates over the 5th selector shaft and the shoulder faces the bearing.

66 Refitting is a reversal of removal, with reference also to paragraphs 26 and 27. Fit a new gasket to the rear cover.

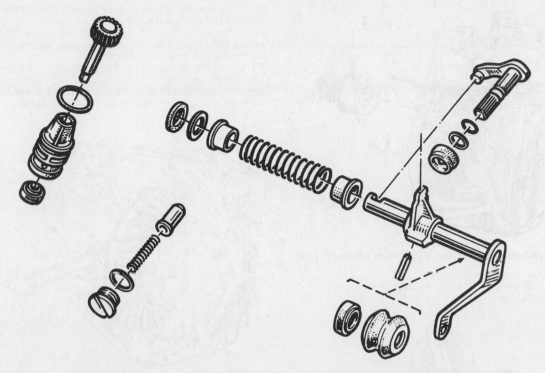

Fig. 13.79 Exploded view of the rear cover components on the NGO gearbox (Sec 8)

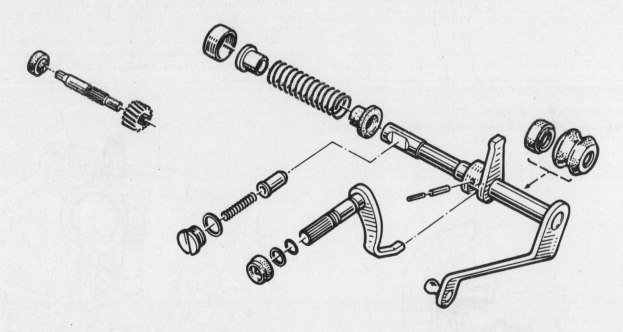

Fig. 13.80 Exploded view of the rear cover components on the NG1 and NG3 gearboxes (Sec 8)

Differential oil seals (NGO, NG1 and NG3 gearbox) – renewal

67 Lip type oil seals are located in the differential bearing ring nuts and O-rings are located on the outer edges of the ring nuts. To renew the seals, mark the ring nuts in relation to the casings then unbolt the lockplates and unscrew the ring nuts, noting the exact number of turns necessary to remove them.

68 Remove the old seals and O-rings then press in the new seals and locate the O-rings in the grooves (Fig. 13.83).

69 Wrap adhesive tape over the shaft splines and lubricate the seals and O-rings with a little grease.

70 Screw the ring nuts into their original positions and fit the lockplates. Remove the tape.

Differential bearing preload (NGO, NG1 and NG3 gearbox) – adjustment

71 If new differential bearings have been fitted the casing ring nuts must be adjusted to obtain the correct preload. First fit the differential

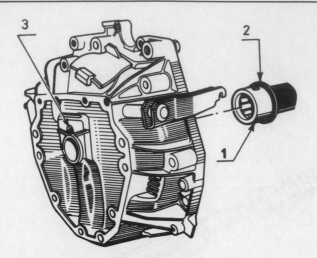

Fig. 13.81 Clutch housing and short clutch shaft bearing (Sec 8)

1   Bearing   3   Lubrication hole
2   O-ring

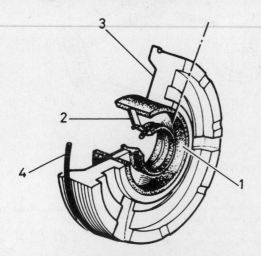

**Fig. 13.82 Clutch release fork retaining pin protrusion dimension
(Sec 8)**

D = 0.04 in (1.0 mm)

**Fig. 13.83 Cutaway section of a differential ring nut and oil seal
(Sec 8)**

1   Oil seal   4   O-ring
2 and 3   Faces to be flush

in the casings without the primary or secondary shafts and tighten the
casing bolts to the correct torque.
72 Screw in the ring nuts turning the one on the differential housing
side slightly more than the opposite side. Continue turning until the
differential unit becomes hard to move.
73 Using a spring balance and cord (Fig. 13.84) check that the
differential turns within the specified loadings, and adjust the ring nuts
accordingly. Do not alter the settings until adjusting the pinion
backlash. Remove the differential from the casings.

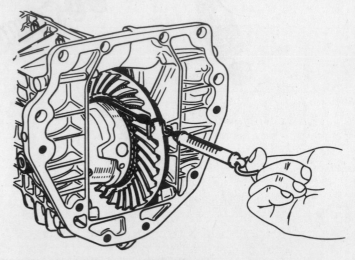

**Fig. 13.84 Checking the differential bearing preload setting
(Sec 8)**

NGO, NG1 and NG3 gearbox — reassembly

74 Locate the primary and secondary shaft assemblies in the right-
hand side casing followed by the differential unit.
75 Apply suitable sealant to the mating faces then fit the left-hand
side casing, insert the bolts and tighten them to the specified torque.
76 Using a dial gauge, as shown in Fig. 13.85, check that the

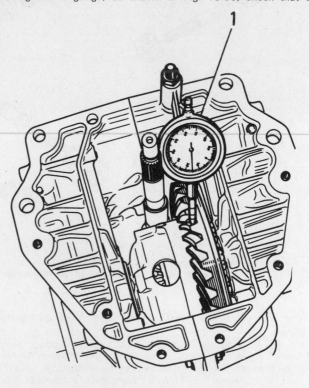

**Fig. 13.85 Using a dial gauge (1) to check the crownwheel and
pinion backlash (Sec 8)**

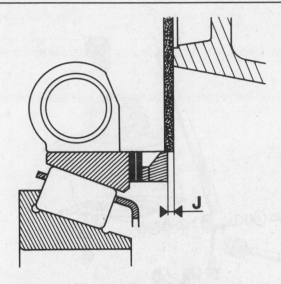

Fig. 13.86 Primary shaft bearing endfloat dimension for the rear cover on the NGO gearbox (Sec 8)

J = 0.001 to 0.005 in (0.02 to 0.12 mm)

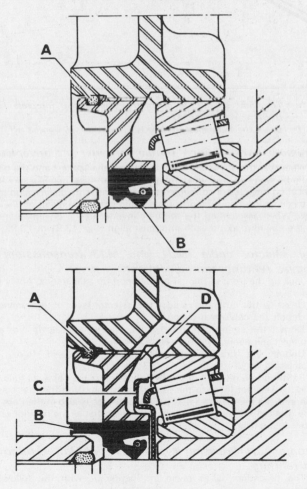

Fig. 13.87 Cross-section of early and late differential oil seal types (Sec 8)

A O-ring   B Oil seal   C Deflector   D Lubrication hole

backlash between the crownwheel and pinion is as given in the Specifications. If not, move the differential unit as necessary by turning the ring nuts by equal amounts. Note that unequal turning of the ring

nuts will result in an incorrect bearing preload. Mark the ring nut positions after making the adjustment.

77 On the NGO gearbox fit the shims and distance piece to the primary shaft and tap towards the bearing using a metal tube. Fit the gasket and measure the dimension J shown in Fig. 13.86. If necessary select different shims to provide the correct dimension.

78 Fit the rear cover, together with a new gasket, then fit the bolts and tighten them to the specified torque.

79 On NG1 and NG3 gearboxes insert the 5th gear detent ball and spring then tighten the plug into the cover.

80 Wrap adhesive tape lightly over the clutch shaft splines and smear with a little grease.

81 Refit the clutch housing using a new gasket and tighten the bolts. Remove the tape.

82 Check that the ring nuts are positioned correctly then fit the lockplates and tighten the bolts.

Gear linkage (NGO, NG1 and NG3 gearboxes) – general
83 The gear linkage is shown in Fig. 13.88 and requires no adjustment (photos).

Molybdenum synchro rings
84 From July 1985, the 3rd, 4th and 5th synchro rings on all models are molybdenum type. They are directly interchangeable with the early type but, when fitting, coat the friction surface with 'Molykote M55 +' or equivalent.

8.83A Gear linkage knuckle

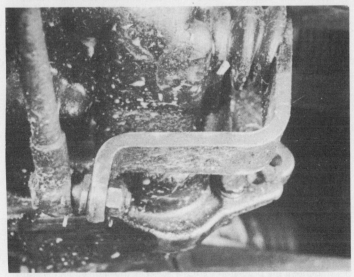

8.83B Gear linkage balljoint

Fig. 13.88 Gear linkage components (Sec 8)

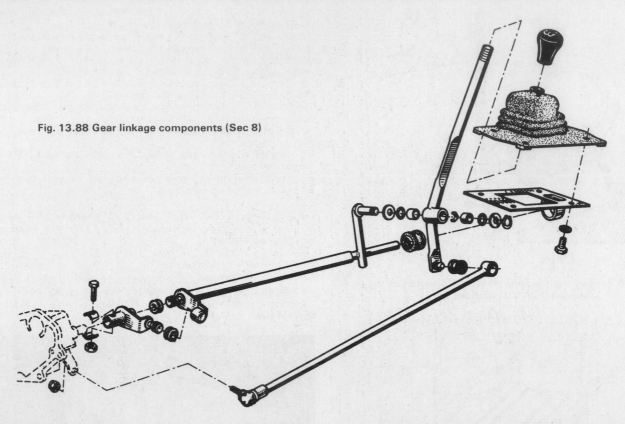

9   Automatic transmission

General description
1   Certain later models are fitted with an MJ1 or MJ3 automatic transmission, both types being identical except for small modifications necessary to fit different engines. The mechanical function of the units is similar to that described in Chapter 7, but the controls are fully computerised.
2   The computer or module is continually supplied with signals from the speed sensor, load potentiometer, multi-purpose switch, kickdown switch and its own control unit, and from this information it transmits signals to the pilot solenoid valves to select the correct gear range.
3   The kickdown switch is located at the bottom of the accelerator pedal travel and a lower ratio is selected when the pedal is fully depressed.
4   The multi-purpose switch is located on the rear of the transmission and is operated by the range selector lever. It controls the engine starting circuit, the reversing light circuit and the pilot solenoid valve circuit.
5   The load potentiometer is operated by the throttle valve and provides a variable voltage to the computer dependent on the throttle position.
6   The speed sensor is mounted on the left-hand side of the transmission and it provides an output signal dependent on the speed of the parking pawl wheel, which is proportional to the speed of the car.
7   A vacuum capsule located on the left-hand side of the transmission is connected direct to the inlet manifold, and its purpose is to regulate the oil pressure within the transmission to suit the engine loading.
8   If either the transmission itself or the control units develop a fault the car should be taken to a Renault dealer who will have the special equipment necessary to diagnose the faulty component.

Automatic transmission fluid – checking, draining and refilling (1985-on)
9   In general the procedures are the same as those given for the earlier types in Section 3 of Chapter 7, but note the following differences:

(a) The fluid should be drained at the revised intervals (see Routine Maintenance)
(b) Drain the transmission fluid when it has fully cooled off

Automatic transmission – torque converter and driveplate
10   Whenever the driveplate is removed both the spacers and the bolts should be renewed, and when refitting the transmission make sure that all locating dowels are correctly fitted, otherwise the transmission may not be centralised and the driveplate could crack as a result of uneven stress. When assembling the torque converter to the driveplate, note that the timing mark on each unit must align (Fig. 13.91 or 13.92).

Gear selector cable (MJ1 and MJ3 transmission) – removal, refitting and adjustment
11   Jack up the front of the car and support on axle stands. Apply the handbrake.
12   Unscrew the nuts and remove the selector lever bottom cover.
13   Unbolt the cable rear clamp.
14   Extract the spring clip, remove the washer and clevis pin, and disconnect the cable from the selector lever.
15   Unbolt the bracket and disconnect the cable from the transmission.
16   Refitting is a reversal of removal, but adjust the cable as follows. Move the selector lever to the Park (P) position. With the rear clamp loose, preload the cable so that the transmission is also in the Park (P) mode. Tighten the rear clamp bolts and check that all the positions can be selected easily.

Automatic transmission (types MJ1 and MJ3) – removal and refitting
17   The procedure is as given in Chapter 7, with the following exceptions:

(a) It is not necessary to remove the starter motor
(b) Disconnect the controls with reference to Fig. 13.90, and the selector cable with reference to paragraph 15 of this Section
(c) Note that it is not possible to remove the upper right-hand transmission-to-engine bolt with the engine in position, so this bolt must be located in the transmission casing before lifting the transmission onto the rear of the engine

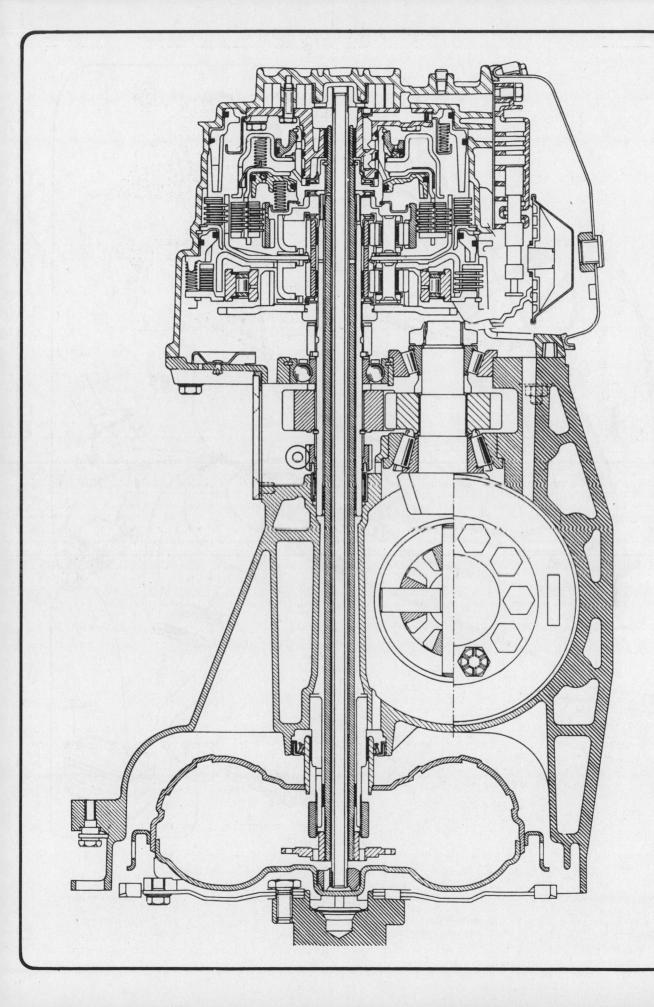

Fig. 13.89 Cross-section of the MJ1 and MJ3 automatic transmission (Sec 9)

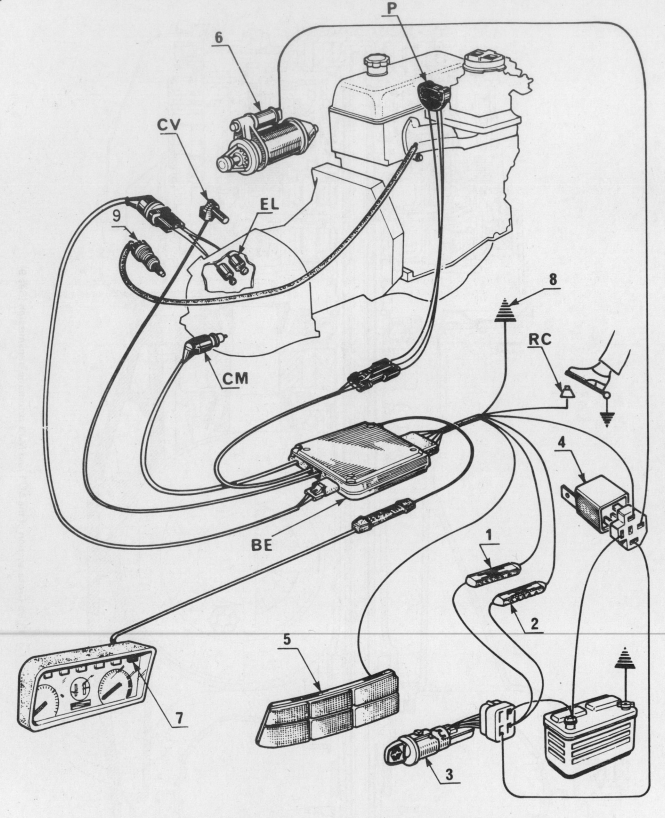

Fig. 13.90 Control units for the MJ1 and MJ3 automatic transmission (Sec 9)

1	Reversing light fuse (5A)	6	Starter
2	Supply fuse (1.5A)	7	Instrument panel
3	Starter switch		(warning light)
4	Starter relay	8	Automatic transmission
5	Reversing lights		earth

9	Vacuum capsule	EL	Pilot solenoid valves
BE	Computer or module	P	Load potentiometer
CM	Multi-purpose switch	RC	Kick-down switch
CV	Speed sensor		

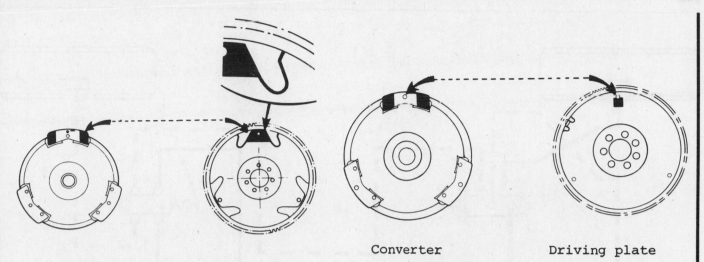

Converter   Driving plate

Fig. 13.91 Ignition mark locations on the torque converter and driving plate on 1647 cc engine models (Sec 9)

Fig. 13.92 Ignition timing mark locations on the torque converter and driving plate on 1995 cc engine models (Sec 9)

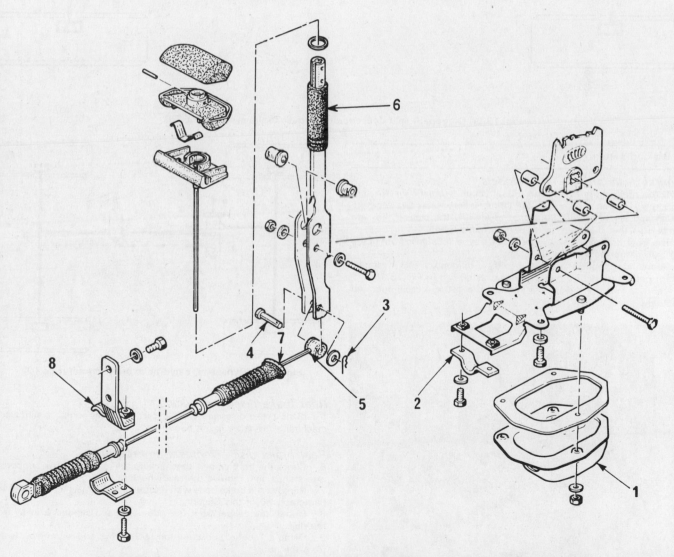

Fig. 13.93 Exploded view of the selector lever and cable for MJ1 and MJ3 transmissions (Sec 9)

1	Cover	3	Clip	5	Cable	7	Guide tube
2	Clamp	4	Clevis pin	6	Lever	8	Bracket

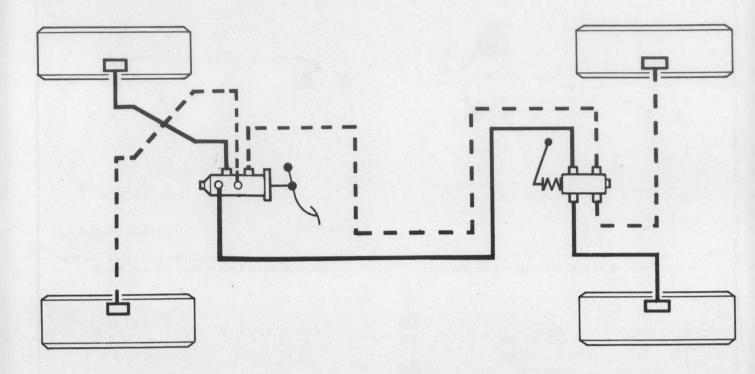

Fig. 13.94 Diagonally split dual circuit hydraulic brake system (Sec 10)

10 Braking system

General description – later models

1   Turbo models are fitted with a dual circuit hydraulic system split diagonally. The rear brake pressure limiter incorporates two separate circuits and it is therefore important to identify the pipes if they are removed so that they can be refitted correctly (photo).

2   The front brakes are of Bendix Series 4 type, located in front of the stub axle carrier on later models.

3   Some later models are fitted with a diagonally split hydraulic system and some with a bypass system as shown in Fig. 13.95.

4   As from 1983 rear discs are fitted incorporating a cable-operated handbrake.

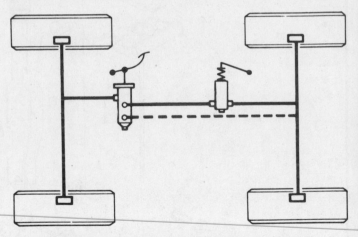

Fig. 13.95 Bypass type hydraulic brake circuit (Sec 10)

Rear brake discs – servicing

5   If the brake discs are worn excessively they must be renewed – machining the faces is not permitted.

Rear brake disc – removal and refitting

6   Chock the front wheels then jack up the rear of the car, support on axle stands, and remove the roadwheel.

7   Remove the brake pads with reference to Chapter 9, Section 9 and paragraph 26 of this Section.

8   Unbolt the caliper from the axle beam and support it away from the disc.

9   Using a Torx key, unscrew the two screws and withdraw the disc from the hub.

10 Refitting is a reversal of removal, but tighten the caliper bolts to the specified torque.

Rear brake caliper – removal and refitting

11 Chock the front wheels then jack up the rear of the car, support on axle stands, and remove the roadwheel.

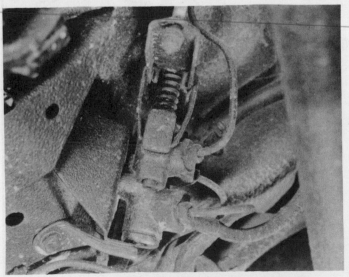

10.1 Rear brake pressure limiter on a Turbo model

12 Clamp the rear brake flexible hose or, alternatively, remove the brake fluid reservoir cap and tighten it down onto a piece of polythene sheet in order to prevent leakage of fluid.
13 Disconnect the handbrake cable.
14 Unscrew the rigid hydraulic pipe union and withdraw the pipe from the bottom of the caliper.
15 Remove the clips, tap out the keys and withdraw the caliper from the bracket. Remove the brake pads (photos).
16 Refitting Is a reversal of removal, but bleed the hydraulic system and adjust the handbrake, as described in Chapter 9.

Rear brake caliper – overhaul
17 With the caliper removed, clean the exterior surfaces then mount it in a soft-jawed vice.
18 Prise off the dust cover and unscrew the piston using a square rod.
19 Place a block of wood in the caliper and, using a tyre pump or air line in the hydraulic pipe aperture, carefully force the piston from the cylinder.
20 Prise the O-ring from the cylinder bore then clean all the components in methylated spirit and allow to dry.
21 Check the piston and cylinder bore for wear or corrosion and, if evident, renew the complete caliper.
22 If the components are in good condition first dip the O-ring in brake fluid and position it in the bore groove.

10.15C ... withdraw tho rear brake callper ...

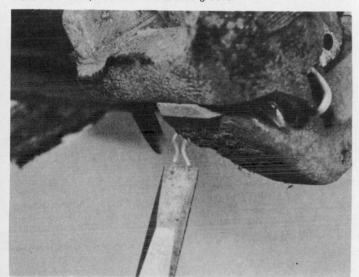

10.15A Remove the clips ...

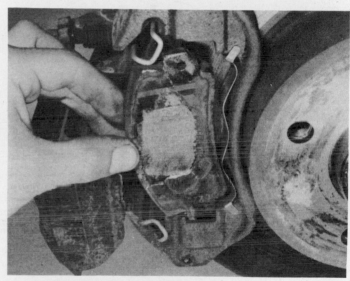

10.15D ... and remove the brake pads

10.15B ... tap out the keys ...

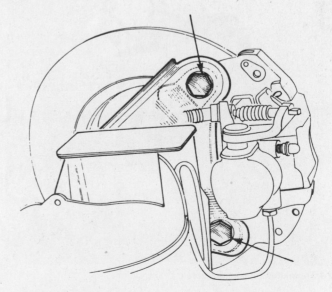

Fig. 13.96 Rear brake caliper mounting bolts (Sec 10)

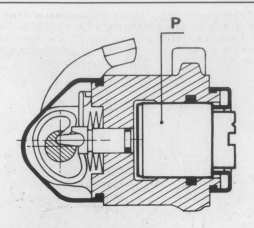

Fig. 13.97 Cross-section of the rear brake caliper (Sec 10)

P   Piston

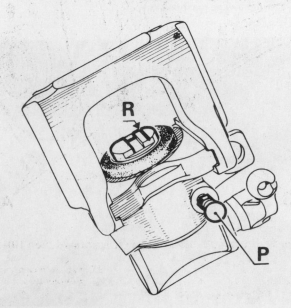

Fig. 13.98 Correct alignment of the rear brake caliper piston recess (R) with the bleed screw (P) (Sec 10)

10.24 The piston recess on the rear brake caliper

23 Dip the piston in brake fluid and press it slowly into the cylinder.
24 Screw in the piston until it turns without entering further into the bore. Turn the piston so that the recess (Fig. 13.98) is aligned with the bleed screw (photo). This will ensure correct positioning of the disc pad.
25 Coat the outer edge of the piston with brake grease then refit the dust cover.

Rear brake pads – removal and refitting
26 The procedure is similar to that described in Chapter 9, Section 9, but before fitting the new pads, screw in the piston and position it as described in paragragh 24.

Front brake pads (offset) – removal and refitting
27 As from early 1984 the Bendix type front brake pads are offset and must be fitted the correct way round. Fig. 13.99 illustrates the old and new type pads. The old type has a central groove and two symmetrical indents, while the new type has an offset groove and only one indent.
28 New calipers are modified to accept the offset pads and ensure that they are fitted correctly, but when fitting the new pads to old calipers make sure that the offset is correct. The outer pads must be fitted with the groove toward the bottom of the caliper and the inner pads with the groove toward the top of the caliper. The pad wear wire must be at the top of the caliper.
29 The offset pads are marked with a white arrow which must point in the forward rotation of the disc.
30 The new calipers incorporate bosses to prevent fitting of old type pads, but when refitting worn pads it is possible to insert them over the bosses, but in this case the pad will not be aligned with the disc.

Brake servo unit – adjustment
31 On some models the pushrod clevis adjustment dimension L as shown in Fig. 9.14 has been changed. Refer to the Specifications for details.

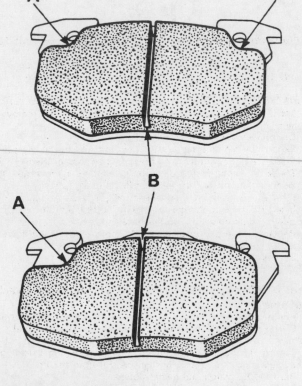

Fig. 13.99 The new offset brake pad (lower) compared with the old symmetrical type (Sec 10)

A   Indents   B   Groove

47 Dip the piston and O-ring in brake fluid and reassemble the caliper in reverse order using a new O-ring and rubbers. Lubricate the rubbers and guides with the special grease supplied in the repair kit.
48 Refitting is a reversal of removal, but bleed the hydraulic system and use locking fluid on the caliper mounting bolts.

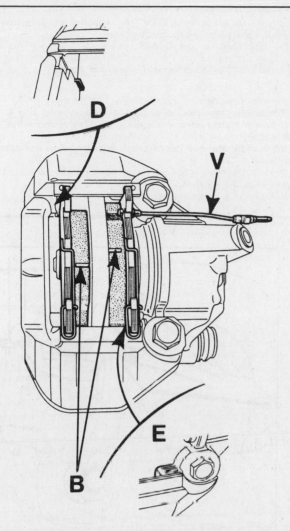

Fig. 13.100 Correct fitting of offset brake pads (Sec 10)

| B | Offset grooves | E | Pad location boss |
| D | Pad location boss | V | Pad wear wire |

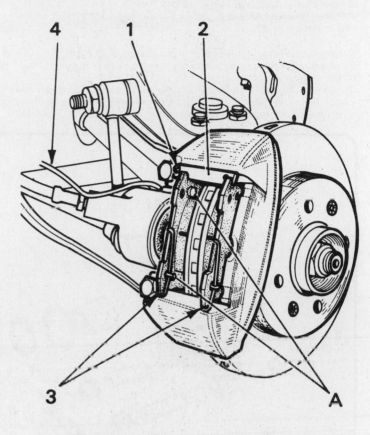

Fig. 13.101 Bendix Series 4 type front brakes (Sec 10)

1	Clip	4	Pad wear wire
2	Key	A	Contact pins
3	Anti-squeal springs		

Front brake pads (Bendix Series 4) – removal and refitting
32 Jack up and support the front of the car and remove the roadwheel. Apply the handbrake.
33 Disconnect the pad wear wire.
34 Extract the clip and tap out the key from the top of the caliper.
35 Press the pads slightly from the disc then extract them.
36 Check the piston and guide rubbers and if necessary renew them, but clean and lubricate their locations with brake grease.
37 If necessary fit the anti-squeal springs on the new pads.
38 Insert the pads in the caliper, tap in the key, and fit the clip. The key may be tapered slightly to facilitate fitting it.
39 Reconnect the pad wear wire, refit the roadwheel and lower the car to the ground.
40 Depress the brake pedal several times to set the pads.

Front brake caliper (Bendix Series 4) – removal, overhaul and refitting
41 Remove the front brake pads, as described in paragraphs 31 to 35.
42 Clamp the flexible hose or alternatively remove the brake fluid reservoir cap and tighten it onto a piece of polythene sheet.
43 Loosen the flexible hose union then unscrew the caliper guide screws, withdraw the caliper and unscrew it from the hose.
44 Remove the piston and guide rubbers.
45 Position a block of wood in the caliper then use a tyre pump or air line in the hose aperture to force out the piston.
46 Clean all components in methylated spirit and check them for wear, damage and corrosion.

Handbrake (rear disc brakes) – adjustment
49 The handbrake adjustment on models fitted with rear disc brakes differs from that described for drum brake models in Chapter 9. Proceed as follows.
50 Loosen the cable tension by rotating the adjuster sleeve (underneath the car). Loosen the cable so that there is a 0.196 in (5 mm) clearance between the cable guide and caliper lever (Fig. 13.103). Fully apply the brake pedal several times then operate the handbrake levers on the calipers a few times. Ensure that they fully return.
51 Retighten the adjuster sleeve to tension the cable and take up the guide to caliper clearance. When correctly adjusted, the handbrake lever travel should be 10 or 11 notches from the rest to fully applied position.
52 If the handbrake lever movement is less than this, the pads will rub and wear out quicker than normal, there will be an increase in the brake pedal travel and the handbrake will be inefficient.

Hydraulic brake hoses – modification
53 Later models may be fitted with brake hoses incorporating taper seat unions not fitted with copper washers. The tapered type union is easily identified by the tapered shoulder (Fig. 13.104) and the union should be tightened to the specified torque only. Make sure that the union components are clean before assembling them.

11 Electrical system

Part A – General equipment

Alternator – general description

1   Later models are fitted with a built-in regulator instead of the earlier remote type.

Alternator voltage regulator – removal and refitting

2   Disconnect the battery and pull the wiring multi-plug from the rear of the alternator.

3   On the Motorola alternator remove the screws, disconnect the wires after noting their location, and withdraw the regulator. Refitting is a reversal of removal, but make sure that the wires are fitted correctly, as shown in Fig. 13.106. The regulator will be permanently damaged if the wires are incorrectly fitted.

4   On SEV Marchal and Ducellier alternators remove the plastic cover from the rear of the alternator then disconnect the wiring and unbolt the regulator. Refitting is a reversal of removal.

Alternator brushes – removal and refitting

5   Remove the regulator, as described in paragraphs 2 to 4.

6   On the Motorola alternator remove the screws and withdraw the brush holder.

7   On the SEV Marchal alternator unsolder the wires and detach the brush holder.

8   On the Ducellier alternator it is not possible to separate the brushes from the regulator.

9   Refitting is a reversal of removal.

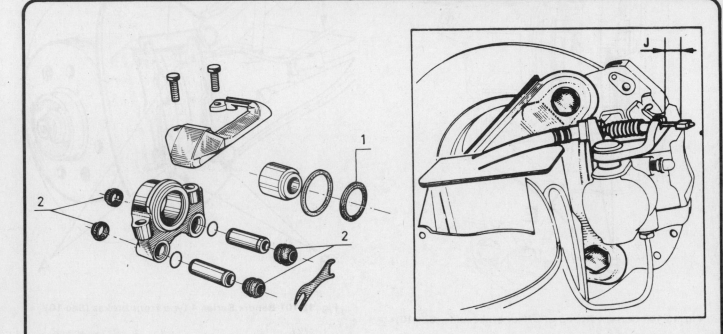

Fig. 13.102 Exploded view of the Bendix Series 4 front brake caliper (Sec 10)

1 Piston dust seal	*2 Guide rubbers*

Fig. 13.103 Handbrake adjustment on rear disc brake models (Sec 10)

J = 0.196 in (5 mm)

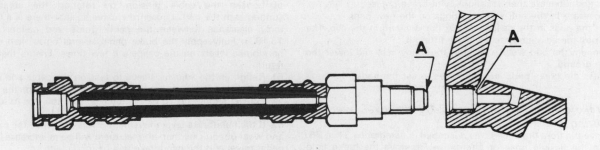

Fig. 13.104 Later tapered type brake hose (Sec 10)

A   Tapered seat surfaces

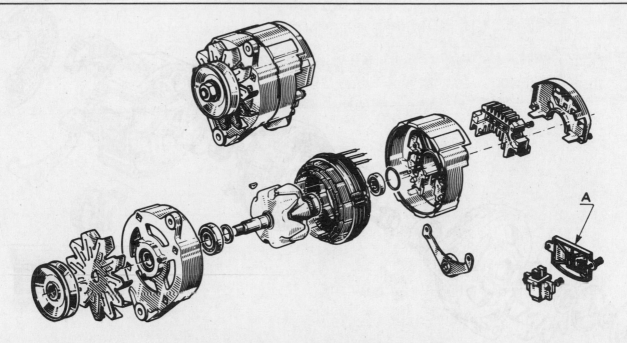

Fig. 13.105 Exploded view of the Motorola alternator (Sec 11)

A   Voltage regulator

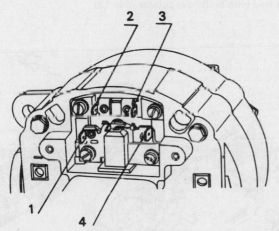

Fig. 13.106 Voltage regulator wire positions on the Motorola alternator (Sec 11)

1	Black	3	Violet
2	Orange	4	Green

Instrument panel (basic version) – removal and refitting
10 Disconnect the battery negative lead.
11 Pull out the switch frame and disconnect the wiring plugs.
12 Pull out the heater control panel.
13 Press in the side clips and withdraw the instrument panel sufficiently to disconnect the speedometer cable and remaining wiring multi-plugs.
14 Refitting is a reversal of removal.

Instrument panel (1984 LHD Turbo models) – removal and refitting
15 Disconnect the battery negative lead.
16 Remove the steering wheel and column shrouds (Chapter 11).

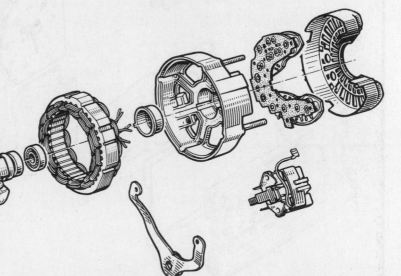

Fig. 13.107 Exploded view of the SEV Marchal alternator fitted with built-in regulator (Sec 11)

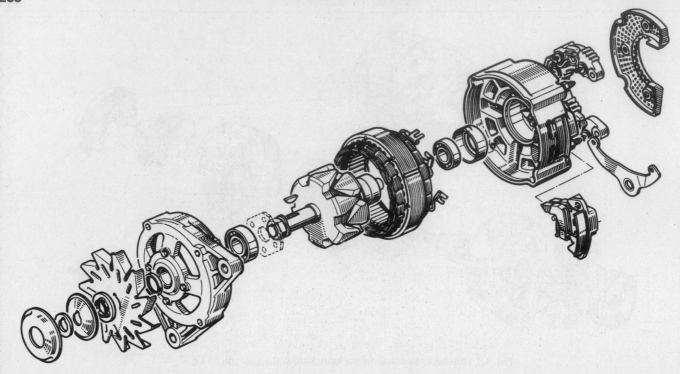

Fig. 13.108 Exploded view of the Ducellier alternator fitted with built-in regulator (Sec 11)

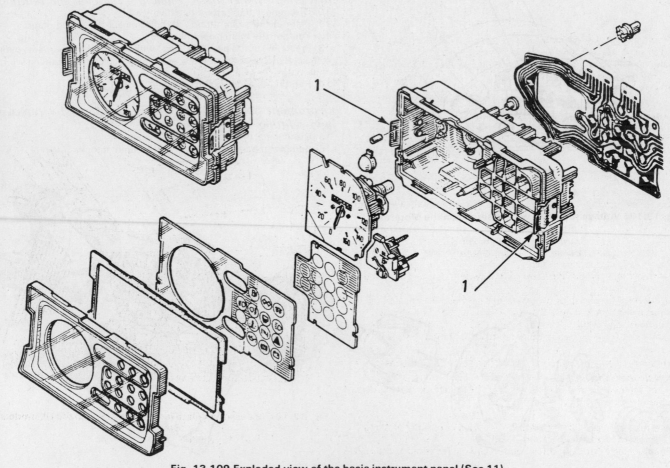

1

1

Fig. 13.109 Exploded view of the basic instrument panel (Sec 11)

1 Spring clips

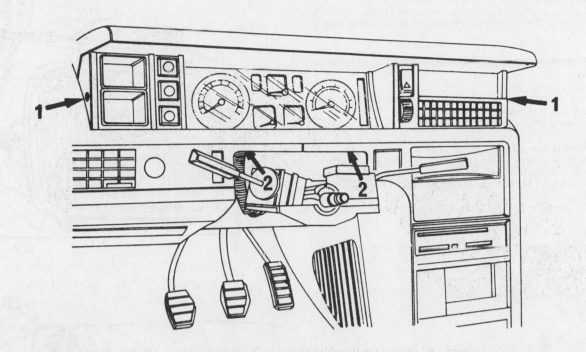

Fig. 13.110 Instrument panel retaining screws (1 and 2) on 1984 LHD Turbo models (Sec 11)

17 Remove the screws shown in Fig. 13.110 and withdraw the surround panel. Disconnect the wiring.
18 Unclip the instrument panel and disconnect the wiring and speedometer cable.
19 Refitting is a reversal of removal.

Tailgate wiper (Estate models) – general
20 As from 1983 the tailgate wiper incoporates a timed delay to operate the wiper intermittently. The delay timer is located on the left-hand rear wheel arch.

Starter motor (US/Canada models) – removal and refitting
21 On R1348 and R1358 models remove the catalytic converter and heat shield, and also the starter motor heat shield.
22 On R1341 and R1351 models remove the air filter and heat shields.
23 Drain the cooling system and disconnect the hoses as necessary.
24 Disconnect the starter wiring.
25 Unscrew and remove the upper left mounting bolt and loosen the upper right bolt.
26 Unscrew the rear support bracket bolt and recover the spacer.
27 Unscrew the mounting bolts and withdraw the starter motor.
28 Refitting is a reversal of removal, but delay tightening the bracket bolts until the main mounting bolts are fully tightened.

Accessories plate/fusebox (US/Canada DL models) – removal and refitting
29 After removal of the screws (Fig. 13.111), release the fusebox from the upper tabs.

Instrument panel (later models) – removal and refitting
30 Disconnect the battery negative lead.
31 Prise off the surround bottom switch covers and remove the two screws.
32 Withdraw the surround sufficiently far to disconnect the switch wiring, then squeeze the clips and withdraw the instrument panel and disconnect the speedometer cable and wiring.
33 Refitting is a reversal of removal.

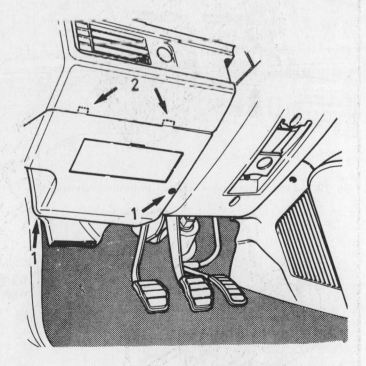

Fig. 13.111 Accessories plate location on US/Canada DL models (Sec 11)
1 Screws   2 Tabs

Headlight sealed beam unit (US/Canada models) – removal, refitting and adjustment
34 With the bonnet open remove the two screws and withdraw the headlight surround.

35 Remove the screws from the sealed beam unit retaining ring then disconnect the wiring plug and withdraw the unit.

36 Refitting is a reversal of removal, but adjust the beam alignment with reference to Chapter 10 using the adjustment knobs shown in Fig. 13.114.

Front direction indicators (US/Canada models) – removal and refitting

37 Remove the radiator grille.

38 Remove the bulbs and if necessary separate the lamps from the grille by removing the screws.

39 Refitting is a reversal of removal.

Windscreen washer pump – removal and refitting

40 The windscreen washer pump is located in front of the battery.

41 Remove the battery (Chapter 10).

42 Lift out the washer bottle and, where the pump is located on the bottle, disconnect the wiring.

43 Detach the pump.

44 Refitting is a reversal of removal.

Headlight wiper blade and arm – removal and refitting

45 Prise up the cover and unscrew the nut from the spindle (photos).

46 Pull the arm from the spindle and disconnect the washer tube (photo).

47 Refitting is a reversal of removal.

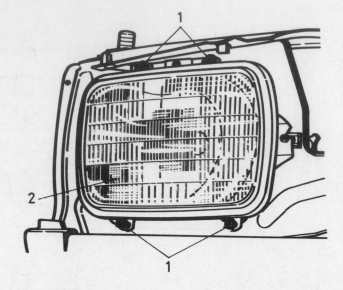

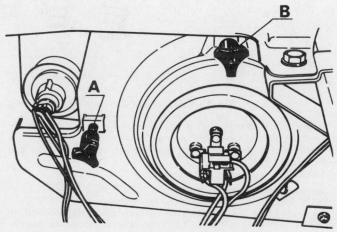

Fig. 13.113 Sealed beam unit fitted to US/Canada models (Sec 11)

1   Retaining ring screws   2   Sealed beam unit
A   Horizontal adjustment knob   B   Vertical adjustment knob

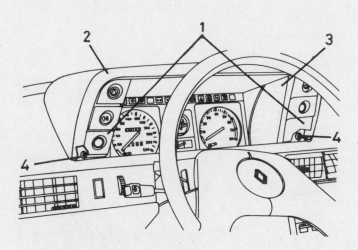

Fig. 13.112 Instrument panel removal on later models (Sec 11)

1   Switch covers   3   Instrument panel
2   Surround   4   Surround retaining screws

11.45A Lift the headlight wiper cover ...

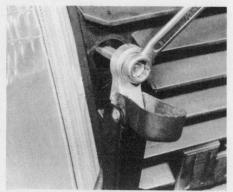

11.45B ... and unscrew the spindle nut

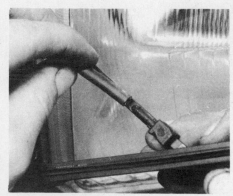

11.46 Disconnecting the washer tube

11.48 Luggage compartment lamp

11.50 Switch location for the luggage compartment lamp

Luggage compartment lamp – removal and refitting
48 Prise out the lamp using a screwdriver (photo).
49 To remove the bulb, ease it from the terminals. Disconnect the wiring if complete lamp removal is required.
50 The lamp switch is located in the boot lid lock and retained by a single screw (photo).
51 Refitting is a reversal of removal.

Trip computer
52 This device was fitted to some models from 1984 on. Its function is to supply the following details as required for a particular journey.

   (a) The time
   (b) The distance covered
   (c) The average speed
   (d) The average fuel consumption
   (e) The actual fuel consumption
   (f) The range
   (g) The amount of fuel left
   (h) The outside air temperature

53 The information is collected by four sensors, these being a fuel flow sensor, a speed sensor (located above the pedal unit and attached to the speedometer cable), a thermistor unit and a special fuel gauge sender unit which supplies fuel information to the instrument panel and the trip computer unit.

54 The system wiring diagram is shown in Fig. 13.114.

55 The operating details for the trip computer are given in the owners handbook supplied with the car.

56 Any faults which may develop in the computer or associated system components must be checked and if necessary, repaired by a Renault dealer.

Wiring diagrams
57 Owing to the numerous Renault 18 model variants it is only possible to include a typical selection of wiring diagrams.

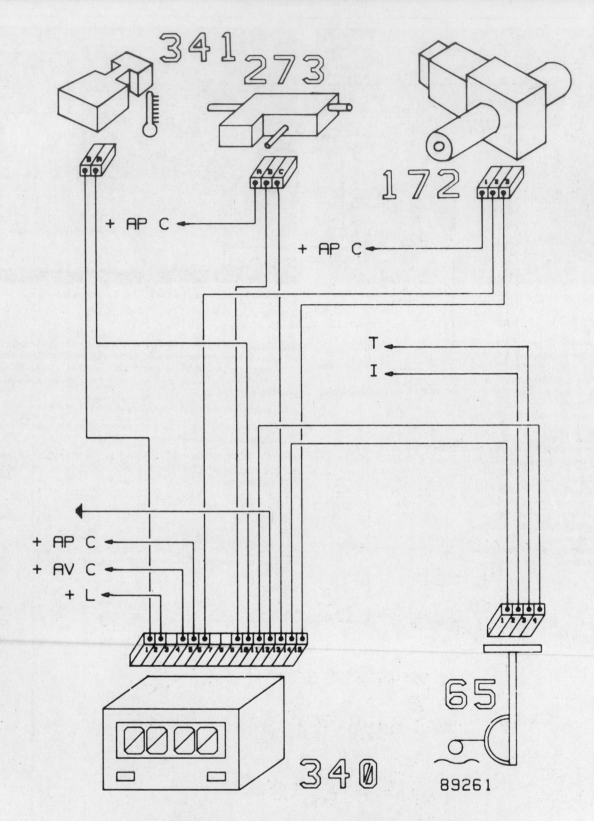

Fig. 13.114 Trip computer system wiring diagram (Sec 11)

65	Fuel gauge	+AVC	+ before ignition switch
172	Speed sensor	+L	Lighting bulbs +
273	Flowmeter		accessories +
340	Trip computer	T	Min fuel level warning
341	Temperature sensor		light bulb
+APC	+ after ignition switch	I	Fuel level gauge

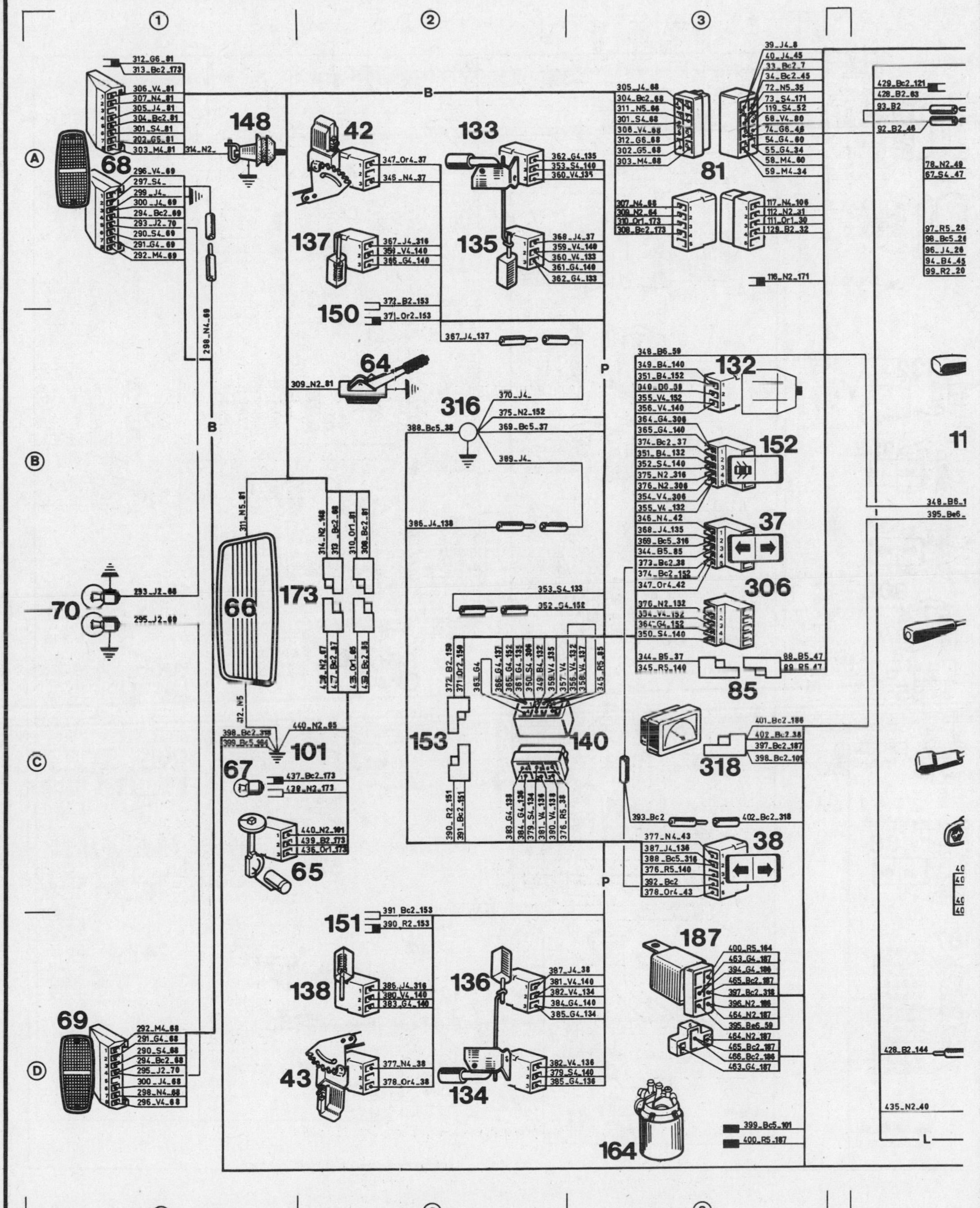

Fig. 13.115 Wiring diagram for 1982 UK Turbo models

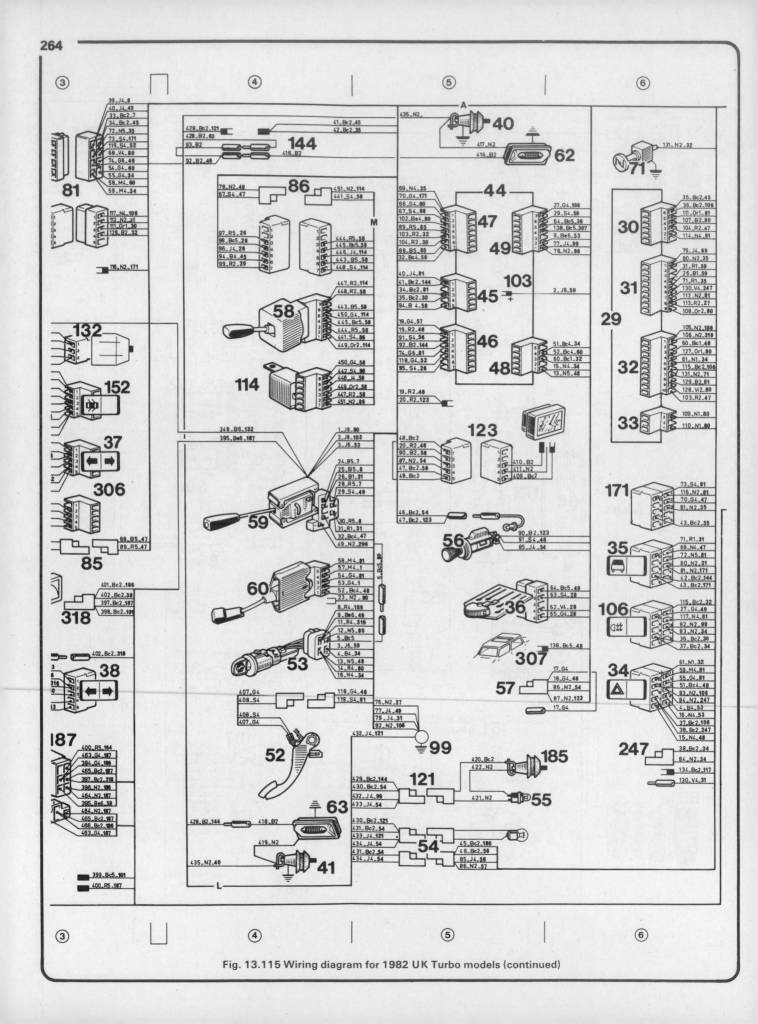

Fig. 13.115 Wiring diagram for 1982 UK Turbo models (continued)

91 24
114.N4.30 423.N4
424.N4.25
131.N2.32
113.R2.31 27
76.N2.99
296 241
403.B5.12
404.N2
405.N2
406.N2
49.N2.59
50.N2.10
123.J4.13 1
53.G4.60
57.M4.60
56.M4.2
33.Bc2.81 7
121.N2
122.Bc5.13
28.R5.59
24.B5.59
338.Or2.178 175
339.Or2.176
341.B2.176
343.N2.175
333.N2.13
343.N2.175
35.Bc2.45
36.Bc2.106
111.Or1.81
107.B2.80
104.R2.47
114.N4.91
7.J4.80 80
136.Bc2.106
6.Bc5.53
108.Or2.31 317.G6.15
107.B2.30 323.S4.21
14.R4.53 322.M4.146
324.R4.22
12.N5.53 326.N5.12
127.Or1.32 325.J4.21
327.N2.209
329.N2.209
109.N1.33
110.N1.33
23.N2.53 320.G4.72
66.G4.47 321.V4.72
102.Bc4.47
68.V4.81
128.Vi2.32
1.J8.59 316.J8.12
8.R4.53 394.G4.187
136.Bc2.80 466.Bc2.187
105.N2.32 396.N2.187
44.Bc2 401.Bc2.318
45.Bc2.54 186
65.G4.38
53.S4.38
75.N5.20
62.V4.36
28
146
322.M4.80
327.N2.80 209
328.N2.80
333.N2.175 13
123.J4.1
122.Bc5.7
133.G4.22
132.G4.22
178
338.Or2.175
334.R2.176
132.G4.110 110
133.G4.110
324.R4.80 R
22
323.S4.80
21
403.B5.296
403.B5.110
320.G4.80 72
321.V4.80
403.B5.12
407.N2.14
406.R4.17
404.N2
405.N2.110
17
406.R4.110
408.Bc4.14
134.Bc2.247 117
135.Bc2.117
318.B5
319.B5
315.J8.15
318.J8.80
326.J4.80 12
326.N5.80
415.J8
10
50.N2.296
135.Bc2.117 117
407.N2.110 14
408.Bc4.17
125.J4.2
125.Bc5.8
331.N2.176
276 425.N10 427.N10
16
426.J8
315.J8.12
317.G6.80
79.J4.99
80.N2.35
31.R1.59
26.B1.59
71.R1.35
130.V4.247
112.N2.81
113.R2.27
108.Or2.80
105.N2.186
106.N2.319
60.Bc1.48
127.Or1.80
61.N1.34
115.Bc2.106
131.N2.71
129.B2.81
128.Vi2.80
103.R2.47
109.N1.80
110.N1.80
73.S4.81
116.N2.81
70.G4.47
81.N2.35
43.Bc2.35
71.R1.31
69.N4.47
70.N16.81
80.N2.31
41.N2.171
42.Bc2.144
43.Bc2.171
115.Bc2.32
27.G4.49
117.N4.81
82.N2.99
83.N2.34
35.Bc2.30
37.Bc2.34
61.N1.32
59.M4.81
55.G4.81
51.Bc4.48
83.N2.106
84.N2.247
4.B4.53
16.N4.53
37.Bc2.106
38.Bc2.247
15.N4.48
38.Bc2.34
84.N2.34
134.Bc2.117
130.V4.31
26
97 97.R5.54
75.N5.28
95.S4.40
99.Bc5.58
56.J4.59
100.N2.20 321
453.N4.321
457.R4.291
455.Bc4.291
452.R4.319
453.N4.97
454.Or4.319
456.R4.291
15 176
337.Or2.174
339.Or2.175
340.B2.174
341.B2.175
329.Bc2.8
334.R2.21
335.R2.179
331.N2.97
332.N2.174
414.Or2.177 179
413.R2.20
336.Or2.174
335.R2.176
332.N2.176
342.N2.174
336.Or2.179
337.Or2.176
340.B2.176
342.N2.174
329.Bc2.176
330.Bc2
8
39.J4.81
124.N2
125.Bc5.14 174
30.R5.59
25.B5.59
100.N2.321 R2
99.R2.58 482
101.N2
413.R2.179
414.Or2.179
291
20 177
456.R4.321 454.Or4.321
457.R4.97 458.R5.319
455.Bc4.321 460.N2.232
106.N2.32
11.R4.53
10.G4
461.N2.319 461.N2.232
460.N2.319 458.R5.319
232 459.R2.319 319
126.J4.14 2
56.M4.1
427.N4.91 25

Fig. 13.115 Wiring diagram for 1982 UK Turbo models (continued)

Key to wiring diagram for 1982 UK Turbo models

The wiring diagram has a grid for ease of unit identification. Figures 1 to 9 run horizontally and letters A to D run vertically

Example: Unit No 1 (LH sidelight and/or direction indicator)
This unit will be found in the rectangle bordered by letter A vertically and figure 9 horizontally

Component number/name	Location	Component number/name	Location
1 LH sidelight and/or direction indicator	A-9	52 Stop-lights switch	D-4
2 RH sidelight and/or direction indicator	D-9	53 Ignition-starter/anti-theft switch	C-4
7 LH headlight	A-9	54 Heater control panel illumination	D-5
8 RH headlight		55 Glove compartment illumination	A-5
10 RH horn	C-9	56 Cigar lighter	C-5
12 Alternator	C-8	57 Feed to car radio	C-5
13 LH front earth	B-9	58 Windscreen wiper/washer switch	B-4
14 RH front earth	C-9	59 Lighting and direction indicators switch	B-4
15 Starter	D-8	60 Direction indicators switch or connector	C-4
16 Battery	C-8	62 LH side or central interior light	A-6
17 Engine cooling fan motor	B-9	63 RH interior light	D-4
20 Windscreen washer pump	D-7	64 Handbrake switch	B-2
21 Oil pressure switch	B-8	65 Fuel gauge tank unit	C-2
22 Thermal switch on radiator (at bottom on		66 Rear screen demister	B-1
Master)	B-8	67 Luggage compartment light	B-1
24 LH front brake	A-8	68 LH rear light assembly	A-1
25 RH front brake	D-8	69 RH rear light assembly	D-1
26 Windscreen wiper motor	C-7	70 Number plate lights	B-1
27 Nivocode or ICP (pressure drop indicator)	A-7	71 Choke 'On' warning light	A-6
28 Heating-ventilating fan motor	B-7	72 Reversing lights switch	C-9
29 Instrument panel	B-6	80 Junction block – front and engine harnesses	A-7
30 Connector No 1 – instrument panel	A-6	81 Junction block – front and rear harnesses	A-3
31 Connector No 2 – instrument panel	B-6	85 Junction block – window or electro-magnetic	
32 Connector No 3 – instrument panel	B-6	lock harnesses	C-3
33 Connector No 4 – instrument panel	B-6	86 Junction block – windscreen wiper time switch	
34 'Hazard' warning lights switch	C-6	relay	A-4
35 Rear screen demister switch	C-6	91 Wire junction – brake pad wear warning light	A-7
36 Heating-ventilating fan motor rheostat or		97 Bodyshell earth	C-7
resistance	C-5	99 Dashboard earth	D-5
37 LH window switch	B-3	101 Fuel tank mounting earth	C-2
38 RH window switch	C-3	103 Feed to accessories plate	B-5
40 LH front door pillar switch	A-5	106 Rear foglight switch	C-6
41 RH front door pillar switch	D-4	110 Engine cooling fan motor relay	B-9
42 LH window motor	A-2	114 Windscreen wiper time switch relay	B-4
43 RH window motor	D-2	117 Wire junction – front foglights	C-9
44 Accessories plate or fusebox	A-5	121 Wire junction – glove compartment light	D-5
45 Junction block – front harness – accessories		123 Clock	B-5
plate	B-5	132 Inertia switch	B-3
46 Junction block – front harness – accessories		133 LH front door switch	A-2
plate	B-5	134 RH front door switch	D-2
47 Junction block – front harness – accessories		135 LH front door solenoid	A-2
plate	A-5	136 RH front door solenoid	D-2
48 Junction block – front harness – accessories		137 LH rear door solenoid	A-2
plate	B-5	138 RH rear door solenoid	D-2
49 Junction block – front harness – accessories		140 Junction block – electro-magnetic lock harness	C-3
plate	A-5	144 Wire junction – interior light	A-4

Key to wiring diagram for 1982 UK Turbo models (continued)

Component number/name	Location	Component number/name	Location
146 Temperature switch or thermal switch	B-8	185 Glove compartment light switch	D-6
148 Tailgate or luggage compartment lid fixed contact	A-1	186 Wire junction – electric pump	B-7
150 LH front door loudspeaker	B-2	187 Rev counter relay	D-3
151 RH front door loudspeaker	D-2	209 Oil level indicator sensor	C-8
152 Electro-magnetic locks centre switch	B-3	232 Boost pressure switch	D-8
153 Loudspeaker wire	C-2	241 Horn compressor	A-8
164 Electric pump	D-3	247 Wire junction – auxiliary lights	D-6
171 Rear screen wiper-washer switch	B-6	276 Engine earth	C-7
173 Wire junction – fuel tank wiring	B-2	291 Pinking detector	D-8
174 RH headlight wiper motor	D-9	296 Horn compressor relay	A-8
175 LH headlight wiper motor	A-9	300 Remote control door unlocking device	B-3
176 Headlight wipers time switch relay	C-8	307 Wire junction – sunroof	C-5
177 Headlight washers pump	D-7	316 Console earth	B-2
178 Headlight washers pump (RHD)	B-8	318 Boost pressure gauge illumination	C-3
179 Wire junction – windscreen washer/headlight washers pump	D-8	319 Ignition cut-out relay	D-9
		321 AEI module	C-7

Wire identification Each wire is identified by a number followed by a letter(s) indicating its colour, a number giving its diameter and finally a number giving the unit destination

Colour code

B	Blue	G	Grey	N	Black	S	Pink
Bc	White	J	Yellow	Or	Orange	V	Green
Be	Beige	M	Maroon	R	Red	Vi	Violet
C	Clear						

Wire diameters

No	mm	No	mm	No	mm	No	mm
1	0.7	4	1.2	7	2.5	10	5.0
2	0.9	5	1.6	8	3.0	11	7.0
3	1.0	6	2.1	9	4.5	12	8.0

Harness identification

A	Front	L	Interior lights – door pillar switches	N	Automatic transmission
B	Rear			P	Door locks
K	Starter	M	Windscreen wiper	R	Engine

Fig. 13.116 Wiring diagram for 1982 UK R1351 models

The wiring diagram has a grid to make it easier to find units. This grid is marked horizontally 1 to 9 and vertically A to D

Example: Unit No 1 (LH sidelight and/or direction indicator)
This unit will be found in the rectangle bounded vertically by letter A and horizontally by number 9

No	Description	Grid
1	LH sidelight and/or direction indicator	A9
2	RH sidelight and/or direction indicator	D9
7	LH headlight	A9
8	RH headlight	D9
9	LH horn	A9
10	RH horn	D9
12	Alternator	B8
13	LH earth	B9
14	RH earth	C9
15	Starter	B7
16	Battery	C7
17	Engine cooling fan motor	C9
20	Windscreen washer pump	B8
21	Oil pressure switch	A8
22	Thermal switch on radiator (bottom on Master)	A8
24	LH front brake	A7
25	RH front brake	D7
26	Windscreen wiper motor	D7
27	Nivocode or ICP (pressure drop indicator)	B9
28	Heating-ventilating fan motor	C4
29	Instrument panel	B6
30	Connector No 1 – instrument panel	A6
31	Connector No 2 – instrument panel	B6
32	Connector No 3 – instrument panel	B6
33	Connector No 4 – instrument panel	B6
34	'Hazard' warning lights switch	C6
35	Rear screen demister switch	C5
36	Heating-ventilating fan motor rheostat or resistance	D4
37	LH window switch	A3
38	RH window switch	D3
40	LH front door pillar switch	A5
41	RH front door pillar switch	D5
42	LH window motor	A2
43	RH window motor	D2
44	Accessories plate or fusebox	B4
45	Junction block – front harness and accessories plate	B4
46	Junction block – front harness and accessories plate	C4
47	Junction block – front harness and accessories plate	B4
48	Junction block – front harness and accessories plate	C4
49	Junction block – front harness and accessories plate	B4
52	Stop-lights switch	D4
53	Ignition-starter/anti-theft switch	B4
54	Heater controls illumination	D5
55	Glove compartment illumination	D5
56	Cigar lighter	B5
57	Feed to car radio	C5
58	Windscreen wiper/washer switch	A4
59	Lighting switch	B5
60	Direction indicators switch or connector	B5
62	LH or front interior light	A6
63	RH interior light	D6
64	Handbrake 'On' warning light switch	C3
65	Fuel gauge tank unit	D1
66	Rear screen demister	B1
68	LH rear light assembly	A1
69	RH rear light assembly	D1
70	Number plate lights	B1
71	Choke 'On' warning light	C6
73	Rear light assembly earth	D1
78	Rear screen wiper motor	B1
79	Rear screen washer pump	B1
80	Junction block – front and engine harnesses	A7
81	Junction block – front and rear harnesses	B2
84	Junction block – front and automatic transmission harnesses	B8
85	Junction block – window or electro-magnetic lock harnesses	B3
86	Junction block – windscreen wiper time switch relay	A5
90	Wire junction – air conditioning compressor	B7
91	Wire junction – brake pad wear warning light	D8
97	Bodyshell earth	C7
99	Dashboard earth	D8
100	Side panel gusset earth	C8
101	Fuel tank earth	D1
103	Feed to 'accessories' plate	C4
104	Wire junction – steering wheel	C5
106	Rear foglight switch	C5
110	Engine cooling fan motor relay	B9
114	Windscreen wiper time switch relay	A4
117	Wire junction – front foglights	C9
121	Wire junction – glove compartment light	D6
123	Clock	B3
124	Automatic transmission	B8
128	Kickdown switch	C8
130	Automatic transmission earth	A9
132	Inertia switch	B2
133	LH front door switch	A3
134	RH front door switch	D3
135	LH front door solenoid	A3
136	RH front door solenoid	D3
137	LH rear door solenoid	A1
138	RH rear door solenoid	D1
140	Junction block – electro-magnetic lock harness	D2
141	Wire junction – rear foglight switch	C5
144	Wire junction – interior light	A6
146	Thermal switch	A8
150	LH front door loudspeaker	A2
151	RH front door loudspeaker	D2
152	Electro-magnetic locks centre switch	A2
153	Loudspeaker wire	D2
155	Rear or LH rear interior light	B2
158	Automatic transmission selector illumination	C4
170	Wire junction – air conditioning harness	C5
171	Rear screen wiper-washer switch	C6
172	Impulse generator	D5
173	Wire junction – fuel tank wiring	C2
177	Headlight washers pump	B8
185	Glove compartment light switch	D6
192	Tailgate earth	C1
204	Starter relay	B3
210	Wire junction – AEI wiring	C7
214	Front foglights relay	B9
219	Wire junction – engine cooling fans relay	C9
252	'Normalur' Cruise Control switch	C4
276	Engine earth	C8
277	Junction block – rear harness and luggage compartment lid wiring	A2
300	Connector No 1 – dashboard and front harnesses	A1
314	Wire junction – Econometer wiring	A5
316	Console strut earth	C3
320	'Normalur' Cruise Control servo motor	C9
321	AEI module	C8
322	Declutching switch	D5
323	'Normalur' Cruise Control computer	D7
336	Connector No 5 – dashboard	B6
337	Wire junction – 'Driving Aid' engine compartment and passenger compartment	D4
344	Junction block – rear harness and rear screen wiper motor feed wires	C1

For wire identification and colour code see the key to Fig. 13.115

68

1
2
3
4
5
6
7
8

300 S4 - 81
298 B2 - 81
293 G4 - 81
306 J4 - 73
290 N2 - 81
295 Bc2 - 81

Ⓐ

Ⓑ

Ⓒ

Ⓓ

137
349 J4 - 316
340 V4 - 140
348 G4 - 140

306
358 N2 - 152
336 V4 - 152
346 G4 - 152
332 S4 - 140

277
381 Bc2 - 155
382 Bc2 - 70
379 N5 - 66

297 Bc2 - 69
305 N5 - 81

381 Bc2 - 277
155

310 R2 - 81
311 N2 - 79
311 N2 - 79
79

379 N5 - 277
380 N2 - 78

382 Bc2 - 277
70
303 Bc2

78
66
192
20 N5

380 N2 - 80
39 R2 - 344
344
39 R2 - 78
309 R2 - 81

73
306 J4 - 88
307 J4 - 89

65
36 Or2 - 173
37 B2 - 173
38 N2 - 101

101
38 N2 - 65

69
296 J4 - 81
297 Bc2 - 277
291 N2 - 81
307 J4 - 73
294 M4 - 81
292 G4 - 81
299 B2 - 81
302 S4 - 81

138
52 J4 - 316
44 V4 - 140
49 G4 - 140

42
329 Or4 - 37
328 N4 - 37

150
353 Or2 - 153
354 B2 - 153

152
346 G4 - 306
347 G4 - 140
356 Bc2 - 37
333 B4 - 132
334 S4 - 140
357 N2 - 316
358 N2 - 306
336 V4 - 306
337 V4 - 132

132
331 B4 - 140
333 B4 - 132
330 Be6 - 59
337 V4 - 152
338 V4 - 140

81
301 S4 - 81
300 S4 - 68
302 S4 - 69
292 G4 - 69
293 G4 - 68
294 M4 - 80
298 B2 - 68
299 B2 - 69
305 N5 - 277
290 N2 - 69
296 J4 - 69
291 N2 - 69
295 Bc2 - 68

151 G4 - 80
152 G4 - 34
155 M4 - 60
156 M4 - 34
165 V4 - 80
171 G6 - 46
169 N5 - 35
170 S4 - 171
136 J4 - 8
137 J4 - 45
216 S4 - 337
130 Bc2 - 7
131 Bc2 - 45

301 S4 - 81
303 N4 - 81
304 Or2 - 173
300 B2 - 173

214 N4 - 106
209 N2 - 31
208 Or1 - 30
207 B4 - 32

310 R2 - 79
309 R2 - 344

217 R2 - 171
305 N2

213 N2 - 171

303 N2 - 81
64

173
37 B2 - 65
38 Or2 - 65

309 B2 - 81
304 Or2 - 81

140
49 G4 - 138
50 G4 - 136
45 V4 - 134
43 S4 - 138
44 V4 - 138
40 R5 - 38

335 S4 - 133
334 S4 - 152
345 G4 - 140
347 G4 - 152
348 G4 - 137
343 G4 - 152
345 G4 - 140
338 V4 - 132
339 V4 - 140
341 V4 - 140
340 V4 - 137
331 B4 - 132
332 S4 - 306
327 R5 - 85

153
354 B2 - 150
353 Or2 - 150

56 R2 - 151
57 Bc2 - 151

151
57 Bc2 - 153
56 R2 - 153

43
41 N4 - 38
42 Or4 - 38

135
350 J4 - 37
341 V4 - 140
342 V4 - 133
343 G4 - 140
344 G4 - 133

133
344 G4 - 135
335 S4 - 140
342 V4 - 135

37
320 N4 - 42
350 J4 - 135
351 Bc5 - 316
326 B5 - 85
355 Bc2 - 38
356 Bc2 - 152
329 Or4 - 42

204
21 B5 - 59
23 Bc2 - 53
24 N2 - 53
22 Bc5 - 53

85
326 B5 - 37
327 R5 - 140

185 B5 - 47
186 R5 - 47

123
145 Bc2 - 123
117 R2 - 46
187 B2 - 56
184 N2 - 57
144 Bc2 - 56
145 Bc2 - 123

62 B2 - 123
63 N2 - 123
61 Bc2 - 123

62 B2 - 123
63 N2 - 123
61 Bc2 - 123

316
357 N2 - 152
352 J4
351 Bc5 - 37

52 J4 - 138
55 J4
54 Bc5 - 38

349 J4 - 137

P

38
141 Bc2 - 150
48 Bc2
47 Bc2 - 38
41 N4 - 43
53 J4 - 136
54 Bc5 - 316
40 R5 - 140
47 Bc2 - 38
42 Or4 - 43

355 Bc2 - 37

134
46 V4 - 138
43 S4 - 140
51 G4 - 136

136
53 J4 - 38
45 V4 - 140
46 V4 - 134
50 G4 - 140
51 G4 - 134

114

5

24 N2 - 204
23 Bc2 - 204
22 Bc5 - 204

166 N4 - 35
167 G4 - 171
163 G4 - 80
164 R2 - 80
199 Be4 - 80
186 R5 - 85
200 R2 - 32
201 R2 - 30
185 B5 - 85
129 Bc4 - 59
137 J4 - 81
138 Bc2 - 144
131 Bc2 - 81
132 Bc2 - 30
191 B4 - 80

117 R2 - 123
116 R2

119 G4 - 57
190 G4 - 56
189 B2 - 144
171 G6 - 81
215 G4 - 337
192 S4 - 26

160 S4 - 36
162 G4 - 36
172 N5 - 26
159 V4 - 36

158
141 Bc2 - 38
142 Bc2 - 54

5
96 G4
97 S4

337
215 G4 - 46
216 S4 - 81

161 Bc5 - 49
160 S4 - 28
159 V4 - 28
162 G4 - 28

Fig. 13.116 Wiring diagram for 1982 UK R1351 models (continued)

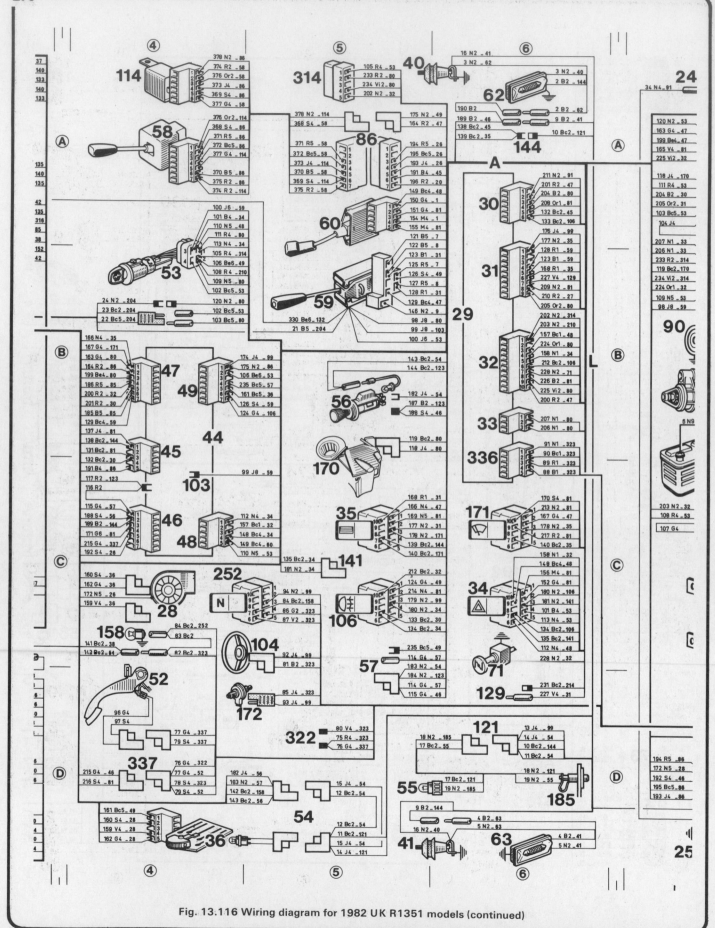

Fig. 13.116 Wiring diagram for 1982 UK R1351 models (continued)

⑥ ⑦ ⑧ ⑨

3 N2 _ 40
2 B2 _ 144
144
2 B2 _ 62
0 B2 _ 41
10 Bc2 _ 121

34 N4 _ 91
24

A

120 N2 _ 53
163 G4 _ 47
199 Be4 _ 47
165 V4 _ 81
225 Vi2 _ 32
118 J4 _ 170
111 R4 _ 53
204 B2 _ 80
205 Or2 _ 31
103 Bc5 _ 53
104 J4
80
207 N1 _ 33
206 N1 _ 33
233 R2 _ 314
119 Bc2 _ 170
234 Vi2 _ 314
224 Or1 _ 32
109 N5 _ 53
98 J8 _ 50
313 J8 _ 12

317 V4 _ 124
315 Be4 _ 84
316 G4 _ 124
325 Vi2 _ 84
321 J4 _ 90
322 R4 _ 22
318 M4 _ 146
319 S4 _ 21
314 G6 _ 15
320 Bc2 _ 12
323 J4 _ 12
324 N5 _ 12

21   319 S4 _ 80
146 318 M4 _ 80

317 V4 _ 80
316 G4 _ 80

27 V4 _ 124
26 G4 _ 124
33 V4 _ 128
29 G4 _ 84
32 S4 _ 130
25 J4 _ 84
26 G4 _ 124
27 V4 _ 124
31 S4 _ 130
28 M4 _ 84
30 S4 _ 130

124
84
325 Vi2 _ 80
315 Be4 _ 80
25 J4 _ 124
28 M4 _ 124
29 G4 _ 124
1 J8 _ 276

22
322 R4 _ 80
229 N4 _ 219
230 N4 _ 22
230 N4 _ 22

30 S4 _ 124
31 S4 _ 124
32 S4 _ 124
130

196 R2 _ 86
197 N2 _ 321
198 N2 _ 177
20
177   198 N2 _ 20

173 N2 _ 99
210 R2 _ 31
27

1
150 G4 _ 60
220 J4 _ 13
153 M4 _ 2
154 M4 _ 60

7
130 Bc2 _ 81
218 N2
219 Bc5 _ 13
125 R5 _ 59
121 B5 _ 59
A

146 N2 _ 59
147 N2 _ 10
9

367 Bc4 _ 17
219 Bc5 _ 7
220 J4 _ 1
366 N2 _ 110
13

231 Bc2 _ 129
232 Bc2 _ 117
214

110
362 B5 _ 12
363 G4 _ 219
364 Bc2 _ 12
365 R4 _ 17
366 N2 _ 13
B

211 N2 _ 91
201 R2 _ 47
204 B2 _ 80
208 Or1 _ 81
132 Bc2 _ 45
133 Bc2 _ 106
176 J4 _ 99
177 N2 _ 35
128 R1 _ 59
123 B1 _ 59
168 R1 _ 35
227 V4 _ 129
209 N2 _ 81
210 R2 _ 27
103 Or2 _ 80
202 N2 _ 314
203 N2 _ 210
157 Bc1 _ 48
224 Or1 _ 80
158 N1 _ 34
212 Bc2 _ 106
228 N2 _ 71
226 B2 _ 81
225 Vi2 _ 80
200 R2 _ 47
207 N1 _ 80
206 N1 _ 80
91 N1 _ 323
00 Be1 _ 323
89 R1 _ 323
88 B1 _ 323

L

B

90
321 J4 _ 80
314 G6 _ 80
312 J8 _ 12
6 N9 _ 16
15
6 N9 _ 15

313 J8 _ 80
312 J8 _ 15
386 B5 _ 12
323 J4 _ 80
324 N5 _ 80
320 Bc2 _ 80
386 B5 _ 12
12

364 Bc2 _ 110
362 B5 _ 110
7 N9 _ 18
1 J8 _ 12
276

16
7 N9 _ 276
9 J8
360 N4 _ 321
97

210
203 N2 _ 22
108 R4 _ 53
361 Or4 _ 321
359 R4 _ 321
107 G4

359 R4 _ 210
360 N4 _ 97
361 Or4 _ 210
321

229 N4 _ 22
219
363 G4 _ 110

367 Bc4 _ 13
365 R4 _ 110
17
C

232 Bc2 _ 214
117

222 Bc5 _ 8
223 J4 _ 2
14

147 N2 _ 9
10

170 S4 _ 81
213 N2 _ 81
167 G4 _ 47
178 N2 _ 35
217 R2 _ 81
140 Bc2 _ 35
150 N1 _ 32
140 Bc4 _ 48
156 M4 _ 81
152 G4 _ 81
180 N2 _ 106
181 N2 _ 141
101 B4 _ 53
113 N4 _ 53
134 Bc2 _ 106
135 Bc2 _ 141
112 N4 _ 48
228 N2 _ 32

C

231 Bc2 _ 214
227 V4 _ 31

82 Bc2 _ 158
81 B2 _ 104
80 V4 _ 322
78 S4 _ 337
75 R4 _ 322
95 N2 _ 99
66 M4 _ 320
67 G4 _ 320
64 R4 _ 320
68 Or3 _ 320
69 B3 _ 320
72 Bc3 _ 320
65 V4 _ 320
88 B1 _ 336
89 R1 _ 336
90 Bc1 _ 336
91 N1 _ 336
85 J4 _ 172
87 V2 _ 252
86 G2 _ 252
323

128
33 V4 _ 184
74 V4 _ 220
73 J4 _ 320
100

71 N3 _ 99
72 Bc3 _ 323
66 M4 _ 323
74 V4 _ 128
67 G4 _ 323
64 R4 _ 323
73 J4 _ 100
68 Or3 _ 323
69 B3 _ 323
70 N3 _ 99
320

136 J4 _ 81
221 N2
222 Bc5 _ 14
127 R5 _ 59
122 B5 _ 59
8
D

70 N3 _ 320
71 N3 _ 320
92 J4 _ 104
93 J4 _ 172
94 N2 _ 252
95 N2 _ 323
173 N2 _ 27
174 N4 _ 49
176 J4 _ 31
179 N2 _ 106
13 J4 _ 121
99

194 R5 _ 86
172 N5 _ 28
192 S4 _ 46
195 Bc5 _ 86
193 J4 _ 86
26

13 J4 _ 99
14 J4 _ 54
10 Bc2 _ 144
11 Bc2 _ 54
18 N2 _ 121
19 N2 _ 55
185
D

4 B2 _ 41
5 N2 _ 41
⑥

34 N4 _ 24
35 N4 _ 91
35 N4 _ 25
91
211 N2 _ 30

25

223 J4 _ 14
153 M4 _ 1
2

⑦ ⑧ ⑨

Fig. 13.116 Wiring diagram for 1982 UK R1351 models (continued)

Key to typical later UK model wiring diagram circuits

	All types	Saloon	Estate	847 engine	841 – A2M engine	843 – A6M engine	829 – J6R engine
AEI ignition	4	–	–	–	–	–	–
Automatic transmission	7	–	–	–	–	–	–
Brake pad wear warning light	1	–	–	–	–	–	–
Charging circuit	8	–	–	–	–	–	–
Choke 'On' warning light	–	–	–	8	–	–	–
Cigar lighter	3	–	–	–	–	–	–
Conventional ignition	–	–	–	8	–	–	–
Coolant temperature switch	7	–	–	–	–	–	–
Cooling fan motor	–	–	–	8	6	6	8
Direction indicators	–	10	10	–	–	–	–
Door locks	6	–	–	–	–	–	–
Front interior lights	3	–	–	–	–	–	–
Fuel gauge	–	1	2	–	–	–	–
Handbrake	1	–	–	–	–	–	–
Headlight dipped beams	–	10	10	–	–	–	–
Headlight main beams	–	10	10	–	–	–	–
Headlight wiper/washers	9	–	–	–	–	–	–
Heating/ventilating	5	–	–	–	–	–	–
Horn	–	10	10	–	–	–	–
Identification plates and switches illumination	5	–	–	–	–	–	–
Idle cut-out	–	–	–	–	–	–	1
Luggage compartment illumination	1	–	–	–	–	–	–
Nivocode	1	–	–	–	–	–	–
Oil pressure switch	7	–	–	–	–	–	–
Radio feed	1	–	–	–	–	–	–
Rear foglight	–	10	10	–	–	–	–
Rear interior light	–	–	4	–	–	–	–
Rear screen demister	–	1	2	–	–	–	–
Rear screen wiper/washer	–	–	2	–	–	–	–
Reversing lights	–	1	2	–	–	–	–
Sidelights	–	10	10	–	–	–	–
Speakers	5	–	–	–	–	–	–
Starter	–	–	–	8	8	8	8
Stop-lights	–	1	2	–	–	–	–
Transistorized ignition	–	–	–	9	–	–	–
Window winders	6	–	–	–	–	–	–
Windscreen wiper/washer	9	–	–	–	–	–	–
Windscreen wiper/washer with timer	8	–	–	–	–	–	–

Key to typical wiring diagram for later UK models

1	LH sidelight and/or direction indicator	84	Junction block – gearbox/automatic transmission harness
2	RH sidelight and/or direction indicator	97	Bodyshell earth
7	LH headlight	99	Dashboard earth
8	RH headlight	101	Fuel tank earth
9	LH horn	103	Feed to accessories plate
10	RH horn	105	Automatic transmission computer
12	Alternator	106	Rear foglight switch
13	LH earth	108	Multi-function switch
14	RH earth	109	Speed sensor
15	Starter	110	Engine cooling fan motor relay
16	Battery	111	Solenoid valves 1 and 2
17	Engine cooling fan motor	114	Windscreen wiper timer relay
18	Ignition coil (or mounting)	121	Junction – glove compartment light
20	Windscreen washer pump	128	Kick-down switch
21	Oil pressure switch	132	Inertia switch
22	Fan motor No 1 activating thermal switch	133	LH front door lock switch
23	Cooling temperature warning thermal switch	134	RH front door lock switch
24	LH front brake	135	LH front door solenoid
25	RH front brake	136	RH front door solenoid
26	Windscreen wiper motor	137	LH rear door solenoid
27	Nivocode or ICP (pressure drop indicator)	138	RH rear door solenoid
28	Heating-ventilating fan motor	140	Junction No 1 – door locking/unlocking harness
29	Instrument panel	146	Temperature or thermal switch
30	Connector No 1 – instrument panel	148	Tailgate or luggage compartment fixed contact
31	Connector No 2 – instrument panel	150	LH front door speaker
32	Connector No 3 – instrument panel	151	RH front door speaker
34	'Hazard' warning lights switch	152	Central door locking switch
35	Rear screen demister switch	153	Radio speaker wires
37	LH window switch	155	Rear or LH rear interior light
38	RH window switch	158	Automatic transmission selector illumination
40	LH front door pillar switch	169	Junction – solenoid valves harness
41	RH front door pillar switch	171	Rear screen wiper/washer switch
42	LH window motor	173	Junction – fuel tank harness
43	RH window motor	174	RH headlight wiper motor
44	Accessories plate or fusebox	175	LH headlight wiper motor
45	Junction block – front harness – accessories plate	176	Headlight wipers timer relay
46	Junction block – front harness – accessories plate	177	Headlight washers pump
47	Junction block – front harness – accessories plate	185	Glove compartment light switch
48	Junction block – front harness – accessories plate	192	Tailgate earth
49	Junction block – front harness – accessories plate	195	Idling cut-out
52	Stop-lights switch	200	Heater plugs
53	Ignition-starter/anti-theft switch	201	Air pre-heating box
54	Heating/ventilating controls illumination	204	Starter relay
55	Glove compartment light	208	Diesel fuel cut-off solenoid
56	Cigar lighter	210	Junction – AEI harness
57	Feed to car radio	272	Throttle spindle switch
58	Windscreen wiper/washer switch	274	Wire junction No 1
59	Lighting and direction indicators switch	276	Engine earth
60	Direction indicator switch or connector	277	Junction – rear and boot lid harnesses
61	Feed terminal before ignition-starter switch	278	Carburettor
62	LH or front centre interior light	283	Advance solenoid valve
63	RH interior light	285	Cold start enrichment relay
64	Handbrake 'On' warning light switch	286	Wire junction No 2
65	Fuel gauge tank unit	289	Wire junction No 3
66	Rear screen demister	293	Junction – windscreen wiper wiring
67	Luggage compartment light	321	AEI module
68	LH rear light assembly	344	Junction block – rear screen wiper motor feed wires
69	RH rear light assembly	353	Thermal switch 15° C
70	Number plate lights	359	Recycling valve solenoid valve
71	Choke 'On' warning light	392	Junction – starter relay harness
72	Reversing lights switch	454	Junction – headlight wipers harness
77	Wire junction – diagnostic socket	455	Rear screen wiper timer relay
78	Rear screen wiper motor	456	Junction – engine cooling fan harness
79	Rear screen washer pump	458	Anti-pollution diode
80	Junction block – engine harness	459	Anti-pollution solenoid valve
81	Junction block – rear harness No 1	477	Junction – headlight wiper/washers intermediate harness

For wire identification and colour code see the key for Fig. 13.115

Not all components are fitted to all models

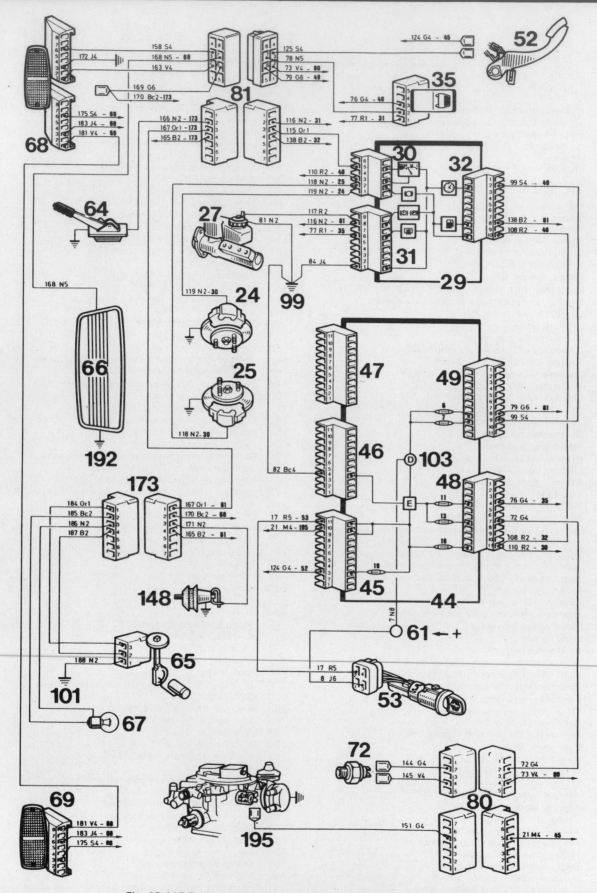

Fig. 13.117 Typical wiring diagram for later UK models – circuit 1

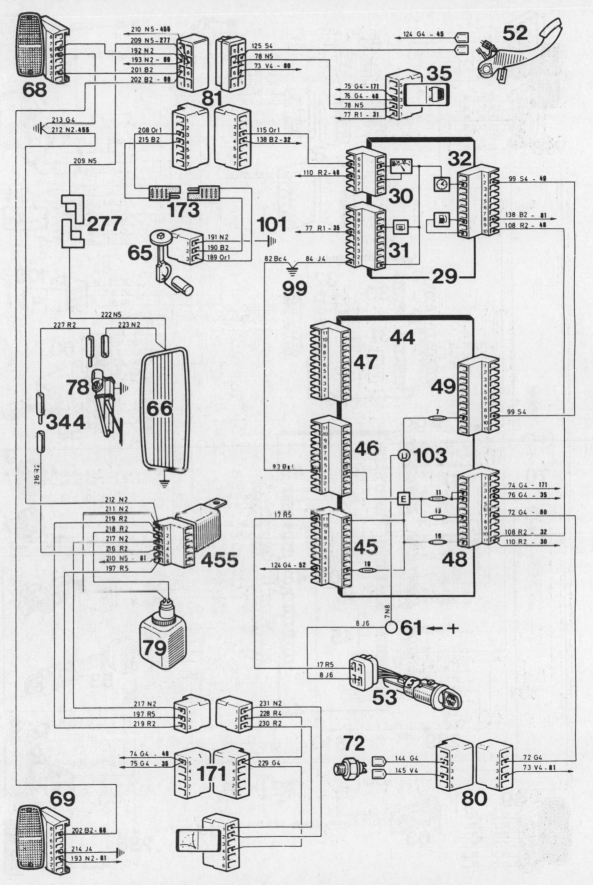

Fig. 13.117 Typical wiring diagram for later UK models – circuit 2

275

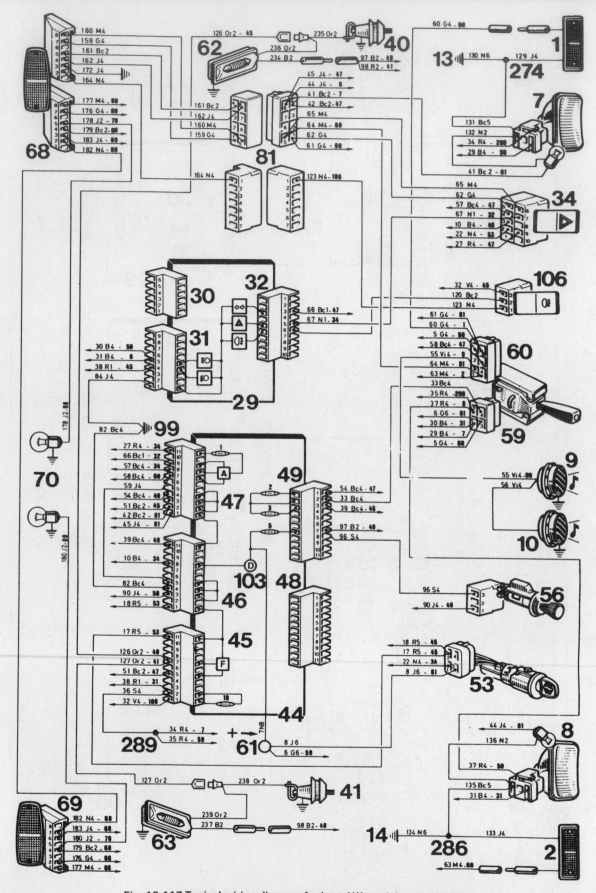

Fig. 13.117 Typical wiring diagram for later UK models – circuit 3

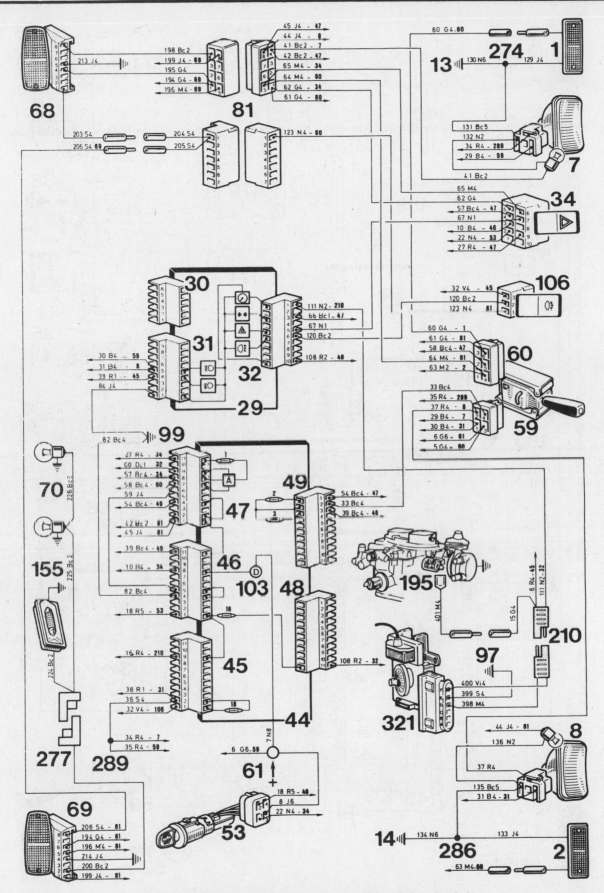

Fig. 13.117 Typical wiring diagram for later UK models – circuit 4

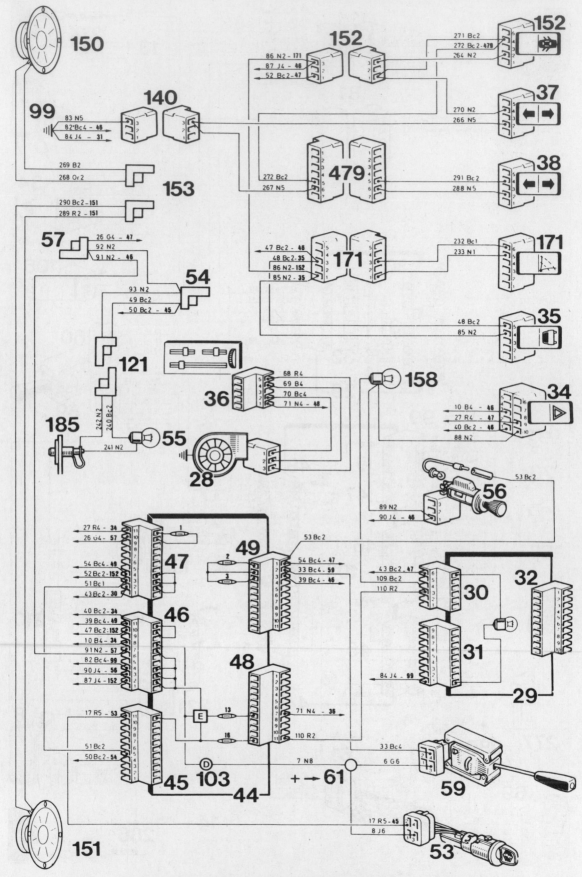

Fig. 13.117 Typical wiring diagram for later UK models – circuit 5

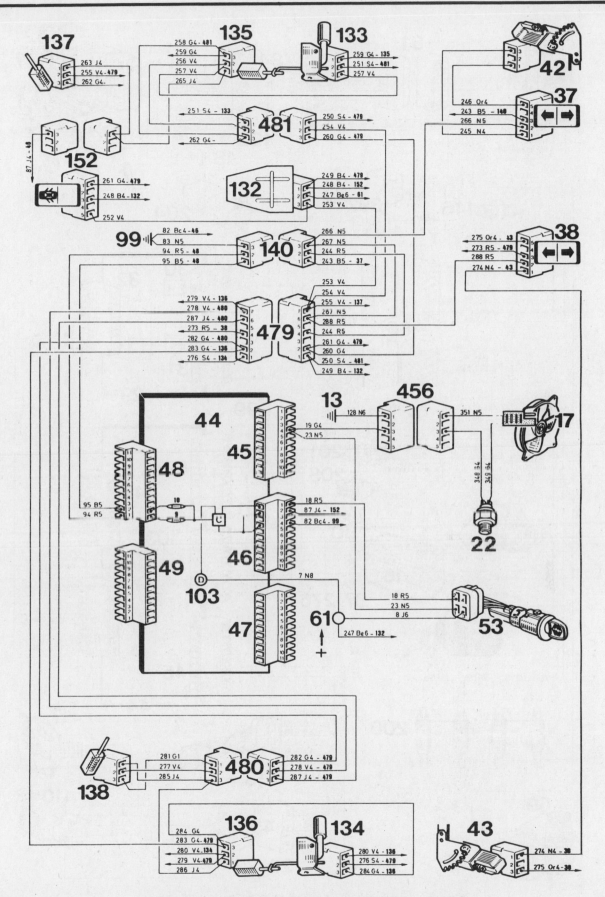

Fig. 13.117 Typical wiring diagram for later UK models – circuit 6

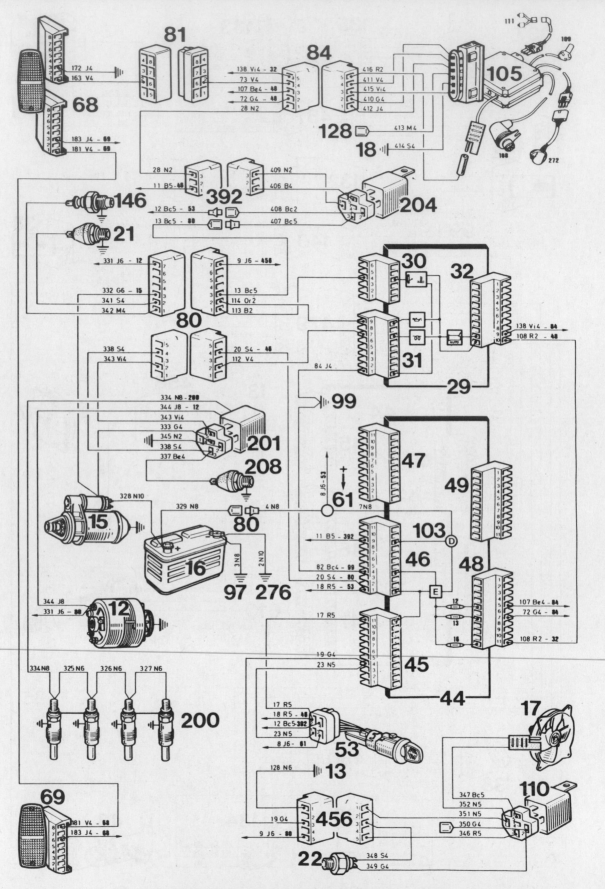

Fig. 13.117 Typical wiring diagram for later UK models – circuit 7

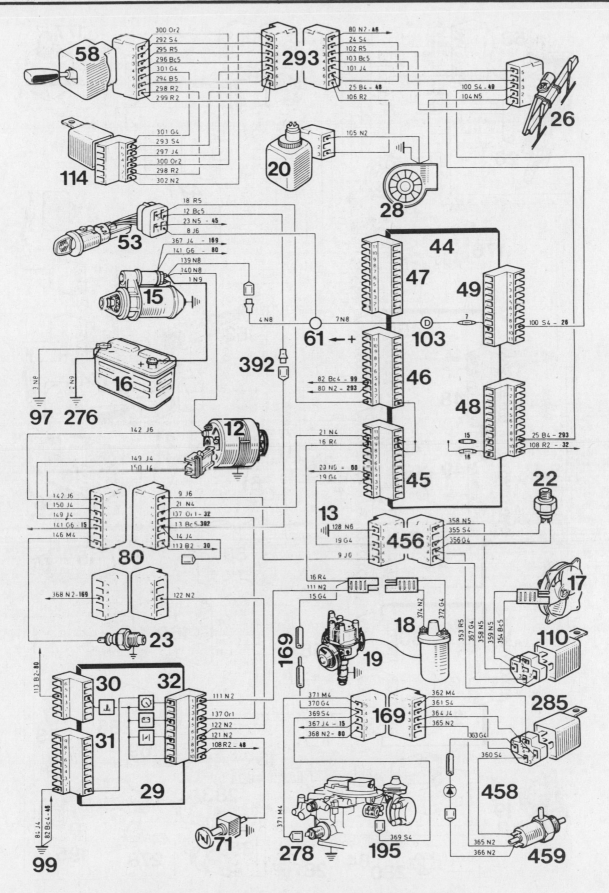

Fig. 13.117 Typical wiring diagram for later UK models – circuit 8

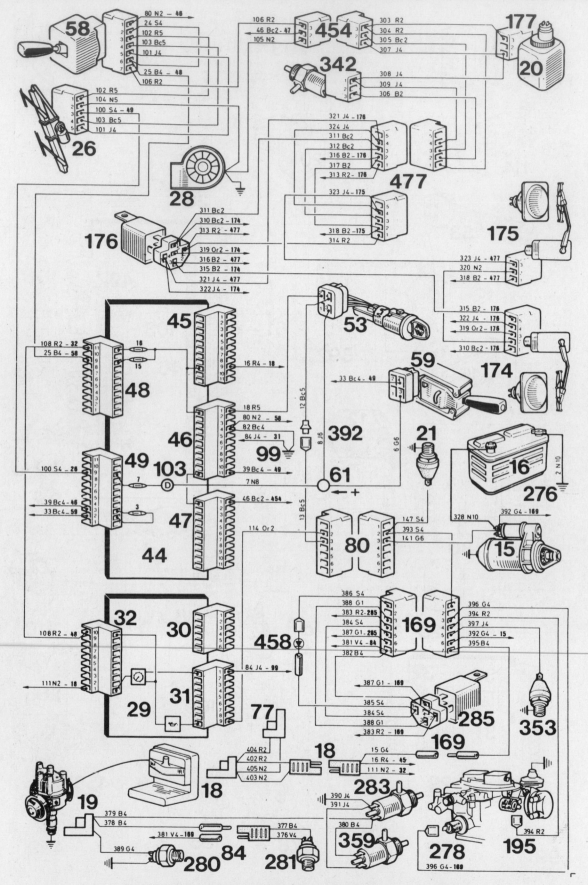

Fig. 13.117 Typical wiring diagram for later UK models – circuit 9

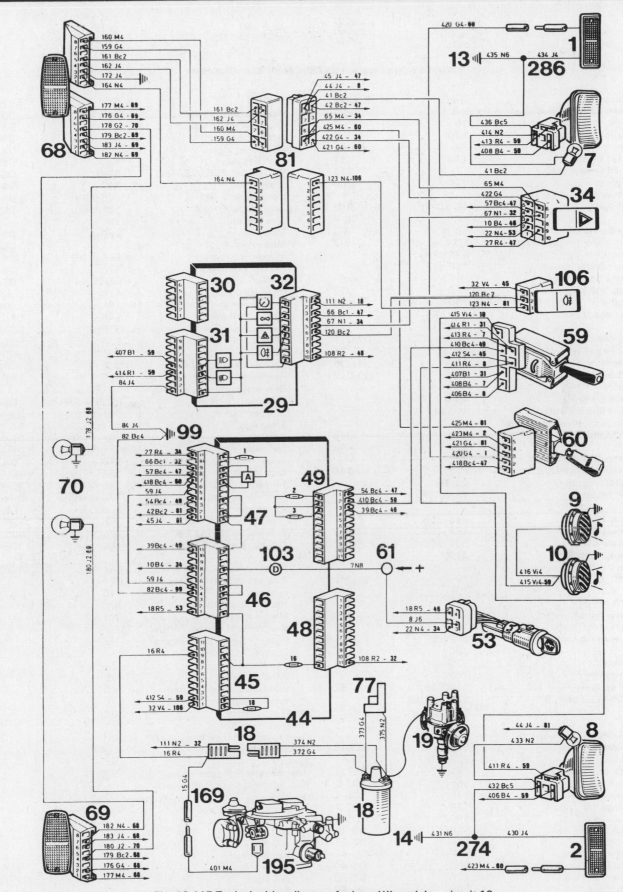

Fig. 13.117 Typical wiring diagram for later UK models – circuit 10

Key to wiring diagram for 1981 and 1982 US models

Component	Grid	Symbol
A/C heater control panel assy.	2-4	
Alternator	2B	
Auto trans. diagnostic conn.	14D	
Battery	3A	Battery
Brake pad sensor, left	17H	S11
Brake pad sensor, right	17H	S12
Cigar lighter	27F	R6
Circuit breakers:		
A/C system	4C-7D-4E	CB-1,CB-2, CB-4
Auxiliary cooling fan motor	8E	B-3
Cooling fan motor (Main)	5D	B-2
Clock assy.	27F,53H	
Computer (auto. trans.)	14E	
Diagnostic connectors:		
Auto. transmission	14D	J17
Engine	13H	J15
Fuel injection	42J-45J	P146
Diodes:		
A/C circuit	8D,7F	D7,D11
A/C heater control panel	3G	D6
Alternator	2B	D1-D6
Instrument cluster PCB	16G,17G	D9,D12
Distributor	10D	
Door lock system	49-51	
Electronic fuel control unit	40F,44F	
Electronic ignition control unit	12F	
Engine diagnostic connector	13H	J15
Flasher	35B	K12
Flowmeter (FI system)	42E	
Fuel injection diagnostic conn.	43J	P146
Fuel injection system	40-45	
Fuel sender assy.	18H,(19H)	
Fuses:		
Panel PCB	34B-26B	F1-F9
Panel PCB	46A-48A	F11,F13
Panel PCB	53C-55C	F10,F12
Speed control	32D	F20
Gauges:		
Fuel	18F	M2
Oil level	20F	M3
Tachometer	15F	M1
Water temperature	15F	M4
Governor (auto. trans.)	14F	
Hazard warning switch assy.	36E,53F	
Horn	56C	
Ignition coil	12D	L1
Ignition sensor	10F	
Ignition sw. assy.	11A,37B	S2
Instrument cluster PCB	15F-21F,35C, 53E,59C	
Lamps:		
A/C-heater control panel (2 Lamps)	52H	DS-13
Back-up, left	23F,(21F)	DS-11
Back-up, right	23F,(22F)	DS-12
Brake ind.	16F	DS-2
Brake pad wear ind.	17F	DS-4
Charge ind.	15F	DS-1
Choke ind.	18F	DS-6
Cigar lighter	53H	DS-14
Clock	53H	DS-27
Dome (Wagon)	(57H)	DS-41
Door lock switch ind.	50D	DS-46
Fasten belt ind.	21F	DS-10
Glove box	58F	DS-40

Component	Grid	Symbol
Hazard warning ind.	35C	DS-21
Hazard warning switch ind.	53F	DS-31
Headlamp, left	59F	DS-43
Headlamp, right	60F	DS-44
High beam ind.	59C	DS-42
Inst. cluster PCB (5 Lamps)	53F	DS-45
Interior, left	30D	DS-16
License plate, left	55F,(56H)	DS-35
License plate, right	56F,(56H)	DS-36
Low fuel ind.	18F	DS-6
Map	29D	DS-15
Marker, left front	54F	DS-32
Marker, left rear	55F,(54H)	DS-33
Marker, right front	57F	DS-39
Marker, right rear	56F,(55H)	DS-38
Oil pressure ind.	17F	DS-3
Oxygen sensor maint. ind.	19F	DS-7
Parking, left front	54F	DS-22B
Parking, right front	57F	DS-23B
PRNDL	52H	DS-26
Radio	30F	DS-28
Rear window defog. ind.	20F	DS-9
Rear window defog. sw. ind.	53F	DS-30
Rear window defog. sw. ind. (Wagon)	(52F)	DS-29
Stop, left	30J,(29J)	DS-18A
Stop, right	31J,(29J)	DS-19A
Tail, left	55F,(54H)	DS-18B
Tail, right	56F,(55H)	DS-19B
Trunk	31D	DS-17
Turn signal, ind.	35C	DS-20
Turn signal, left front	34H,(37H)	DS-22A
Turn signal, right front	35H,(38H)	DS-23A
Turn signal, left rear	36H,(37H)	DS-24
Turn signal, right rear	36H,(39H)	DS-25
Window switch ind. left	47F	DS-47
Window switch ind. right	48F	DS-48
Motors:		
A/C blower	10J	B13
Aux. cooling fan (A/C sys.)	8E	B3
Fuel pump	44C	B8
Heater blower (with A/C)	10G	B11
Main cooling fan	5D	B2
Rear window washer (Wagon)	(20E)	B4
Rear window wiper (Wagon)	(18E)	B5
Starter	6A	B1
Window, left	45E	B9
Window, right	49E	B10
Windshield washer pump	24D	B6
Windshield wiper	27C	B7
Noise suppressor	10C	C1
Oil pressure sender	16J	S10
Oxygen sensor (FI system)	40H	
Power windows system	45-49	
Radio	29F	
Rear window defogger sw. assy.	17C,53F	
Rear window wiper motor assy. (Wagon)	(18E)	
Rear window wiper sw. assy. (Wagon)	(20C,52F)	
Relays, solenoids, etc.:		
A/C No. 1	6G	K4
A/C No. 2	9G	K32
A/C No. 3	9H	K33
A/C No. 4	9G	K31
A/C No. 5	6H	K26
A/C No. 6	6J	K27

Key to wiring diagram for 1981 and 1982 US models (continued)

Component	Grid	Symbol
A/C compressor clutch	12H	K27
A/C fast idle solenoid	11G	K24
Auto transmission (EL1)	14E	K29
Auto transmission (EL2)	14E	K28
Auxiliary air regulator (FI sys.)	43E	K18
Auxiliary cooling fan motor (with A/C only)	6E	K3
Cold start injector	44E	K19
Door lock solenoid, left front	51F	K23
Door lock solenoid, left rear	49G	K21
Door lock solenoid, right front	50F	K20
Door lock solenoid, right rear	50G	K22
Flasher	35B	K12
Fuel injector 1	40E	K14
Fuel injector 2	41E	K15
Fuel injector 3	41E	K16
Fuel injector 4	42E	K17
Fuel pump	42C	K13
Ignition on	16B	K7
Key-in buzzer	30C	K10
Main cooling fan motor	3D	K2
Power window	15B	K6
Rear window defogger	24B	K25
Rear window wiper motor (Wagon)	18E	K30
Seat belt buzzer	24E	K8
Speed sensor	31H	K11
Starter (auto. trans.)	9B	K5
Starter solenoid	4A	K1
Windshield wiper timer	25F	K9
Resistances:		
Cigar lighter	27F	R6
Coolant temp. sender	15J	R2
Coolant temp. sensor (FI system)	41G	R7
Fuel sender	17H,(18H)	R3
Inst. cluster PCB lamps rheostat	53E	R10
Low fuel ind. sender	18H,(19H)	R4
Oil level sender	20H	R5
Rear window defogger	17E	R1
Switch ID lamps rheostat	62E	R11
Thermostat (A/C system)	9E	R8
Senders & sensors:		
Brake pad sensors (left & right)	17H	S11,S12
Coolant temp. sender	15J	R2
Coolant temp. sensor (FI System)	41G	R7
Fuel sender	17H,(18H)	R3
Ignition sensor	10D	
Low fuel ind. sender	18H,(19H)	R4
Oil fuel sender	20H	R5
Oil pressure sender	16J	S10
Oxygen sensor (FI system)	40H	
Speed sensor	31H	K11
Top dead center sensor	13G	L2
Solenoids: (see Relays)		
Spark plugs	9D	
Speakers	27J	
Speed control regulator	31F	
Speed control servo	30H	
Speed control system	31-33	
Speed sensor	31H	K11
Starter motor assy.	5A	B1

Component	Grid	Symbol
Switches:		
A/C-heater blower motor	4G	S37
A/C-heater control panel	2H	S39
A/C high pressure (high arrest)	11H	S43
A/C low pressure	11G	S42
Back-up light	23D	S7
Brake fluid level ind.	16H	S9
Brake pad sensor, left	17H	S11
Brake pad sensor, right	17H	S12
Changeover, left (door lock system)	51D	S31
Changeover, right (door lock system)	50D	S30
Combination	55B-60B	S33
Coolant thermal	6C	S1
Dimmer	59B	S33C
Dome lamp (Wagon)	(57H)	S35
Door jam, left	29E	S19
Door jam, right	29E	S18
Door lock	50C	S33
Glove box lamp	58H	S34
Handbrake	16J	S8
Hazard warning	36E,(39E)	S25
Horn	56B	S33B
Ignition	10A,37A	S2
Inertia (door lock system)	51C	S32
Interior lamp, left	30D	S17
Key-in	30E	S20
Kickdown (auto. trans.)	15C	S3
Light	55B	S33A
Map lamp	29D	S16
Microswitch (A/C system)	6F	S36
Multi-function (auto. trans.)	13E,21D	S4
Oil pressure sender	16J	S10
Oxygen sensor maint. ind	19J	S13
Radiator thermal	6C	S1
Rear window defogger	17C	S5
Rear window wiper/washer (Wagon)	(19C)	S6
Seat belt	24H	S14
Speed control	32D	S23
Stop lamp	31C	S22
Thermal coolant time (FI system)	44E	S26
Throttle position (FI system)	42H	S27
Trunk lamp	30E	S21
Turn signal	34G	S24
Window, left	46E	S28
Window, right	48E	S29
Windshield wiper/washer	25C	S15
Tail lamp assy, left	23F,30J, 36H,55G	DS11,DS18, DS24,DS18
Tail lamp assy, right	23F,31J, 36H,56G	DS12,DS19, DS25,DS19
Tail lamp assy, left (Wagon)	22F,29J, 37H,54H	DS11,DS18, DS24,DS18
Tail lamp assy, right (Wagon)	22F,29J, 39H,55H	DS12,DS19, DS25,DS19
Thermostat (A/C system)	9E	R8
Top dead center system	13G	L2
Voltage regulator	3C	
Water temp. sender	15J	R2

All components listed are not fitted to every model

Wiring harnesses

A	Front main	H	Seat belt
B	Rear (Sedan)	J	Fuel tank (Sedan)
C	Fuel pump and relay	K	Fuel tank (Wagon)
D	Glove box lamp	L	Interior lamp/glove box
E	Tail lamp	M	Diagnostic connector (auto. trans.)
F	Door locks/power windows, left	N	Automatic transmission
G	Door locks/power windows, right		

O	Starter relay	U	Tailgate (Wagon)
P	Brake pad wear warning	W	Auxiliary cooling fan motor
Q	Diagnostic connector (engine)	X	Rear (Wagon)
R	Engine	AA	Speed control
T	Interior lamp or map reading lamp		

Colour codes

BE	Beige	GN	Green	S	Salmon
BK	Black	O	Orange	V	Violet
BL	Blue	P	Pink	W	White
BN	Brown	R	Red	Y	Yellow

13.118 Wiring diagram for 1981 and 1982 US models

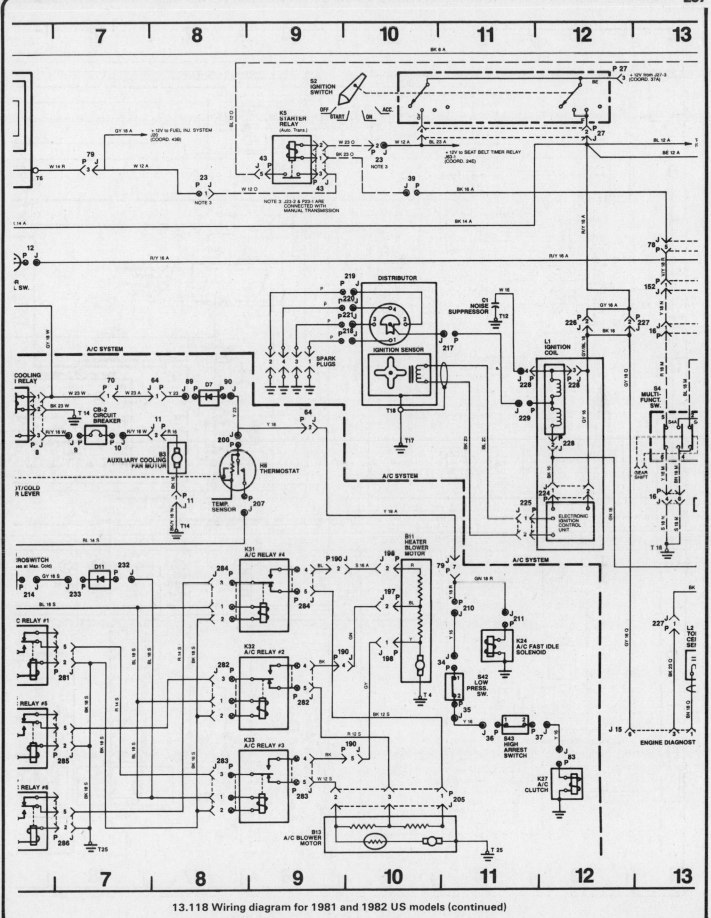

13.118 Wiring diagram for 1981 and 1982 US models (continued)

12　13　14　15　16　17　18

BK 6 A
BL 12 O
W 6 A

FUSE BOX P.C.B.

+12V from J27-3 (COORD. 37A)
P 27
BE
J 27

+12V to FUEL PUMP RELAY J62-5 (COORD. 40B)

+12V to FUSES F11 & F13 (COORD. 46A)

50　50　J51P

44

K6 POWER WINDOW RELAY

K7 IGNITION ON RELAY
51

BL 12 A　To J46A-2 (COORD. 21C)
BE 12 A

45

F 16 1.5a　F 14 5a　F 18 16a　F 17 16a　F 19 5a

R/Y 16 A

47　45　47　GY 18 A

GND to K9 J135-6 (COORD. 25G)

78

S3 KICK-DOWN SW.
40

REAR WINDOW DEFOG. SW. ASSY.
41

S5 REAR WINDOW DEFOGGER SWITCH (Spring Loaded)
ON　OFF

S6 REAR WIPER SWITCH

152

41

AUTO. TRANS. DIAGNOSTIC CONNECTOR J17

GY 16 A
BK 16
227

S4 MULTI-FUNCT. SW.

S4A　S4B

K28　K29
EL-2　EL-1
COMPUTOR

GEAR SHIFT

(STATION WAGON)
REAR WINDOW WIPER MOTOR ASSY

81
BK 12 B
194
67　P 67 J
GY 16

R1 REAR WINDOW DEFOGGER
53
54

K30 REAR WINDOW WIPER MOTOR RELAY

B5 REAR WINDOW WIPER MOTOR
T 26　T27

R 23 A

INSTRUMENT CLUSTER P.C.B.
10　32

GOVERNOR

+12V from VOLTAGE REGULATOR J78-2 (COORD. 3C)

W/GN 16 A

DS-1 CHARGE IND.
M1 TACH.　M4 TEMP. GAUGE
DS-2 BRAKE IND.　DS-3 OIL PRESSURE IND.　DS-4 BRAKE PAD WEAR IND.　M2 FUEL GAUGE　DS-6 LOW FUEL IND.　DS-8 CHOKE IND.

BK 18 A
32

TIMING PULSES to FUEL INJ. SYSTEM J62-3 (COORD. 40D)

D9　D12

227
L2 TOP DEAD CENTER SENSOR

(STATION WAGON)

30　31　30　32　N.C.

103

BN 18 Q　W 18 Q　Y 18 Q
T 16

J 15

ENGINE DIAGNOSTIC CONNECTOR

100

S9 BRAKE FLUID LEVEL IND. SW.
101

104　105
S11 LEFT　S12 RIGHT

R3 FUEL SENDER　R4 LOW FUEL IND. SENDER

FUEL SENDER ASSY

187　107

R3　R4

FUEL SENDER ASSY.

BRAKE PAD SENSORS

79　82

106
R2 COOLANT TEMP. SENDER

99
S8 HAND-BRAKE SWITCH

102
S10 OIL PRESSURE SENDER

109

109

12　13　14　15　16　17　18

13.118 Wiring diagram for 1981 and 1982 US models (continued)

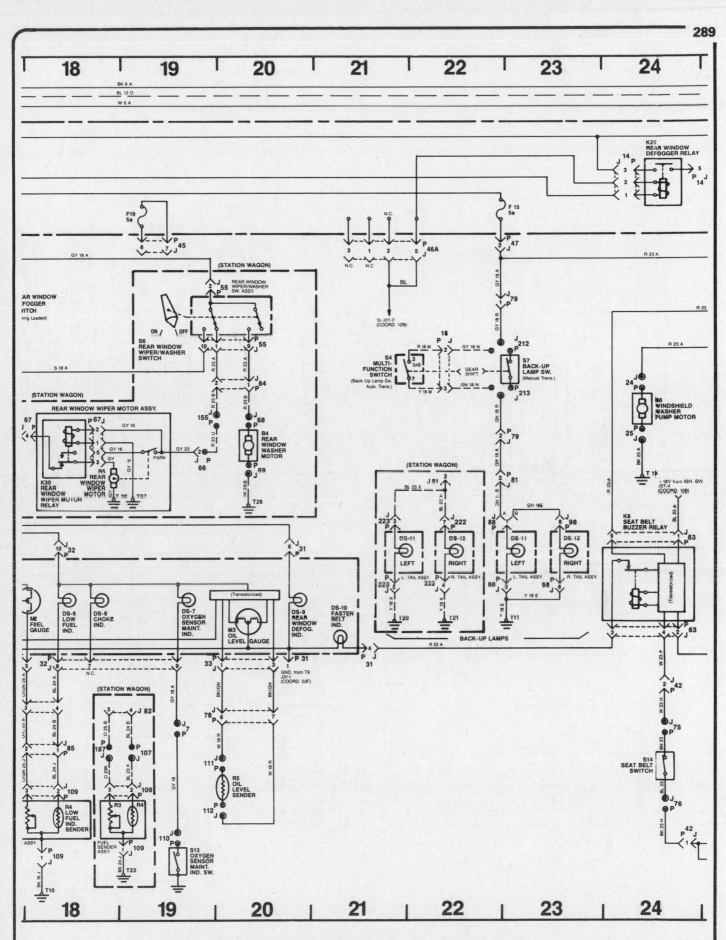

13.118 Wiring diagram for 1981 and 1982 US models (continued)

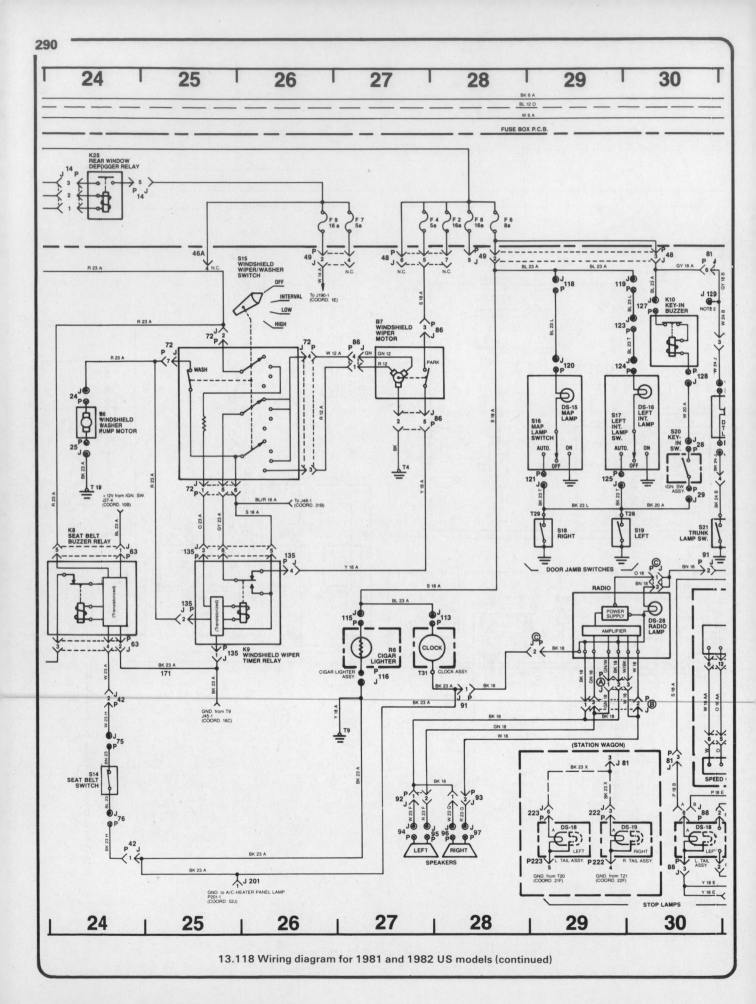

13.118 Wiring diagram for 1981 and 1982 US models (continued)

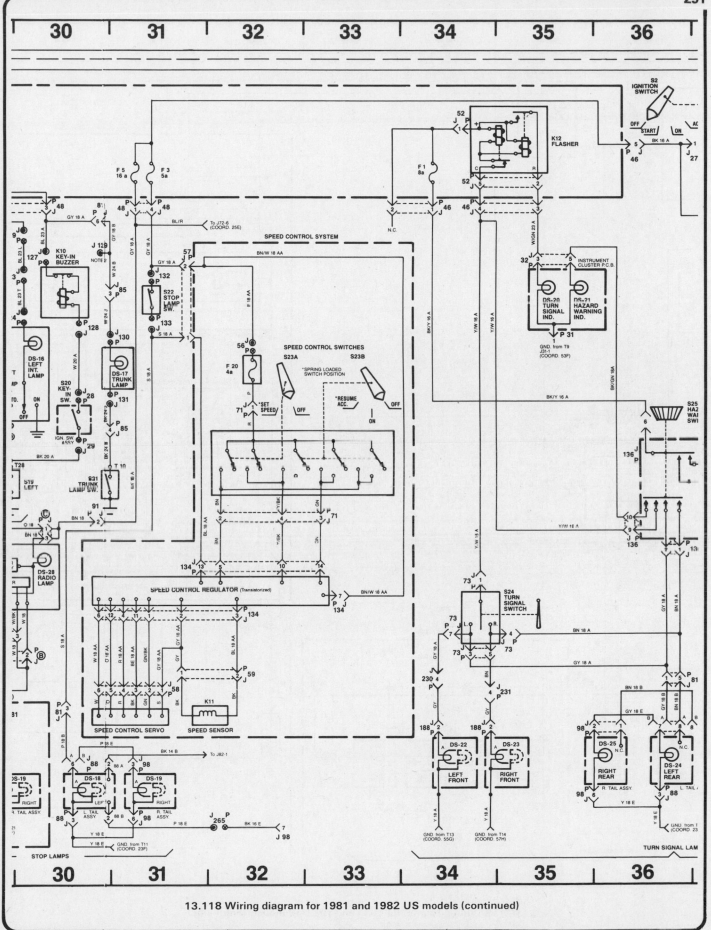

13.118 Wiring diagram for 1981 and 1982 US models (continued)

36 37 38 39 40 41 42

BL 12 O
W 6 A
Y 9 A
BL 12 A

S2 IGNITION SWITCH

OFF START ON ACC.

BE
P J 27
3

BK 16 A
P 46
5
1 J 27

BL/R 16 A

+12V from STARTING CIRCU
P19-4
(COORD. 5B)

+12V from IGNITION SWITCH
J45-3
(COORD. 14B)

FUEL INJECTION SYSTEM

R/Y 16 A
BL 12 A
J 62
P 5
1

R/Y 16 C
BL 12 C
S 17 C
J 60
P 9
2
6

86c 88z 88y

K13A K13B

85 88b 88e 88a 86

P 61 J 8 5

RUMENT
TER P.C.B.

RD
ING

BK/GN 18A

S25 HAZARD WARNING SWITCH

HAZARD WARNING SW. ASSY.

136 J P
P136 J
5
4
BL/R 16 A
10 J P 136

2 1 136
J

GY 18 A
BN 18 A

TIMING PULSES from IGN. COIL J32-1 (COORD. 14F)

BK 23 A
J 62 P 3

Y 10 I

FUEL INJECTORS

Y 21 I Y 21 I Y 21 I Y 21 I V 21 I W 21 I

BL/R 21 I
BK 12 C

2 J 137
K14 CYL.1
1 J 137

2 P 138
K15 CYL.2
1 J 138

2 P 139
K16 CYL.3
1 J 139

2 P 140
K17 CYL.4
1 J 140

144 39 36

144 9 7

J 65 P 1

Y 21 I Y/R 21 I W/R 21 I W/BK 21 I Y/BK 21 I GY/Y 18 I O 21 I S 21 I GN 21 I R 21 I

J 141 P 28 15 33 32 14 29 10 9 8 7

ELECTRONIC FUEL CONTROL UNIT (Transistorize

J 141 24 23 16 17 13 5 2 1

(STATION WAGON)

230 J 2 4 J 231

GY 18 I
P 142 J
BK 18

W/BL 15 I W/BL 15 I W 21 I V 21 I V 21 I V 21 I BL 21 I GY/R 21 I

J 143 P 1
R7 COOLANT TEMP. SENSOR
J 143 P 2

2 3 P 145
IDLE FULL
S27 THROTTLE POSITION SWITCH
1 P

P 146 2 7

P 81 J
5
1
BN 18 B

P 81 1 J
GY 18 X

5

BN 18 B
GY 18 E
A B
J 88 P

223 4 P
GY 18 A
L. TAIL ASSY.

189 J 2 P
GY 18 A

222 J 5 P
GY 18 X
R. TAIL ASSY.

6 J 222

OXYGEN SENSOR

T24

DS-25
N.C.
RIGHT REAR
R. TAIL ASSY.

DS-24
N.C.
LEFT REAR
P 88 L. TAIL ASSY.

DS-24
LEFT REAR
P 223

DS-22
LEFT FRONT

DS-23
RIGHT FRONT

DS-25
RIGHT REAR
P 222

Y 18 E 3 P 88

Y 18 E GND. from T11 (COORD. 23F)

Y 18 A GND. from T20 (COORD. 21F)
GND. from T13 (COORD. 55G)

Y 18 A GND. from T14 (COORD. 57H)
GND. from T21 (COORD. 22F)

TURN SIGNAL LAMPS

36 37 38 39 40 41 42

13.118 Wiring diagram for 1981 and 1982 US models (continued)

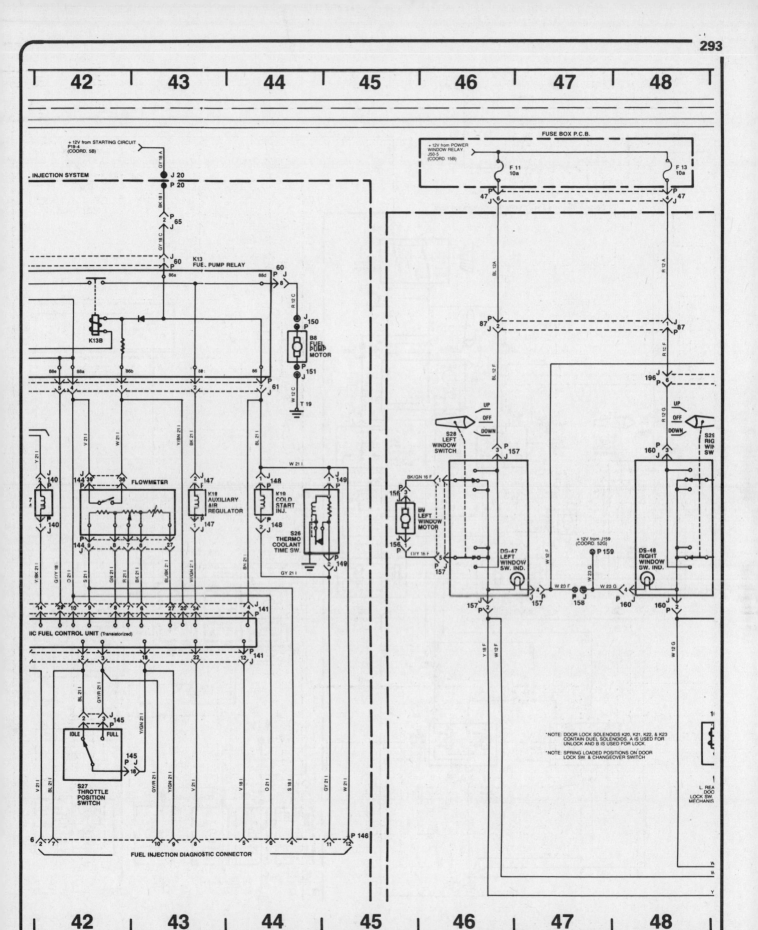

13.118 Wiring diagram for 1981 and 1982 US models (continued)

48   49   50   51   52   53   54

BK 6 A
W 6 A
Y 9 A
BL 12 O
W 8 A
BE 10 F

T35

S33A
LIGHT SWITCH

OFF   FLASH
PARK   ON

F 13
10a

+ 12V to SPLICE 3E

Y 8 A

J 47
P

POWER WINDOWS & ELECTRONIC DOOR LOCKS SYSTEMS

P 162
J 162

R 12 A

S32
INERTIA
SWITCH

OFF/
RESET

J 162
P 162

J 87

R 12 F

S33
DOOR
LOCK SW.

FUSE BOX P.C.B.

F 10
5a

LOCK   UNLOCK

BU/R 18 F

S 18 F

J 169
P

J 49
P
W 23 A

W 23 F

196
J
P 6

DS-46
DOOR
LOCK
SW. IND.

DOOR LOCK
SWITCH

S31
LEFT
CHANGE-
OVER
SWITCH

R 12 G

UP
OFF
DOWN

S29
RIGHT
WINDOW
SWITCH

S30
RIGHT
CHANGEOVER
SWITCH

DOOR LOCK
SWITCH

P 170
J

160
P
J 3

BK 16 G

P 161

B10
RIGHT
WINDOW
MOTOR

GN 18 G

161
J

R. FRONT
DOOR LOCK
MECHANISM

GN 18 G

L. FRONT
DOOR LOCK
MECHANISM

W 23 A

J 172
P

R11
SWITCH ID.
LAMPS
RHEOSTAT

INSTRUMENT
CLUSTER P.C.B.

R10
INST.
CLUSTER
P.C.B.
LAMPS
RHEOSTAT

DS-45
INST. CLUSTER
P.C.B. LAMPS
(5 Lamps)

P 230

DS-48
RIGHT
WINDOW
SW. IND.

160
J
P

J 160

164
J

165
J

K20
RIGHT FRONT
DOOR LOCK
SOLENOID

K23
LEFT
FRONT
DOOR
LOCK
SOLENOID

J 173

W 23 A   W 23 A

DS-29
REAR
WINDOW
WIPER
SW. IND.

DS-30
REAR
WINDOW
DEFOG.
SW. IND.

DS-31
HAZARD
WARNING
SW. IND.

DS-32
LEFT
FRONT
MARKER
LAMP

DS-22
L. FRONT
PARKING
LAMP

REAR
WINDOW
DEFOGGER
SW. ASSY.

W 23 A

55
P

41
P

136
P

31

175
P

188
P

88
P

Y 18 G

W 12 G

GY 18 F

GN 18 F

196
J   P 196

K21
LEFT
REAR
DOOR
LOCK
SOLENOID

K22
RIGHT
REAR
DOOR
LOCK
SOLENOID

J 166

J 168

L. REAR
DOOR
LOCK SW. &
MECHANISM

R. REAR
DOOR
LOCK SW. &
MECHANISM

+ 12V to P159
(COORD. 47E)

J 159

201
P

J 174

CIGAR
LIGHTER
ASSY.

J 117   114

CLOCK ASSY.

J 81

223
P

L. REAR
MARKER
LAMP

LEFT
TAIL
LAMP

DS-33

DS-18

W 23 A

DS-13
(2 Lamps)

DS-26

DS-14

DS-27

A/C-HTR.
PANEL
LAMPS

PRNDL
LAMP

CIGAR
LIGHTER
LAMP

CLOCK
LAMP

A/C-HTR.
CONTROL
PANEL ASSY.

P 201

P 116

P 31

T 19

GND. from
CIGAR LIGHTER
J116
(COORD. 27G)

GND. from
RADIO
J91-1
(COORD. 29G)

LEFT TAIL
ASSY.

GND. from
J201-1
(COORD. 28G)

GND. from T20
(COORD. 21F)

GND. from T21
(COORD. 22F)

48   49   50   51   52   53   54

13.118 Wiring diagram for 1981 and 1982 US models (continued)

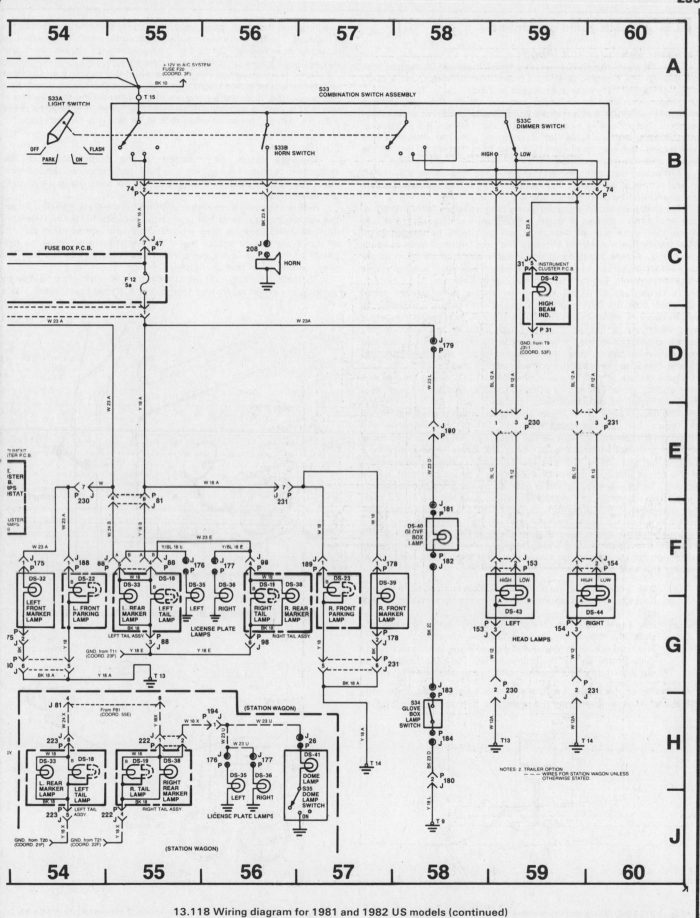

13.118 Wiring diagram for 1981 and 1982 US models (continued)

Part B – Mobile radio equipment

Aerials – selection and fitting

The choice of aerials is now very wide. It should be realised that the quality has a profound effect on radio performance, and a poor, inefficient aerial can make suppression difficult.

A wing-mounted aerial is regarded as probably the most efficient for signal collection, but a roof aerial is usually better for suppression purposes because it is away from most interference fields. Stick-on wire aerials are available for attachment to the inside of the windscreen, but are not always free from the interference field of the engine and some accessories.

Motorised automatic aerials rise when the equipment is switched on and retract at switch-off. They require more fitting space and supply leads, and can be a source of trouble.

There is no merit in choosing a very long aerial as, for example, the type about three metres in length which hooks or clips on to the rear of the car, since part of this aerial will inevitably be located in an interference field. For VHF/FM radios the best length of aerial is about one metre. Active aerials have a transistor amplifier mounted at the base and this serves to boost the received signal. The aerial rod is sometimes rather shorter than normal passive types.

A large loss of signal can occur in the aerial feeder cable, especially over the Very High Frequency (VHF) bands. The design of feeder cable is invariably in the co-axial form, ie a centre conductor surrounded by a flexible copper braid forming the outer (earth) conductor. Between the inner and outer conductors is an insulator material which can be in solid or stranded form. Apart from insulation, its purpose is to maintain the correct spacing and concentricity. Loss of signal occurs in this insulator, the loss usually being greater in a poor quality cable. The quality of cable used is reflected in the price of the aerial with the attached feeder cable.

The capacitance of the feeder should be within the range 65 to 75 picofarads (pF) approximately (95 to 100 pF for Japanese and American equipment), otherwise the adjustment of the car radio aerial trimmer may not be possible. An extension cable is necessary for a long run between aerial and receiver. If this adds capacitance in excess of the above limits, a connector containing a series capacitor will be required, or an extension which is labelled as 'capacity-compensated'.

Fitting the aerial will normally involve making a $\frac{7}{8}$ in (22 mm) diameter hole in the bodywork, but read the instructions that come with the aerial kit. Once the hole position has been selected, use a centre punch to guide the drill. Use sticky masking tape around the area for this helps with marking out and drill location, and gives protection to the paintwork should the drill slip. Three methods of making the hole are in use:

(a) Use a hole saw in the electric drill. This is, in effect, a circular hacksaw blade wrapped round a former with a centre pilot drill.

(b) Use a tank cutter which also has cutting teeth, but is made to shear the metal by tightening with an Allen key.

(c) The hard way of drilling out the circle is using a small drill, say $\frac{1}{8}$ in (3 mm), so that the holes overlap. The centre metal drops out and the hole is finished with round and half-round files.

Whichever method is used, the burr is removed from the body metal and paint removed from the underside. The aerial is fitted tightly

ensuring that the earth fixing, usually a serrated washer, ring or clamp, is making a solid connection. *This earth connection is important in reducing interference.* Cover any bare metal with primer paint and topcoat, and follow by underseal if desired.

Aerial feeder cable routing should avoid the engine compartment and areas where stress might occur, eg under the carpet where feet will be located. Roof aerials require that the headlining be pulled back and that a path is available down the door pillar. It is wise to check with the vehicle dealer whether roof aerial fitting is recommended.

Loudspeakers

Speakers should be matched to the output stage of the equipment, particularly as regards the recommended impedance. Power transistors used for driving speakers are sensitive to the loading placed on them.

Before choosing a mounting position for speakers, check whether the vehicle manufacturer has provided a location for them. Generally door-mounted speakers give good stereophonic reproduction, but not all doors are able to accept them. The next best position is the rear parcel shelf, and in this case speaker apertures can be cut into the shelf, or pod units may be mounted.

For door mounting, first remove the trim, which is often held on by 'poppers' or press studs, and then select a suitable gap in the inside

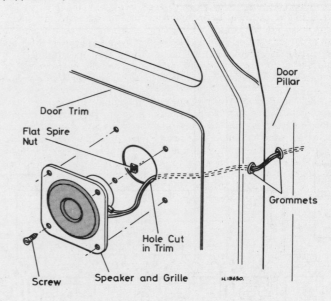

Fig. 13.120 Door-mounted speaker installation (Sec 11)

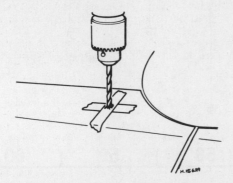

Fig. 13.119 Drilling the bodywork for aerial mounting (Sec 11)

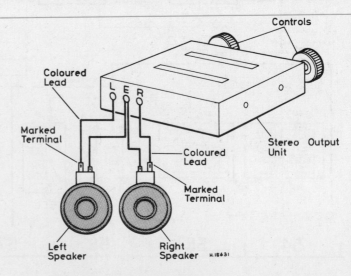

Fig. 13.121 Speaker connections must be correctly made as shown (Sec 11)

door assembly. Check that the speaker would not obstruct glass or winder mechanism by winding the window up and down. A template is often provided for marking out the trim panel hole, and then the four fixing holes must be drilled through. Mark out with chalk and cut cleanly with a sharp knife or keyhole saw. Speaker leads are then threaded through the door and door pillar, if necessary drilling 10 mm diameter holes. Fit grommets in the holes and connect to the radio or tape unit correctly. Do not omit a waterproofing cover, usually supplied with door speakers. If the speaker has to be fixed into the metal of the door itself, use self-tapping screws, and if the fixing is to the door trim use self-tapping screws and flat spire nuts.

Rear shelf mounting is somewhat simpler but it is necessary to find gaps in the metalwork underneath the parcel shelf. However, remember that the speakers should be as far apart as possible to give a good stereo effect. Pod-mounted speakers can be screwed into position through the parcel shelf material, but it is worth testing for the best position. Sometimes good results are found by reflecting sound off the rear window.

Unit installation

Many vehicles have a dash panel aperture to take a radio/audio unit, a recognised international standard being 189.5 mm x 60 mm. Alternatively a console may be a feature of the car interior design and this, mounted below the dashboard, gives more room. If neither facility is available a unit may be mounted on the underside of the parcel shelf; these are frequently non-metallic and an earth wire from the case to a good earth point is necessary. A three-sided cover in the form of a cradle is obtainable from car radio dealers and this gives a professional appearance to the installation; in this case choose a position where the controls can be reached by a driver with his seat belt on.

Installation of the radio/audio unit is basically the same in all cases, and consists of offering it into the aperture after removal of the knobs *(not* push buttons) and the trim plate. In some cases a special mounting plate is required to which the unit is attached. It is worthwhile supporting the rear end in cases where sag or strain may occur, and it is usually possible to use a length of perforated metal strip attached between the unit and a good support point nearby. In general it is recommended that tape equipment should be installed at or nearly horizontal.

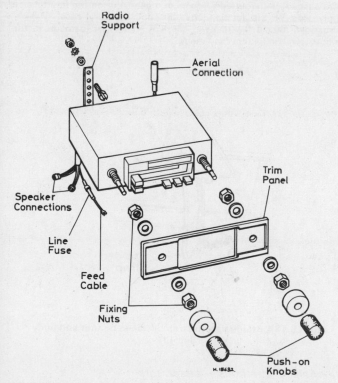

Fig. 13.122 Mounting component details for radio/cassette unit (Sec 11)

Connections to the aerial socket are simply by the standard plug terminating the aerial downlead or its extension cable. Speakers for a stereo system must be matched and correctly connected, as outlined previously.

Note: *While all work is carried out on the power side, it is wise to disconnect the battery earth lead.* Before connection is made to the vehicle electrical system, check that the polarity of the unit is correct. Most vehicles use a negative earth system, but radio/audio units often have a reversible plug to convert the set to either + or − earth. *Incorrect connection may cause serious damage.*

The power lead is often permanently connected inside the unit and terminates with one half of an in-line fuse carrier. The other half is fitted with a suitable fuse (3 or 5 amperes) and a wire which should go to a power point in the electrical system. This may be the accessory terminal on the ignition switch, giving the advantage of power feed with ignition or with the ignition key at the 'accessory' position. Power to the unit stops when the ignition key is removed. Alternatively, the lead may be taken to a live point at the fusebox with the consequence of having to remember to switch off at the unit before leaving the vehicle.

Before switching on for initial test, be sure that the speaker connections have been made, for running without load can damage the output transistors. Switch on next and tune through the bands to ensure that all sections are working, and check the tape unit if applicable. The aerial trimmer should be adjusted to give the strongest reception on a weak signal in the medium wave band, at say 200 metres.

Interference

In general, when electric current changes abruptly, unwanted electrical noise is produced. The motor vehicle is filled with electrical devices which change electric current rapidly, the most obvious being the contact breaker.

When the spark plugs operate, the sudden pulse of spark current causes the associated wiring to radiate. Since early radio transmitters used sparks as a basis of operation, it is not surprising that the car radio will pick up ignition spark noise unless steps are taken to reduce it to acceptable levels.

Interference reaches the car radio in two ways:

(a) by conduction through the wiring.
(b) by radiation to the receiving aerial.

Initial checks presuppose that the bonnet is down and fastened, the radio unit has a good earth connection (not through the aerial downlead outer), no fluorescent tubes are working near the car, the aerial trimmer has been adjusted, and the vehicle is in a position to receive radio signals, ie not in a metal-clad building.

Switch on the radio and tune it to the middle of the medium wave (MW) band off-station with the volume (gain) control set fairly high. Switch on the ignition (but do not start the engine) and wait to see if irregular clicks or hash noise occurs. Tapping the facia panel may also produce the effects. If so, this will be due to the voltage stabiliser, which is an on-off thermal switch to control instrument voltage. It is located usually on the back of the instrument panel, often attached to the speedometer. Correction is by attachment of a capacitor and, if still troublesome, chokes in the supply wires.

Switch on the engine and listen for interference on the MW band. Depending on the type of interference, the indications are as follows.

A harsh crackle that drops out abruptly at low engine speed or when the headlights are switched on is probably due to a voltage regulator.

A whine varying with engine speed is due to the dynamo or alternator. Try temporarily taking off the fan belt – if the noise goes this is confirmation.

Regular ticking or crackle that varies in rate with the engine speed is due to the ignition system. With this trouble in particular and others in general, check to see if the noise is entering the receiver from the wiring or by radiation. To do this, pull out the aerial plug, (preferably shorting out the input socket or connecting a 62 pF capacitor across it). If the noise disappears it is coming in through the aerial and is *radiation noise.* If the noise persists it is reaching the receiver through the wiring and is said to be *line-borne.*

Interference from wipers, washers, heater blowers, turn-indicators, stop lamps, etc is usually taken to the receiver by wiring, and simple treatment using capacitors and possibly chokes will solve the problem. Switch on each one in turn (wet the screen first for running wipers!)

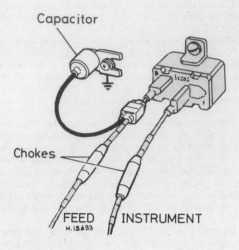

Fig. 13.123 Voltage stabiliser interference suppression (Sec 11)

and listen for possible interference with the aerial plug in place and again when removed.

Electric petrol pumps are now finding application again and give rise to an irregular clicking, often giving a burst of clicks when the ignition is on but the engine has not yet been started. It is also possible to receive whining or crackling from the pump.

Note that if most of the vehicle accessories are found to be creating interference all together, the probability is that poor aerial earthing is to blame.

Component terminal markings

Throughout the following sub-sections reference will be found to various terminal markings. These will vary depending on the manufacturer of the relevant component. If terminal markings differ from those mentioned, reference should be made to the following table, where the most commonly encountered variations are listed.

Alternator	Alternator terminal (thick lead)	Exciting winding terminal
DIN/Bosch	B+	DF
Delco Remy	+	EXC
Ducellier	+	EXC
Ford (US)	+	DF
Lucas	+	F
Marelli	+B	F

Ignition coil	Ignition switch terminal	Contact breaker terminal
DIN/Bosch	15	1
Delco Remy	+	−
Ducellier	BAT	RUP
Ford (US)	B/+	CB/−
Lucas	SW/+	−
Marelli	BAT/+B	D

Voltage regulator	Voltage input terminal	Exciting winding terminal
DIN/Bosch	B+/D+	DF
Delco Remy	BAT/+	EXC
Ducellier	BOB/BAT	EXC
Ford (US)	BAT	DF
Lucas	+/A	F
Marelli		F

Suppression methods – ignition

Suppressed HT cables are supplied as original equipment by manufacturers and will meet regulations as far as interference to neighbouring equipment is concerned. It is illegal to remove such suppression unless an alternative is provided, and this may take the form of resistive spark plug caps in conjunction with plain copper HT

cable. For VHF purposes, these and 'in-line' resistors may not be effective, and resistive HT cable is preferred. Check that suppressed cables are actually fitted by observing cable identity lettering, or measuring with an ohmmeter – the value of each plug lead should be 5000 to 10 000 ohms.

A 1 microfarad capacitor connected from the LT supply side of the ignition coil to a good nearby earth point will complete basic ignition interference treatment. *NEVER fit a capacitor to the coil terminal to the contact breaker – the result would be burnt out points in a short time.*

If ignition noise persists despite the treatment above, the following sequence should be followed:

(a) Check the earthing of the ignition coil; remove paint from fixing clamp.

(b) If this does not work, lift the bonnet. Should there be no change in interference level, this may indicate that the bonnet is not electrically connected to the car body. Use a proprietary braided strap across a bonnet hinge ensuring a first class electrical connection. If, however, lifting the bonnet increases the interference, then fit resistive HT cables of a higher ohms-per-metre value.

(c) If all these measures fail, it is probable that re-radiation from metallic components is taking place. Using a braided strap between metallic points, go round the vehicle systematically – try the following: engine to body, exhaust system to body, front suspension to engine and to body, steering column to body (especially French and Italian cars), gear lever to engine and to body (again especially French and Italian cars), Bowden cable to body, metal parcel shelf to body. When an offending component is located it should be bonded with the strap permanently.

(d) As a next step, the fitting of distributor suppressors to each lead at the distributor end may help.

(e) Beyond this point is involved the possible screening of the distributor and fitting resistive spark plugs, but such advanced treatment is not usually required for vehicles with entertainment equipment.

Electronic ignition systems have built-in suppression components, but this does not relieve the need for using suppressed HT leads. In some cases it is permitted to connect a capacitor on the low tension supply side of the ignition coil, but not in every case. Makers' instructions should be followed carefully, otherwise damage to the ignition semiconductors may result.

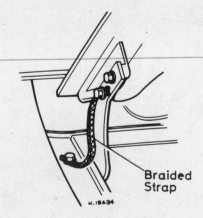

Fig. 13.124 Braided earth strap between bonnet and body (Sec 11)

Suppression methods – generators

For older vehicles with dynamos a 1 microfarad capacitor from the D (larger) terminal to earth will usually cure dynamo whine. Alternators should be fitted with a 3 microfarad capacitor from the B+ main output terminal (thick cable) to earth. Additional suppression may be

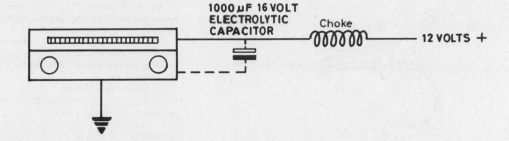

Fig. 13.125 Line-borne interference suppression (Sec 11)

obtained by the use of a filter in the supply line to the radio receiver.
It is most important that:

(a) *Capacitors are never connected to the field terminals of either a dynamo or alternator.*
(b) *Alternators must not be run without connection to the battery.*

Suppression methods — voltage regulators

Voltage regulators used with DC dynamos should be suppressed by connecting a 1 microfarad capacitor from the control box D terminal to earth.

Alternator regulators come in three types:

(a) *Vibrating contact regulators separate from the alternator. Used extensively on continental vehicles.*
(b) *Electronic regulators separate from the alternator.*
(c) *Electronic regulators built-in to the alternator.*

In case (a) interference may be generated on the AM and FM (VHF) bands. For some cars a replacement suppressed regulator is available. Filter boxes may be used with non-suppressed regulators. But if not available, then for AM equipment a 2 microfarad or 3 microfarad capacitor may be mounted at the voltage terminal marked D+ or B+ of the regulator. FM bands may be treated by a feed-through capacitor of 2 or 3 microfarad.

Electronic voltage regulators are not always troublesome, but where necessary, a 1 microfarad capacitor from the regulator + terminal will help.

Integral electronic voltage regulators do not normally generate much interference, but when encountered this is in combination with alternator noise. A 1 microfarad or 2 microfarad capacitor from the warning lamp (IND) terminal to earth for Lucas ACR alternators and Femsa, Delco and Bosch equivalents should cure the problem.

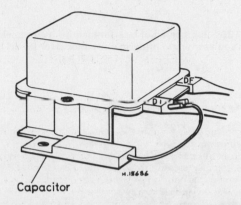

Capacitor

Fig. 13.127 Suppression of AM interference by vibrating contact voltage regulator (alternator equipment) (Sec 11)

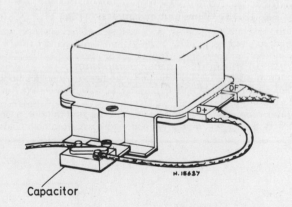

Capacitor

Fig. 12.128 Suppression of FM interference by vibrating contact voltage regulator (alternator equipment) (Sec 11)

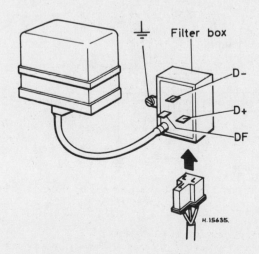

Filter box

D—
D+
DF

Fig. 13.126 Typical filter box for vibrating contact voltage regulator (alternator equipment) (Sec 11)

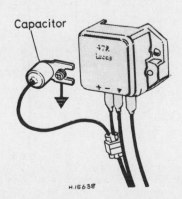

Capacitor

Fig. 13.129 Electronic voltage regulator suppression (Sec 11)

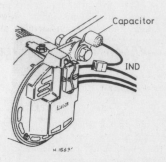

Fig. 13.130 Suppression of interference from electronic voltage regulator when integral with alternator (Sec 11)

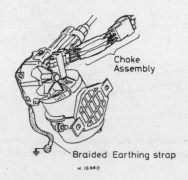

Fig. 13.131 Wiper motor suppression (Sec 11)

Suppression methods – other equipment

Wiper motors – Connect the wiper body to earth with a bonding strap. For all motors use a 7 ampere choke assembly inserted in the leads to the motor.

Heater motors – Fit 7 ampere line chokes in both leads, assisted if necessary by a 1 microfarad capacitor to earth from both leads.

Electronic tachometer – The tachometer is a possible source of ignition noise – check by disconnecting at the ignition coil CB terminal. It usually feeds from ignition coil LT pulses at the contact breaker terminal. A 3 ampere line choke should be fitted in the tachometer lead at the coil CB terminal.

Horn – A capacitor and choke combination is effective if the horn is directly connected to the 12 volt supply. The use of a relay is an alternative remedy, as this will reduce the length of the interference-carrying leads.

Electrostatic noise – Characteristics are erratic crackling at the receiver, with disappearance of symptoms in wet weather. Often shocks may be given when touching bodywork. Part of the problem is the build-up of static electricity in non-driven wheels and the acquisition of charge on the body shell. It is possible to fit spring-loaded contacts at the wheels to give good conduction between the rotary wheel parts and the vehicle frame. Changing a tyre sometimes helps – because of tyres' varying resistances. In difficult cases a trailing flex which touches the ground will cure the problem. If this is not acceptable it is worth trying conductive paint on the tyre walls.

Fuel pump – Suppression requires a 1 microfarad capacitor between the supply wire to the pump and a nearby earth point. If this is insufficient a 7 ampere line choke connected in the supply wire near the pump is required.

Fluorescent tubes – Vehicles used for camping/caravanning frequently have fluorescent tube lighting. These tubes require a relatively high voltage for operation and this is provided by an inverter (a form of oscillator) which steps up the vehicle supply voltage. This can give rise to serious interference to radio reception, and the tubes themselves can contribute to this interference by the pulsating nature of the lamp discharge. In such situations it is important to mount the aerial as far away from a fluorescent tube as possible. The interference problem may be alleviated by screening the tube with fine wire turns spaced an inch (25 mm) apart and earthed to the chassis. Suitable chokes should be fitted in both supply wires close to the inverter.

Radio/cassette case breakthrough

Magnetic radiation from dashboard wiring may be sufficiently intense to break through the metal case of the radio/cassette player. Often this is due to a particular cable routed too close and shows up as ignition interference on AM and cassette play and/or alternator whine on cassette play.

The first point to check is that the clips and/or screws are fixing all parts of the radio/cassette case together properly. Assuming good earthing of the case, see if it is possible to re-route the offending cable – the chances of this are not good, however, in most cars.

Next release the radio/cassette player and locate it in different positions with temporary leads. If a point of low interference is found,

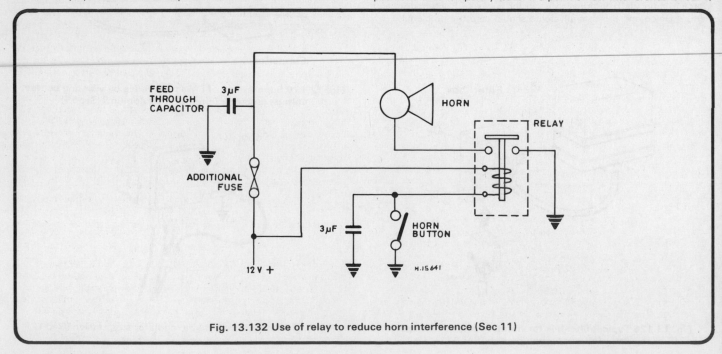

Fig. 13.132 Use of relay to reduce horn interference (Sec 11)

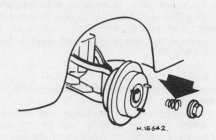

Fig. 13.133 Use of spring contacts at wheels (Sec 11)

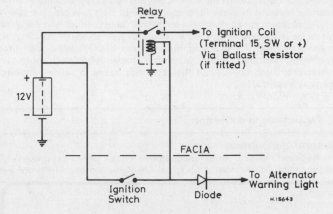

Fig 13.134 Use of ignition coil relay to suppress case breakthrough (Sec 11)

then if possible fix the equipment in that area. This also confirms that local radiation is causing the trouble. If re-location is not feasible, fit the radio/cassette player back in the original position.

Alternator interference on cassette play is now caused by radiation from the main charging cable which goes from the battery to the output terminal of the alternator, usually via the + terminal of the starter motor relay. In some vehicles this cable is routed under the dashboard, so the solution is to provide a direct cable route. Detach the original cable from the alternator output terminal and make up a new cable of at least 6 mm² cross-sectional area to go from alternator to battery with the shortest possible route. Remember – do not run the engine with the alternator disconnected from the battery.

Ignition breakthrough on AM and/or cassette play can be a difficult problem. It is worth wrapping earthed foil round the offending cable run near the equipment, or making up a deflector plate well screwed down to a good earth. Another possibility is the use of a suitable relay to switch on the ignition coil. The relay should be mounted close to the ignition coil; with this arrangement the ignition coil primary current is not taken into the dashboard area and does not flow through the ignition switch. A suitable diode should be used since it is possible that at ignition switch-off the output from the warning lamp alternator terminal could hold the relay on.

Connectors for suppression components

Capacitors are usually supplied with tags on the end of the lead, while the capacitor body has a flange with a slot or hole to fit under a nut or screw with washer.

Connections to feed wires are best achieved by self-stripping connectors. These connectors employ a blade which, when squeezed down by pliers, cuts through cable insulation and makes connection to the copper conductors beneath.

Chokes sometimes come with bullet snap-in connectors fitted to the wires, and also with just bare copper wire. With connectors, suitable female cable connectors may be purchased from an auto-accessory shop together with any extra connectors required for the cable ends after being cut for the choke insertion. For chokes with bare wires, similar connectors may be employed together with insulation sleeving as required.

VHF/FM broadcasts

Reception of VHF/FM in an automobile is more prone to problems than the medium and long wavebands. Medium/long wave transmitters are capable of covering considerable distances, but VHF transmitters are restricted to line of sight, meaning ranges of 10 to 50 miles, depending upon the terrain, the effects of buildings and the transmitter power.

Because of the limited range it is necessary to retune on a long journey, and it may be better for those habitually travelling long distances or living in areas of poor provision of transmitters to use an AM radio working on medium/long wavebands.

When conditions are poor, interference can arise, and some of the suppression devices described previously fall off in performance at very high frequencies unless specifically designed for the VHF band. Available suppression devices include reactive HT cable, resistive distributor caps, screened plug caps, screened leads and resistive spark plugs.

For VHF/FM receiver installation the following points should be particularly noted:

(a) Earthing of the receiver chassis and the aerial mounting is important. Use a separate earthing wire at the radio, and scrape paint away at the aerial mounting.

(b) If possible, use a good quality roof aerial to obtain maximum height and distance from interference generating devices on the vehicle.

(c) Use of a high quality aerial downlead is important, since losses in cheap cable can be significant.

(d) The polarisation of FM transmissions may be horizontal, vertical, circular or slanted. Because of this the optimum mounting angle is at 45° to the vehicle roof.

Citizens' Band radio (CB)

In the UK, CB transmitter/receivers work within the 27 MHz and 934 MHz bands, using the FM mode. At present interest is concentrated on 27 MHz where the design and manufacture of equipment is less difficult. Maximum transmitted power is 4 watts, and 40 channels spaced 10 kHz apart within the range 27.60125 to 27.99125 MHz are available.

Aerials are the key to effective transmission and reception. Regulations limit the aerial length to 1.65 metres including the loading coil and any associated circuitry, so tuning the aerial is necessary to obtain optimum results. The choice of a CB aerial is dependent on whether it is to be permanently installed or removable, and the performance will hinge on correct tuning and the location point on the vehicle. Common practice is to clip the aerial to the roof gutter or to employ wing mounting where the aerial can be rapidly unscrewed. An alternative is to use the boot rim to render the aerial theftproof, but a popular solution is to use the 'magmount' – a type of mounting having a strong magnetic base clamping to the vehicle at any point, usually the roof.

Aerial location determines the signal distribution for both transmission and reception, but it is wise to choose a point away from the engine compartment to minimise interference from vehicle electrical equipment.

The aerial is subject to considerable wind and acceleration forces. Cheaper units will whip backwards and forwards and in so doing will alter the relationship with the metal surface of the vehicle with which it forms a ground plane aerial system. The radiation pattern will change correspondingly, giving rise to break-up of both incoming and outgoing signals.

Interference problems on the vehicle carrying CB equipment fall into two categories:

(a) Interference to nearby TV and radio receivers when transmitting.

(b) Interference to CB set reception due to electrical equipment on the vehicle.

Problems of break-through to TV and radio are not frequent, but can be difficult to solve. Mostly trouble is not detected or reported because the vehicle is moving and the symptoms rapidly disappear at the TV/radio receiver, but when the CB set is used as a base station any trouble with nearby receivers will soon result in a complaint.

It must not be assumed by the CB operator that his equipment is faultless, for much depends upon the design. Harmonics (that is, multiples) of 27 MHz may be transmitted unknowingly and these can fall into other user's bands. Where trouble of this nature occurs, low pass filters in the aerial or supply leads can help, and should be fitted in base station aerials as a matter of course. In stubborn cases it may be necessary to call for assistance from the licensing authority, or, if possible, to have the equipment checked by the manufacturers.

Interference received on the CB set from the vehicle equipment is, fortunately, not usually a severe problem. The precautions outlined previously for radio/cassette units apply, but there are some extra points worth noting.

It is common practice to use a slide-mount on CB equipment enabling the set to be easily removed for use as a base station, for example. Care must be taken that the slide mount fittings are properly earthed and that first class connection occurs between the set and slide-mount.

Vehicle manufacturers in the UK are required to provide suppression of electrical equipment to cover 40 to 250 MHz to protect TV and VHF radio bands. Such suppression appears to be adequately effective at 27 MHz, but suppression of individual items such as alternators/dynamos, clocks, stabilisers, flashers, wiper motors, etc, may still be necessary. The suppression capacitors and chokes available from auto-electrical suppliers for entertainment receivers will usually give the required results with CB equipment.

Other vehicle radio transmitters

Besides CB radio already mentioned, a considerable increase in the use of transceivers (ie combined transmitter and receiver units) has taken place in the last decade. Previously this type of equipment was fitted mainly to military, fire, ambulance and police vehicles, but a large business radio and radio telephone usage has developed.

Generally the suppression techniques described previously will suffice, with only a few difficult cases arising. Suppression is carried out to satisfy the 'receive mode', but care must be taken to use heavy duty chokes in the equipment supply cables since the loading on 'transmit' is relatively high.

Glass-fibre bodied vehicles

Such vehicles do not have the advantage of a metal box surrounding the engine as is the case, in effect, of conventional vehicles. It is usually necessary to line the bonnet, bulkhead and wing valances with metal foil, which could well be the aluminium foil available from builders merchants. Bonding of sheets one to another and the whole down to the chassis is essential.

Wiring harness may have to be wrapped in metal foil which again should be earthed to the vehicle chassis. The aerial base and radio chassis must be taken to the vehicle chassis by heavy metal braid. VHF radio suppression in glass-fibre cars may not be a feasible operation.

In addition to all the above, normal suppression components should be employed, but special attention paid to earth bonding. A screen enclosing the entire ignition system usually gives good improvement, and fabrication from fine mesh perforated metal is convenient. Good bonding of the screening boxes to several chassis points is essential.

---

12 Suspension and steering

General description

1 Negative offset front suspension has been progressively introduced on UK models, although all US/Canada models incorporate

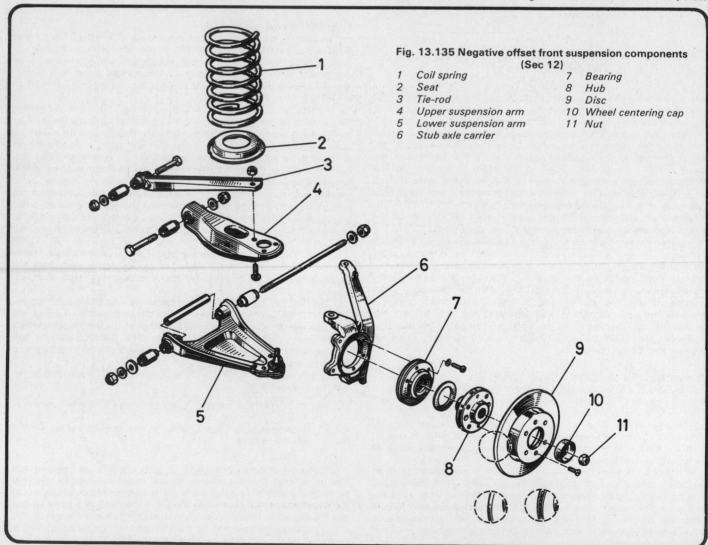

Fig. 13.135 Negative offset front suspension components (Sec 12)

1 Coil spring
2 Seat
3 Tie-rod
4 Upper suspension arm
5 Lower suspension arm
6 Stub axle carrier
7 Bearing
8 Hub
9 Disc
10 Wheel centering cap
11 Nut

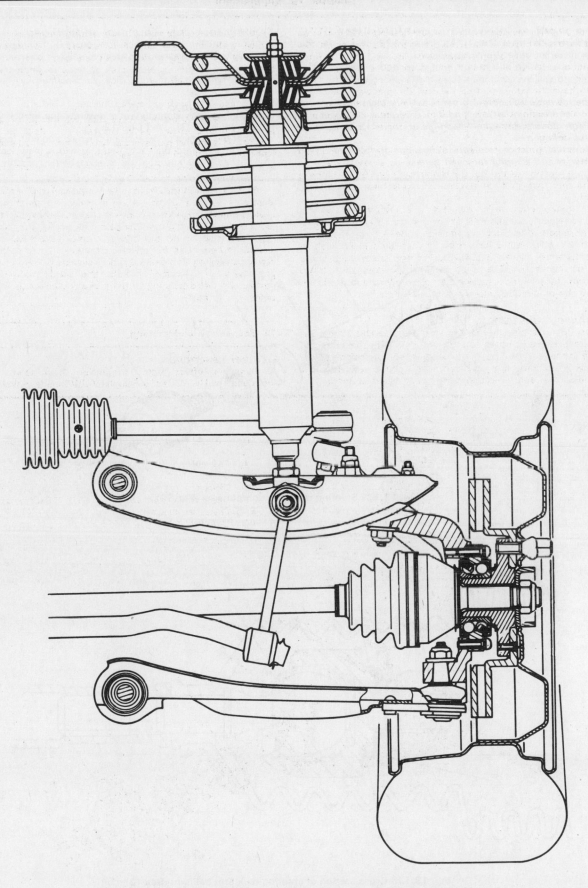

Fig. 13.136 Cross-section of the negative offset front suspension (Sec 12)

it. UK Turbo models are fitted with it, as are R1342, R1352, R1343 and R1353 models from 1981. As from 1983 all UK models incorporate negative offset front suspension.

2   The later suspension differs from that described in Chapter 11 as follows:

(a) *The lower suspension arm outer balljoint faces upwards*
(b) *Top and bottom balljoints now incorporate a 20° taper*
(c) *Single double row wheel bearings are bolted to an adapted stub axle carrier*
(d) *Driveshafts incorporate spider couplings at each end*
(e) *New design steering rack and arms*

Steering rack arm – removal and refitting

3   Note that the steering rack arm must be discarded after removal as the special lockwasher damages the balljoint body when the joint is loosened.

4   First remove the steering rack, as described in Chapter 11, and mount it in a vice.

5   Remove the track rod ends (Chapter 11).

6   Hold the rack stationary and unscrew the arm. If available use Renault tools Dir 811 and Dir 812.

7   Coat the threads of the joint on the new arm with locking fluid and screw the arm onto the rack, together with a new lockwasher and stopwasher. Make sure that the stopwasher is correctly aligned with

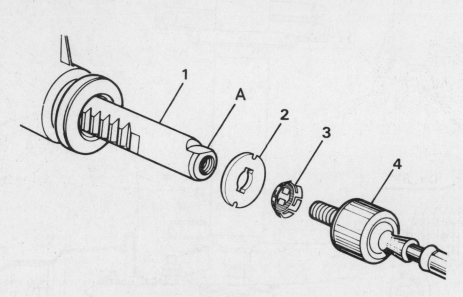

Fig. 13.137 Steering rack arm components (Sec 12)

1   Rack   4   Balljoint
2   Stopwasher   A   Locking flat
3   Lockwasher

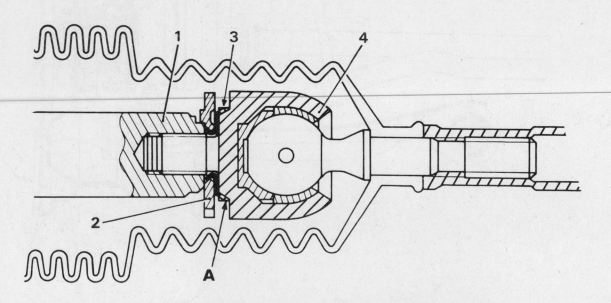

Fig. 13.138 Cross-section of steering rack arm balljoint (Sec 12)

1   Rack   4   Balljoint
2   Stopwasher   A   Shoulder
3   Lockwasher

10 Remove the brake disc, as described in Chapter 9, and unscrew the driveshaft nut.
11 Pull the hub from the bearing and over the driveshaft. To do this, insert two metal bars behind the hub flange and tighten two wheel bolts evenly onto them.
12 Using a Torx key, unscrew the bearing retaining bolts, then withdraw the bearing and remove the loose inner race from the driveshaft.

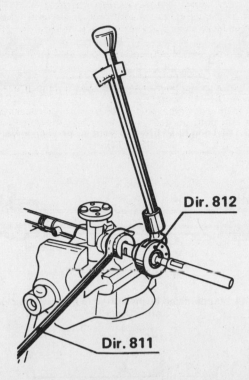

Fig. 13.139 Using the special tools to unscrew the steering rack arm and balljoint (Sec 12)

the flat on the end of the rack before tightening the joint. If necessary have the tightening torque checked by a Renault dealer using tool Dir 812.

Front hub bearings (negative offset) – removal and refitting
8 With the car on the ground and the handbrake applied, remove the hub cap and loosen the driveshaft nut.
9 Jack up the front of the car and support on axle stands. Apply the handbrake and remove the roadwheel.

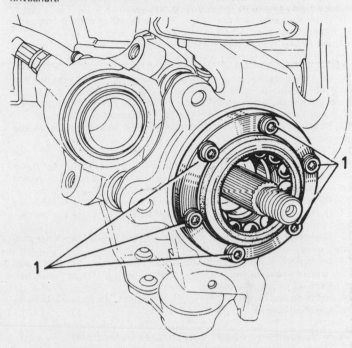

Fig. 13.141 Front hub bearing retaining bolts (1) (Sec 12)

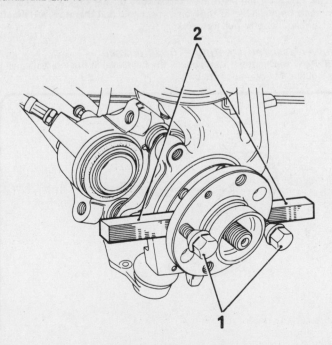

Fig. 13.140 Using metal bars (2) and two wheel bolts (1) to pull the hub from the bearing (Sec 12)

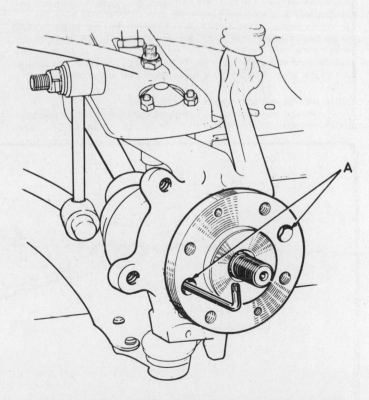

Fig. 13.142 Using a Torx key through the holes in the hub flange (A) to unscrew the bearing bolts (Sec 12)

13 Pull the outer race from the hub using an extractor tool. A clamp type tool having a fine edge is necessary in order to engage the outer shoulder of the race.

14 Clean the bearing, races and seating then inspect the bearing for wear. The bearing should be renewed if there are any signs of pitting on the races or balls.

15 To fit the bearing, locate the inner race on the driveshaft then insert the bearing in the stub axle carrier. Insert the retaining bolts and tighten them evenly to the specified torque.

16 Drive the outer race on the hub using a length of metal tubing with an inner diameter of 1.576 in (40.0 mm).

17 Smear the races and balls with multi-purpose grease then engage the hub over the driveshaft, using a wooden mallet to tap it into the bearing until the driveshaft threads are visible.

18 Fit the wheel locating cup and driveshaft nut, then hold the hub stationary using a metal bar bolted to it and tighten the nut to the specified torque. If necessary, in the interests of safety, final tightening of the nut can be delayed until the car is lowered to the ground.

19 Refit the brake disc and roadwheel and lower the car to the ground.

Stub axle carrier (negative offset) – removal and refitting

20 The procedure is identical to that described in Chapter 11, except for removal of the hub and bearing.

21 With the roadwheel, brake disc, and driveshaft nut removed, use a Torx key through the holes in the hub flange to unscrew and remove the bearing bolts.

22 Withdraw the hub flange, together with the bearing, from the stub axle carrier.

23 The balljoints on the stub axle carrier incorporate 20° tapers and they should separate without using a separator tool. **Do not** use a pair of hammers to free the joints.

24 Refitting is a reversal of removal, but tighten the nuts and bolts to the specified torque.

Front suspension lower arm bushes – renewal

25 The front suspension lower arm bushes have been modified a number of times. On early models a shim was inserted at one end of the pivot to obtain the correct castor angle for the particular model. On 1982 models the shim was omitted and a longer bush fitted in the arm; however, as from late 1983 models, the castor angle has been standardized and identical bushes fitted at each end of the pivot.

26 When renewing the early type bushes on negative offset models the bushes must be positioned as shown in Fig. 13.145. If any doubt exists concerning the correct position of the bushes in the arm, the inner face of the rear bush must be flush with the inner edge of the bore in the arm, then the front bush positioned accordingly.

Front wheel castor angle – adjustment

27 The front suspension upper suspension arm tie-rod on certain later models is adjustable in length to facilitate adjustment of the castor

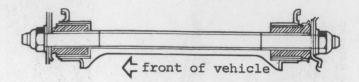

Fig. 13.143 Long bush fitted to front suspension lower arm (Sec 12)

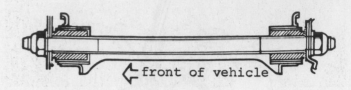

Fig. 13.144 Standardized bushes fitted to front suspension lower arm (Sec 12)

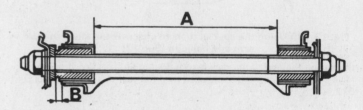

Fig. 13.145 Lower suspension arm bush setting dimension for negative offset models (Sec 12)

A   7.126 in (181.0 mm)
B   Checking dimension before removal if bushes are damaged

angle. However, the instruments necessary to check the angle will not normally be available to the home mechanic and therefore a Renault dealer should carry out the work.

Power-assisted steering – maintenance

28 Refer to Chapter 11 and check the fluid level at the intervals given in Routine Maintenance (photos).

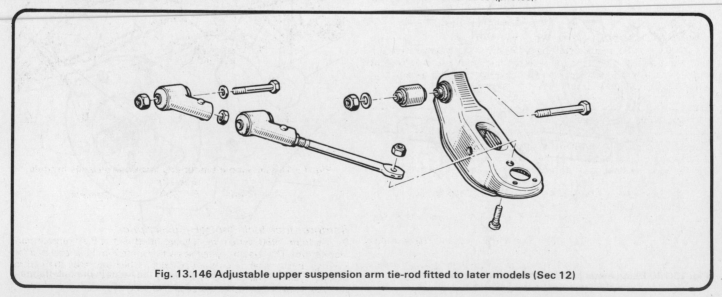

Fig. 13.146 Adjustable upper suspension arm tie-rod fitted to later models (Sec 12)

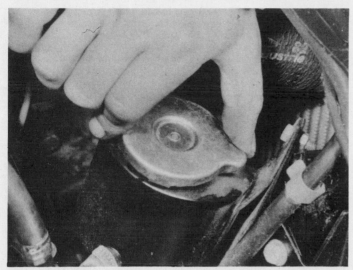

12.28A Remove the cap from the power steering fluid reservoir to check the level

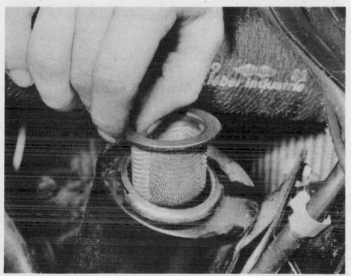

12.28B The strainer located in the power steering fluid reservoir

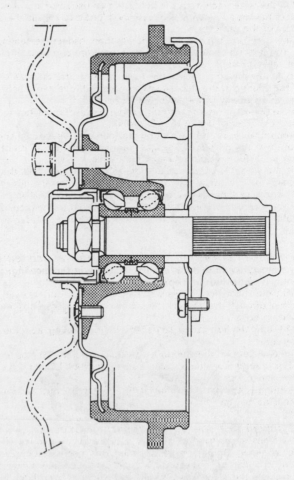

Fig. 13.147 Cross-section of the 1983 on rear hub (Sec 12)

Rear hub bearings – removal and refitting

29 As from 1983 models with negative offset front suspension the rear hub bearings are integrated with the hubs and it is therefore not possible to remove them separately. Renew the complete assembly in the event of worn bearings.

13 Bodywork and fittings

Dashboard (basic) – removal and refitting
1 Disconnect the battery negative lead.
2 Remove the steering wheel and steering column shrouds (Chapter 11).
3 Remove the accessories plate and bracket (Chapter 10).
4 Detach the ashtray.
5 Unscrew the three dashboard mounting bolts shown in Fig. 13.148.
6 Disconnect the speedometer cable and instrument panel multi-plugs.
7 Unclip the dashboard end furthest from the steering column then unclip the other end and withdraw the dashboard.
8 Refitting is a reversal of removal.

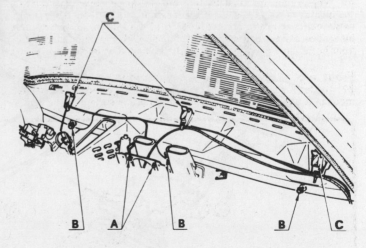

Fig. 13.148 Dashboard mounting locations on basic models (Sec 13)

A Screw   B Bolt   C Brackets

Remote door lock control – description
9 As from 1983 certain models are fitted with a PLIP remote door lock control. The system comprises a transmitter in the shape of a fob keyring, and a receiver on the dashboard which operates the central locking system. The key-operated system remains, so that there are two methods of operating the central locking system.

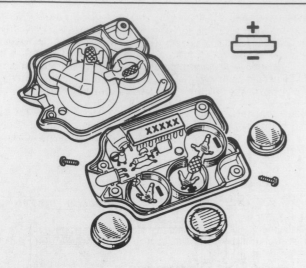

Fig. 13.149 Remote door lock (PLIP) transmitter showing batteries and code location (Sec 13)

10 The transmitter emits infra red coding which is processed by the receiver. Three small batteries are fitted in the transmitter.

11 When obtaining a new transmitter it is necessary to quote the code stamped on the chip inside the transmitter. On some models the code also appears on the front of the receiver beneath the embellisher.

Dashboard (1984 LHD Turbo) – removal and refitting

12 On 1984 LHD Turbo models the wiring harness is fixed inside the dashboard and must therefore be disconnected from the main harness before the dashboard can be removed.

Dashboard (US/Canada DL models) – removal and refitting

13 Disconnect the battery negative lead.

14 Remove the instrument panel (Chapter 10) and the steering wheel and steering column shrouds (Chapter 11).

15 Remove the accessories plate (fusebox), but do not disconnect the wiring.

16 Pull out the top of the console and disconnect the clock wiring.

17 Pull out the bottom of the console and disconnect the cigarette lighter wiring.

18 Remove the glove compartment and unscrew the dashboard rear mounting bolt now accessible.

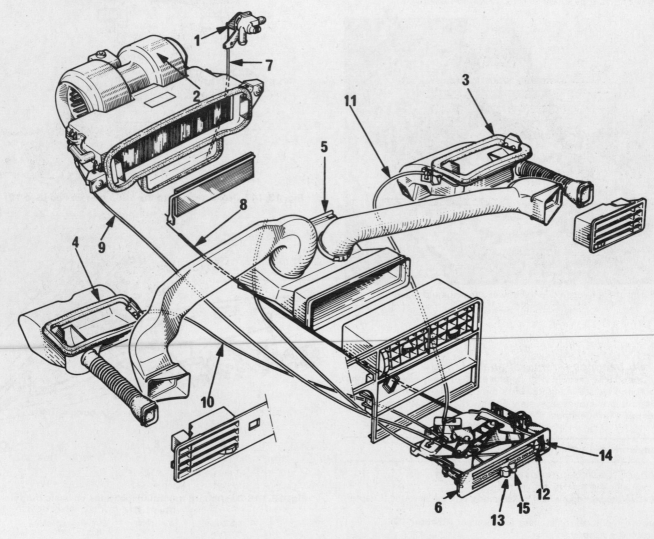

Fig. 13.150 heater controls on US/Canada DL models (Sec 13)

1 Heater valve	4 Left heater duct	8 Ventilation flap cable	12 Distribution lever (high-low)
2 Heater-ventilator assembly (with blower motor and heater core)	5 Ventilation box (fresh air only)	9 Airflow flap cable	13 Heater valve lever
	6 Control panel	10 Left heater duct cable	14 Fan motor dial
3 Right heater duct	7 Heater valve cable	11 Right heater duct cable	15 Fresh air intake lever

19 Remove the console side panels, heater control panel and if applicable the choke cable.
20 Disconnect the glove compartment light wiring.
21 Unscrew the side mounting bolts and the bolts in front of the gearstick, then unclip the dashboard and remove it, at the same time disconnecting the air ducts.
22 Refitting is a reversal of removal.

Heater control panel (US/Canada DL models) – removal and refitting

23 Disconnect the battery negative lead.
24 Remove the console side panels.
25 Unscrew the heater control panel and disconnect the wiring from the fan motor rheostat and lighting connector.
26 Unclip the control cables and withdraw the control panel.
27 Refitting is a reversal of removal, but adjust the cables as described in Chapter 12.

Air conditioning system (US/Canada models) – general

28 The air conditioning components are shown in Fig. 13.151, the system uses a refrigerant to lower the temperature of the air in the passenger compartment.
29 **Never** disconnect any part of the air conditioning circuit unless it has been evacuated by a qualified refrigeration engineer.

30 Where the system components obstruct other mechanical operations, then it is permissible to unbolt their mountings and move them to the limit of their flexible hoses without disconnecting the hoses. If there is still insufficient room to carry out the required work then the system must be evacuated before disconnecting and removing the components. On completion of the work, have the system recharged.
31 Regularly check the condenser for clogging with flies or dirt, and if necessary clear it with a water hose or air line.
32 Regularly check the tension of the compressor drivebelt. The belt deflection should be approximately 0.14 to 0.18 in (3.5 to 4.5 mm) at the centre of its longest run. Adjust the compressor on its bracket if necessary.

Air conditioning fan motor (US/Canada models) – removal and refitting

33 Disconnect the battery negative lead.
34 Remove the console side panels and unscrew the console mounting bolts.
35 Disconnect the motor wiring.
36 Unscrew the mounting bolts then lift the console and withdraw the fan motor.
37 Refitting is a reversal of removal.

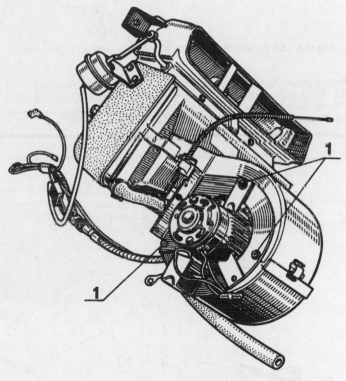

Fig. 13.152 Air conditioning fan motor mounting bolts (1) (Sec 13)

Air conditioning thermostat (US/Canada models) – removal, refitting and adjustment

38 Disconnect the battery negative lead.
39 Remove the left-hand side console panel.
40 Disconnect the control cable and wiring, and detach the thermostat.
41 Refitting is a reversal of removal, but adjust the cable as follows. Move the lever on the control panel to the maximum cold position, then set the thermostat also to the maximum cold position and lock the outer cable with the spring clip. Check the adjustment by turning the rheostat to the lowest position, depressing the air conditioning button, and moving the control panel lever to the maximum cold position – the fan should switch to the high speed as the lever reaches the maximum cold position.

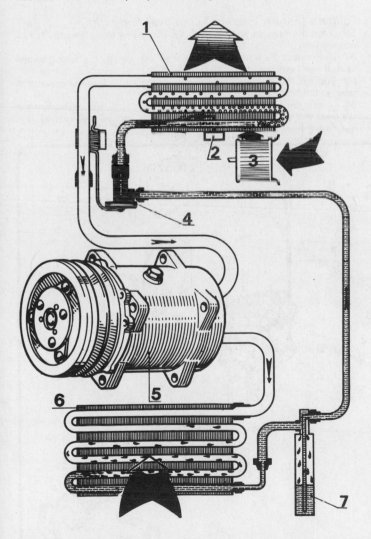

Fig. 13.151 The air conditioning system components (Sec 13)

1	Evaporator	5	Compressor
2	Thermostat	6	Condenser
3	Blower	7	Receiver/drier
4	Thermal expansion valve		

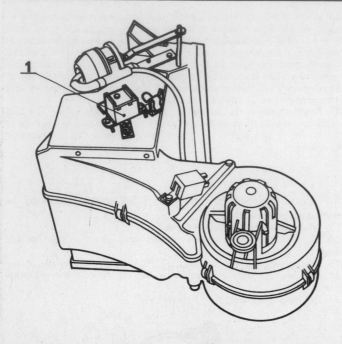

Fig. 13.153 Air conditioning thermostat location (1) (Sec 13)

Front door lock (mechanical) – removal and refitting

42 Close the window and remove the trim panel and plastic sheet.
43 Remove the clip and withdraw the exterior lock barrel.
44 Disconnect the remote control rod from the handle.

45 Remove the lock screws and the locking knob, and extract the lock from inside the door.
46 Refitting is a reversal of removal, but remove the locking knob sleeve, while the lock is being positioned on the door, by turning it.

Front door windows – removal and refitting

47 When fitting the channel to the bottom of the window, position it as shown in Fig. 13.155.

Rear door windows – removal and refitting

48 When fitting the channel to the bottom of the window, position it as shown in Fig. 13.156.

Rear door lock (mechanical) – removal and refitting

49 Close the window and remove the trim panel and plastic sheet.
50 Disconnect the remote control rod from the handle.
51 Remove the lock screws and unclip the remote control rod.
52 Unscrew the locking knob then turn the clip through 90° and remove the locking knob. Unclip and remove the long locking rod.
53 Release the lockplate from the exterior handle and withdraw it from inside the door.
54 Remove the exterior handle and locking knob sleeve.
55 Refitting is a reversal of removal, but make sure that the exterior door handle control arm is located as shown in Fig. 13.157.

Tailgate (Estate models) – removal and refitting

56 Open the tailgate and prise off the trim panel using a wide-bladed screwdriver.
57 Disconnect the battery negative lead followed by the wiring to the heated rear window and wiper.
58 Support the tailgate and disconnect the struts by prising out the tabs at the body end.
59 Unclip the headlining, unscrew the hinge nuts and withdraw the tailgate. Access to the nuts can also be gained by cutting the headlining near the rear crossmember apertures.

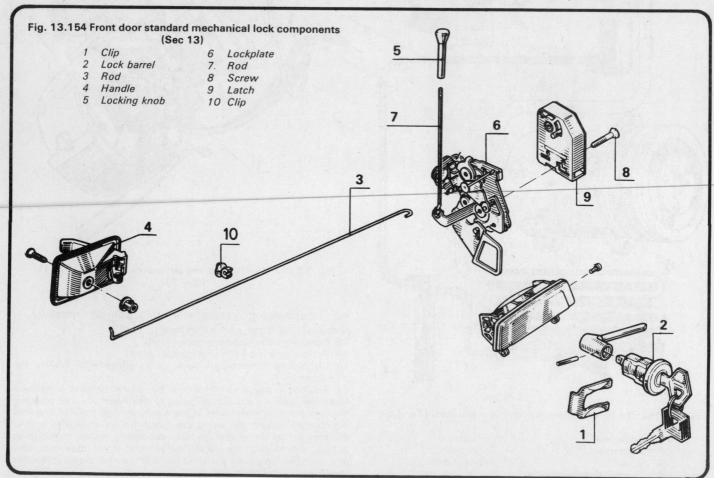

Fig. 13.154 Front door standard mechanical lock components (Sec 13)

1	Clip	6	Lockplate
2	Lock barrel	7	Rod
3	Rod	8	Screw
4	Handle	9	Latch
5	Locking knob	10	Clip

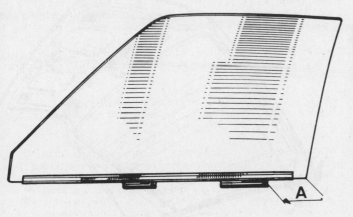

Fig. 13.155 Channel fitting dimension on the front door window (Sec 13)

A = 2.165 in (55.0 mm)

Fig. 13.156 Channel fitting dimension on the rear door window (Sec 13)

A = 2.598 in (66.0 mm)

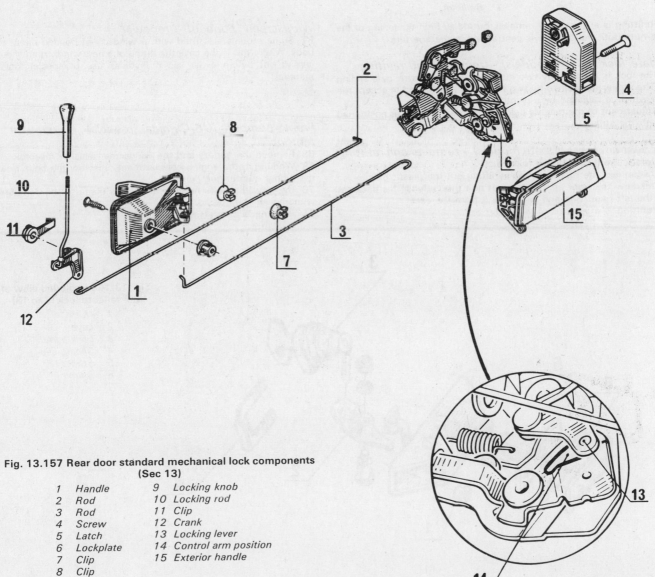

Fig. 13.157 Rear door standard mechanical lock components (Sec 13)

1	Handle	9	Locking knob
2	Rod	10	Locking rod
3	Rod	11	Clip
4	Screw	12	Crank
5	Latch	13	Locking lever
6	Lockplate	14	Control arm position
7	Clip	15	Exterior handle
8	Clip		

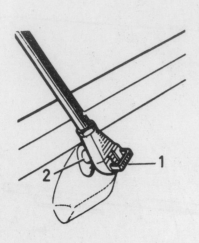

Fig. 13.158 Tailgate strut fixing (Sec 13)

1   Tab   2   Balljoint

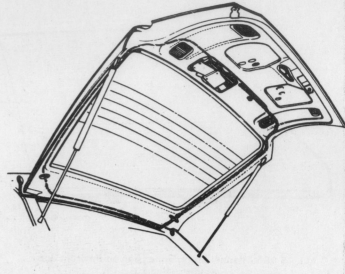

Fig. 13.159 Tailgate and struts (Sec 13)

60 Refitting is a reversal of removal, but delay final tightening of the hinge nuts until the tailgate is central within the aperture.

Tailgate lock (Estate models) – removal and refitting
61 The lock is retained by two nuts and the lock barrel by a sliding clip. The striker is retained by two bolts and it is slotted to accept the catch which is released by a pivoting lever.
62 Adjustment of the tailgate height is by means of shims positioned beneath the striker.

Rear seat backrest (Estate models) – removal and refitting
63 Unlock the backrest and fold it flat.
64 Disconnect the rubber straps by lifting out the plastic pins.
65 Unscrew the right-hand side pivot bolt then release the backrest from the left-hand side and withdraw it from the car.
66 Refitting is a reversal of removal.

Windscreen (bonded) – renewal
67 Some models are fitted with a windscreen bonded direct to the body. Where this is the case the fitting of a new windscreen should be carried out by an expert as it involves use of special tools and materials.

Front bumper (US/Canada models) – removal and refitting
68 Remove the battery and the windscreen washer reservoir.
69 Working in the engine compartment, unscrew the bolts retaining the rubber quarter bumpers.
70 Unscrew the bolts holding the centre blade to the dampers and withdraw the bumper.
71 Refitting is a reversal of removal.

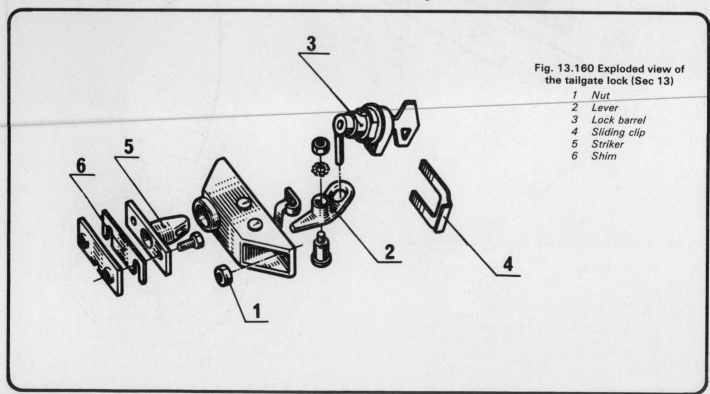

Fig. 13.160 Exploded view of the tailgate lock (Sec 13)

1   Nut
2   Lever
3   Lock barrel
4   Sliding clip
5   Striker
6   Shim

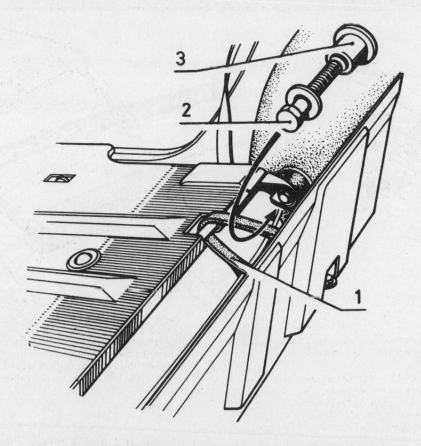

Fig. 13.161 Rear seat backrest showing right-hand side pivot bolt (Sec 13)

1   Rubber strap   2   Bolt   3   Sleeve

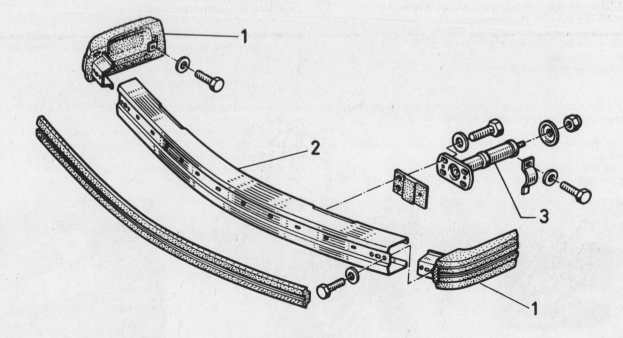

Fig. 13.162 Front bumper components (US/Canada models) (Sec 13)

1   Quarter bumpers   2   Centre blade   3   Damper

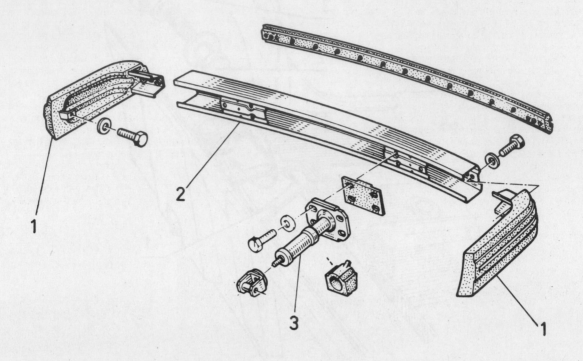

Fig. 13.163 Rear bumper components (US/Canada models) (Sec 13)

1   *Quarter bumpers*   2   *Centre blade*   3   *Damper*

Rear bumper (US/Canada models) – removal and refitting

72 Working in the luggage compartment, pull back the trim panels sufficiently far to be able to unscrew the bolts retaining the rubber quarter bumpers.
73 Unscrew the bolts holding the centre blade to the dampers and withdraw the bumper.
74 Refitting is a reversal of removal.

Remote-controlled exterior mirror – removal and refitting

75 Unscrew the cross-head screw holding the cover to the door (photo).
76 Pull out the control knob then remove the screw and separate the cover (photos).
77 Remove the interior trim panel.
78 Prise off the outer cover and unscrew the mounting screws.

Withdraw the mirror and recover the mounting plate.
79 Refitting is a reversal of removal.

Door trim panel (later models) – removal and refitting

80 Using a Torx key remove the screws and withdraw the armrest (photo).
81 Remove the screws and lift off the pocket (photos).
82 Remove the screw and release the finger plate from the interior door handle (photos).
83 Remove the screw retaining the exterior mirror control cover to the door.
84 Remove the remaining screws and use a wide-bladed screwdriver to release the trim panel clips (photos). Remove the trim panel at the same time feeding the exterior mirror control through the hole.
85 Peel off the protective plastic sheet (photo).
86 Refitting is a reversal of removal.

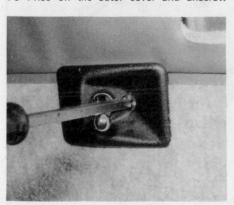

13.75 Remove the crosshead screw ...

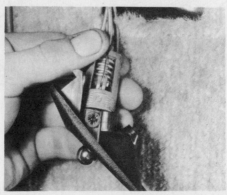

13.76A ... and pull out the remote exterior mirror control

13.76B Exterior mirror control with cover removed

Plastic components

With the use of more and more plastic body components by the vehicle manufacturers (eg bumpers, spoilers, and in some cases major body panels), rectification of more serious damage to such items has become a matter of either entrusting repair work to a specialist in this field, or renewing complete components. Repair of such damage by the DIY owner is not really feasible owing to the cost of the equipment and materials required for effecting such repairs. The basic technique involves making a groove along the line of the crack in the plastic using a rotary burr in a power drill. The damaged part is then welded back together by using a hot air gun to heat up and fuse a plastic filler rod into the groove. Any excess plastic is then removed and the area rubbed down to a smooth finish. It is important that a filler rod of the correct plastic is used, as body components can be made of a variety of different types (eg polycarbonate, ABS, polypropylene).

Damage of a less serious nature (abrasions, minor cracks etc) can be repaired by the DIY owner using a two-part epoxy filler repair material like Holts Body + Plus or Holts No Mix which can be used directly from the tube. Once mixed in equal proportions (or applied direct from the tube in the case of Holts No Mix), this is used in similar fashion to the bodywork filler used on metal panels. The filler is usually cured in twenty to thirty minutes, ready for sanding and painting.

If the owner is renewing a complete component himself, or if he has repaired it with epoxy filler, he will be left with the problem of finding a suitable paint for finishing which is compatible with the type of plastic used. At one time the use of a universal paint was not possible owing to the complex range of plastics encountered in body component applications. Standard paints, generally speaking, will not bond to plastic or rubber satisfactorily, but Holts Professional Spraymatch paints to match any plastic or rubber finish can be obtained from dealers. However, it is now possible to obtain a plastic body parts finishing kit which consists of a pre-primer treatment, a primer and coloured top coat. Full instructions are normally supplied with a kit, but basically the method of use is to first apply the pre-primer to the component concerned and allow it to dry for up to 30 minutes. Then the primer is applied and left to dry for about an hour before finally applying the special coloured top coat. The result is a correctly coloured component where the paint will flex with the plastic or rubber, a property that standard paint does not normally possess.

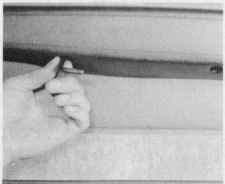

13.80 Removing the armrest

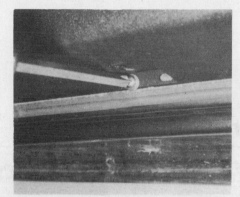

13.81A Remove the screws ...

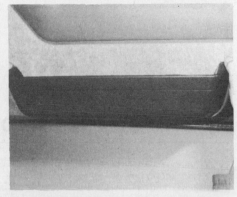

13.81B ... and lift off the pocket

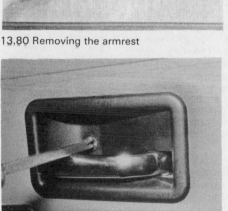

13.82A Remove the screw ...

13.82B ... and remove the interior door handle finger plate

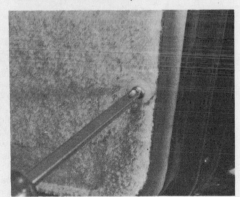

13.84A Remove the screws ...

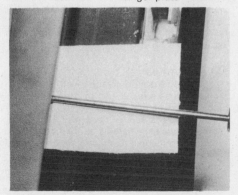

13.84B ... and use a wide-bladed screwdriver and protective card to prise off the trim panel

13.85 Peel off the plastic sheet

General repair procedures

Whenever servicing, repair or overhaul work is carried out on the car or its components, it is necessary to observe the following procedures and instructions. This will assist in carrying out the operation efficiently and to a professional standard of workmanship.

Joint mating faces and gaskets

Where a gasket is used between the mating faces of two components, ensure that it is renewed on reassembly, and fit it dry unless otherwise stated in the repair procedure. Make sure that the mating faces are clean and dry with all traces of old gasket removed. When cleaning a joint face, use a tool which is not likely to score or damage the face, and remove any burrs or nicks with an oilstone or fine file.

Make sure that tapped holes are cleaned with a pipe cleaner, and keep them free of jointing compound if this is being used unless specifically instructed otherwise.

Ensure that all orifices, channels or pipes are clear and blow through them, preferably using compressed air.

Oil seals

Whenever an oil seal is removed from its working location, either individually or as part of an assembly, it should be renewed.

The very fine sealing lip of the seal is easily damaged and will not seal if the surface it contacts is not completely clean and free from scratches, nicks or grooves. If the original sealing surface of the component cannot be restored, the component should be renewed.

Protect the lips of the seal from any surface which may damage them in the course of fitting. Use tape or a conical sleeve where possible. Lubricate the seal lips with oil before fitting and, on dual lipped seals, fill the space between the lips with grease.

Unless otherwise stated, oil seals must be fitted with their sealing lips toward the lubricant to be sealed.

Use a tubular drift or block of wood of the appropriate size to install the seal and, if the seal housing is shouldered, drive the seal down to the shoulder. If the seal housing is unshouldered, the seal should be fitted with its face flush with the housing top face.

Screw threads and fastenings

Always ensure that a blind tapped hole is completely free from oil, grease, water or other fluid before installing the bolt or stud. Failure to do this could cause the housing to crack due to the hydraulic action of the bolt or stud as it is screwed in.

When tightening a castellated nut to accept a split pin, tighten the nut to the specified torque, where applicable, and then tighten further to the next split pin hole. Never slacken the nut to align a split pin hole unless stated in the repair procedure.

When checking or retightening a nut or bolt to a specified torque setting, slacken the nut or bolt by a quarter of a turn, and then retighten to the specified setting.

Locknuts, locktabs and washers

Any fastening which will rotate against a component or housing in the course of tightening should always have a washer between it and the relevant component or housing.

Spring or split washers should always be renewed when they are used to lock a critical component such as a big-end bearing retaining nut or bolt.

Locktabs which are folded over to retain a nut or bolt should always be renewed.

Self-locking nuts can be reused in non-critical areas, providing resistance can be felt when the locking portion passes over the bolt or stud thread.

Split pins must always be replaced with new ones of the correct size for the hole.

Special tools

Some repair procedures in this manual entail the use of special tools such as a press, two or three-legged pullers, spring compressors etc. Wherever possible, suitable readily available alternatives to the manufacturer's special tools are described, and are shown in use. In some instances, where no alternative is possible, it has been necessary to resort to the use of a manufacturer's tool and this has been done for reasons of safety as well as the efficient completion of the repair operation. Unless you are highly skilled and have a thorough understanding of the procedure described, never attempt to bypass the use of any special tool when the procedure described specifies its use. Not only is there a very great risk of personal injury, but expensive damage could be caused to the components involved.

Use of English

As this book has been written in England, it uses the appropriate English component names, phrases, and spelling. Some of these differ from those used in America. Normally, these cause no difficulty, but to make sure, a glossary is printed below. In ordering spare parts remember the parts list may use some of these words:

English	American	English	American
Accelerator	Gas pedal	Locks	Latches
Aerial	Antenna	Methylated spirit	Denatured alcohol
Anti-roll bar	Stabiliser or sway bar	Motorway	Freeway, turnpike etc
Big end bearing	Rod bearing	Number plate	License plate
Bonnet (engine cover)	Hood	Paraffin	Kerosene
Boot (luggage compartment)	Trunk	Petrol	Gasoline (gas)
Bulkhead	Firewall	Petrol tank	Gas tank
Bush	Bushing	'Pinking'	'Pinging'
Cam follower or tappet	Valve lifter or tappet	Prise (force apart)	Pry
Carburettor	Carburetor	Propeller shaft	Driveshaft
Catch	Latch	Quarterlight	Quarter window
Choke/venturi	Barrel	Retread	Recap
Circlip	Snap-ring	Reverse	Back-up
Clearance	Lash	Rocker cover	Valve cover
Crownwheel	Ring gear (of differential)	Saloon	Sedan
Damper	Shock absorber, shock	Seized	Frozen
Disc (brake)	Rotor/disk	Sidelight	Parking light
Distance piece	Spacer	Silencer	Muffler
Drop arm	Pitman arm	Sill panel (beneath doors)	Rocker panel
Drop head coupe	Convertible	Small end, little end	Piston pin or wrist pin
Dynamo	Generator (DC)	Spanner	Wrench
Earth (electrical)	Ground	Split cotter (for valve spring cap)	Lock (for valve spring retainer)
Engineer's blue	Prussian blue	Split pin	Cotter pin
Estate car	Station wagon	Steering arm	Spindle arm
Exhaust manifold	Header	Sump	Oil pan
Fault finding/diagnosis	Troubleshooting	Swarf	Metal chips or debris
Float chamber	Float bowl	Tab washer	Tang or lock
Free-play	Lash	Tappet	Valve lifter
Freewheel	Coast	Thrust bearing	Throw-out bearing
Gearbox	Transmission	Top gear	High
Gearchange	Shift	Torch	Flashlight
Grub screw	Setscrew, Allen screw	Trackrod (of steering)	Tie-rod (or connecting rod)
Gudgeon pin	Piston pin or wrist pin	Trailing shoe (of brake)	Secondary shoe
Halfshaft	Axleshaft	Transmission	Whole drive line
Handbrake	Parking brake	Tyre	Tire
Hood	Soft top	Van	Panel wagon/van
Hot spot	Heat riser	Vice	Vise
Indicator	Turn signal	Wheel nut	Lug nut
Interior light	Dome lamp	Windscreen	Windshield
Layshaft (of gearbox)	Countershaft	Wing/mudguard	Fender
Leading shoe (of brake)	Primary shoe		

Conversion factors

Length (distance)
Inches (in)	X	25.4	= Millimetres (mm)	X 0.0394	= Inches (in)
Feet (ft)	X	0.305	= Metres (m)	X 3.281	= Feet (ft)
Miles	X	1.609	= Kilometres (km)	X 0.621	= Miles

Volume (capacity)
Cubic inches (cu in; in³)	X	16.387	= Cubic centimetres (cc; cm³)	X 0.061	= Cubic inches (cu in; in³)
Imperial pints (Imp pt)	X	0.568	= Litres (l)	X 1.76	= Imperial pints (Imp pt)
Imperial quarts (Imp qt)	X	1.137	= Litres (l)	X 0.88	= Imperial quarts (Imp qt)
Imperial quarts (Imp qt)	X	1.201	= US quarts (US qt)	X 0.833	= Imperial quarts (Imp qt)
US quarts (US qt)	X	0.946	= Litres (l)	X 1.057	= US quarts (US qt)
Imperial gallons (Imp gal)	X	4.546	= Litres (l)	X 0.22	= Imperial gallons (Imp gal)
Imperial gallons (Imp gal)	X	1.201	= US gallons (US gal)	X 0.833	= Imperial gallons (Imp gal)
US gallons (US gal)	X	3.785	= Litres (l)	X 0.264	= US gallons (US gal)

Mass (weight)
Ounces (oz)	X	28.35	= Grams (g)	X 0.035	= Ounces (oz)
Pounds (lb)	X	0.454	= Kilograms (kg)	X 2.205	= Pounds (lb)

Force
Ounces-force (ozf; oz)	X	0.278	= Newtons (N)	X 3.6	= Ounces-force (ozf; oz)
Pounds-force (lbf; lb)	X	4.448	= Newtons (N)	X 0.225	= Pounds-force (lbf; lb)
Newtons (N)	X	0.1	= Kilograms-force (kgf; kg)	X 9.81	= Newtons (N)

Pressure
Pounds-force per square inch (psi; lbf/in²; lb/in²)	X	0.070	= Kilograms-force per square centimetre (kgf/cm²; kg/cm²)	X 14.223	= Pounds-force per square inch (psi; lbf/in²; lb/in²)
Pounds-force per square inch (psi; lbf/in²; lb/in²)	X	0.068	= Atmospheres (atm)	X 14.696	= Pounds-force per square inch (psi; lbf/in²; lb/in²)
Pounds-force per square inch (psi; lbf/in²; lb/in²)	X	0.069	= Bars	X 14.5	= Pounds-force per square inch (psi; lbf/in²; lb/in²)
Pounds-force per square inch (psi; lbf/in²; lb/in²)	X	6.895	= Kilopascals (kPa)	X 0.145	= Pounds-force per square inch (psi; lbf/in²; lb/in²)
Kilopascals (kPa)	X	0.01	= Kilograms-force per square centimetre (kgf/cm²; kg/cm²)	X 98.1	= Kilopascals (kPa)
Millibar (mbar)	X	100	= Pascals (Pa)	X 0.01	= Millibar (mbar)
Millibar (mbar)	X	0.0145	= Pounds-force per square inch (psi; lbf/in²; lb/in²)	X 68.947	= Millibar (mbar)
Millibar (mbar)	X	0.75	= Millimetres of mercury (mmHg)	X 1.333	= Millibar (mbar)
Millibar (mbar)	X	0.401	= Inches of water (inH₂O)	X 2.491	= Millibar (mbar)
Millimetres of mercury (mmHg)	X	0.535	= Inches of water (inH₂O)	X 1.868	= Millimetres of mercury (mmHg)
Inches of water (inH₂O)	X	0.036	= Pounds-force per square inch (psi; lbf/in²; lb/in²)	X 27.68	= Inches of water (inH₂O)

Torque (moment of force)
Pounds-force inches (lbf in; lb in)	X	1.152	= Kilograms-force centimetre (kgf cm; kg cm)	X 0.868	= Pounds-force inches (lbf in; lb in)
Pounds-force inches (lbf in; lb in)	X	0.113	= Newton metres (Nm)	X 8.85	= Pounds-force inches (lbf in; lb in)
Pounds-force inches (lbf in; lb in)	X	0.083	= Pounds-force feet (lbf ft; lb ft)	X 12	= Pounds-force inches (lbf in; lb in)
Pounds-force feet (lbf ft; lb ft)	X	0.138	= Kilograms-force metres (kgf m; kg m)	X 7.233	= Pounds-force feet (lbf ft; lb ft)
Pounds-force feet (lbf ft; lb ft)	X	1.356	= Newton metres (Nm)	X 0.738	= Pounds-force feet (lbf ft; lb ft)
Newton metres (Nm)	X	0.102	= Kilograms-force metres (kgf m; kg m)	X 9.804	= Newton metres (Nm)

Power
Horsepower (hp)	X	745.7	= Watts (W)	X 0.0013	= Horsepower (hp)

Velocity (speed)
Miles per hour (miles/hr; mph)	X	1.609	= Kilometres per hour (km/hr; kph)	X 0.621	= Miles per hour (miles/hr; mph)

Fuel consumption*
Miles per gallon, Imperial (mpg)	X	0.354	= Kilometres per litre (km/l)	X 2.825	= Miles per gallon, Imperial (mpg)
Miles per gallon, US (mpg)	X	0.425	= Kilometres per litre (km/l)	X 2.352	= Miles per gallon, US (mpg)

Temperature

Degrees Fahrenheit = (°C x 1.8) + 32

Degrees Celsius (Degrees Centigrade; °C) = (°F - 32) x 0.56

*It is common practice to convert from miles per gallon (mpg) to litres/100 kilometres (l/100km), where mpg (Imperial) x l/100 km = 282 and mpg (US) x l/100 km = 235

Index